Annual Reports on Fermentation Processes

Volume 6

ACADEMIC PRESS RAPID MANUSCRIPT REPRODUCTION

Annual Reports on Fermentation Processes

VOLUME 6

EDITED BY

GEORGE T. TSAO

Laboratory of Renewable Resources Engineering
Purdue University
A. A. Potter Engineering Center
West Lafayette, Indiana

ASSOCIATE EDITORS

M. C. FLICKINGER

National Cancer Institute
Frederick Cancer Research Facility
Frederick, Maryland

ROBERT K. FINN

School of Chemical Engineering
Cornell University
Ithaca, New York

1983

ACADEMIC PRESS
A Subsidiary of Harcourt Brace Jovanovich, Publishers
New York London
Paris San Diego San Francisco São Paulo Sydney Tokyo Toronto

ACADEMIC PRESS, INC.
111 Fifth Avenue, New York, New York 10003

United Kingdom Edition published by
ACADEMIC PRESS, INC. (LONDON) LTD.
24/28 Oval Road, London NW1 7DX

Library of Congress Catalog Card Number: 0-275-3790

ISBN 0-12-040306-4

PRINTED IN THE UNITED STATES OF AMERICA

83 84 85 86 9 8 7 6 5 4 3 2 1

Contents

Contributors

Numbers in parentheses indicate the pages on which the authors' contributions begin.

K. Aunstrup (175), *NOVO Industries A/S, Bagsvaerd, Denmark*
Daniel N. Bull (359), *New Brunswick Scientific Co., Inc., Edison, New Jersey*
H. R. Bungay (149), *Chemical and Environmental Engineering, Rensselaer Polytechnic Institute, Troy, New York*
M. L. Bungay (149), *Chemical and Environmental Engineering, Rensselaer Polytechnic Institute, Troy, New York*
Ichiro Chibata (1), *Research Laboratory of Applied Biochemistry, Tenabe Seiyaku Co. Ltd., Osaka, Japan*
Alex Ciegler (299), *United States Department of Agriculture, Southern Regional Research Center, New Orleans, Louisiana*
Joseph Feder (35), *Monsanto Company, St. Louis, Missouri*
Motoki Fujimura (1), *Research Laboratory of Applied Biochemistry, Tenabe Seiyaku Co. Ltd., Osaka, Japan*
Ivan G. Gilbert (323), *Laboratory of Renewable Resources Engineering, Purdue University, West Lafayette, Indiana*
Cheng S. Gong (253), *Laboratory of Renewable Resources Engineering, Purdue University, West Lafayette, Indiana*
C. N. Haas (149), *Illinois Institute of Technology, Chicago, Illinois*
Youn W. Han (299), *United States Department of Agriculture, Southern Regional Research Center, New Orleans, Louisiana*
Isao Karube (203), *Research Laboratory of Applied Biochemistry, Tenabe Seiyaku Co., Ltd., Osaka, Japan*
Young H. Lee (101), *Department of Chemical Engineering, Drexel University, Philadelphia, Pennsylvania*
Sydney Luk (101), *Department of Chemical Engineering, Drexel University, Philadelphia, Pennsylvania*
Ralph A. Messing (23), *Research and Development Division, Corning Glass Works, Corning, New York*
James Y. Oldshue (75), *Mixing Equipment Company, Rochester, New York*
Henry J. Peppler (237), *Universal Foods Corporation, Whitefish Bay, Wisconsin*
Shuichi Suzuki (203), *Tokyo Institute of Technology, Yokohama, Japan*
William R. Tolbert (35), *Monsanto Company, St. Louis, Missouri*

Tetsuya Tosa (1), *Research Laboratory of Applied Biochemistry, Tenabe Seiyaku Co. Ltd., Osaka, Japan*

George T. Tsao (323), *Laboratory of Renewable Resources Engineering, Purdue University, West Lafayette, Indiana*

Preface

A collection of informative reviews on a timely subject by outstanding authors is published in this volume. We shall continue to seek out and add new and important topics for review in future volumes. Feedback from readers has been very valuable to us and is always welcome.

The annual lectureship in honor of David Perlman was presented at the Annual Meeting of the Microbial and Biochemical Technology Division of the American Chemical Society, which was held in Kansas City. The third annual address, which was presented by Dr. Henry Peppler, is reprinted here in Chapter 9.

CHAPTER 1

IMMOBILIZED LIVING MICROBIAL CELLS

Ichiro Chibata
Tetsuya Tosa
Motoki Fujimura

Tenabe Seiyaku Co., Ltd.,
Osaka, Japan

I. HISTORICAL BACKGROUND

Recently, in order to produce useful compounds by enzymic or fermentative method, microbial cells have come to be often used in the field of chemical industry. The following advantages using microbial cells may be mentioned. Enzymes which are biocatalysts have higher substrate specificity than common chemical catalysts. And enzymes can catalyze various reactions under mild conditions, and can efficiently produce useful compounds of specific complicated structures which cannot be easily synthesized by chemical process. However, enzymes are generally not so stable as to maintain their catalytic activity for a long period. And it is very difficult to recover enzymes from reaction mixture for reuse and to use them for continuous reaction. The studies on immobilized enzymes started as the means of overcoming these disadvantages of using enzymes.

This immobilized ensyme system has made rapid progress in its industrial application. In 1969, we succeeded in the industrialization of optical resolution of DL-amino acids by immobilized aminoacylase. This is the first industrial application of immobilized enzyme. Following this, production of 6-aminopenicillanic acid (6-APA) by immobilized penicillin amidase, production of high fructose syrup by immobilized glucose isomerase, and degradation of lactose by immobilized β-galactosidase were successively industrialized.

For immobilization of intracellular enzymes, it is necessary to extract the enzymes from the cells. Then, in order to avoid this procedure, the technique for the direct immobilization of whole microbial cells was studied.

ISBN 0-12-040306-4

The advantages of immobilized microbial cells are as follows. It is not necessary to extract the enzymes from the cells. Their heat stability and operational stability in continuous enzyme reaction are the same or superior compared with those of an immobilized enzyme system. On the other hand, there are some disadvantages that many kinds of enzymes in the cells may cause side-reactions or degradation of the product besides the main reaction. However, these unfavorable reactions can be often avoided by heat-treatment, acid-treatment, and chemical-treatment before or after immobilization of microbial cells. On the applications of immobilized microbial cell system, we also first succeeded in industrialization of the system for the continuous production of L-aspartic acid in 1973. Succeeding to this we industrialized the production of L-malic acid and L-alanine. Besides these systems now the production systems for 6-APA and high fructose syrups are industrialized.

So far immobilized microbial cells have been primarily used for a single-step enzyme reaction.

In this system, the immobilized microbial cells are dead, though the enzymes are active and stable. However, many useful compounds especially produced by fermentation are usually formed by multi-step reactions catalyzed with many kinds of enzymes in living microbial cells. Also, these reactions often require generation of ATP and coenzymes such as NAD, NADP, coenzyme A and so on. When immobilized microbial cells are kept in a living state, they may be applied to such multi-enzyme reactions. Such immobilized microbial cells are called "immobilized living cells." And especially when the immobilized living cells are in a growing state, they are called "immobilized growing cells."

Therefore, when immobilized microbial cells are used for multi-enzyme system, they can be classified to three physiological states--dead, living, and growing states. So we must choose the state more suitable for application purpose.

In this review, we describe the studies on application of the immobilized living cell system for the production of useful compounds.

II. IMMOBILIZATION OF MICROBIAL CELLS IN A LIVING STATE

Immobilization of microbial cells should be carried out under mild conditions in order to maintain the activity of the cells. Methods for immobilization of microbial cells can be classified into three categories, that is, carrier-binding, cross-linking and entrapping methods. Among these methods, the entrapping method has been most extensively investigated.

A. Carrier-Binding Method

The carrier-binding method is based on direct binding of cells to water-insoluble carrier by physical adsorption, ionic, and/or covalent bonds. As carriers, water-insoluble polysaccharides (cellulose, dextrane and agarose derivatives), proteins (gelatin and albumin), synthetic polymers (ion-exchange resins and polyvinylchloride), inorganic materials (silica, ceramics, brick, and sand), and metal oxides (zirconium oxide and titanium oxide) are used.

However, this method is not considered to be advantageous, because cells may leak out from the carrier due to autolysis during the enzyme reaction.

B. Cross-Linking Method

Microbial cells are immobilized by cross-linking with bifunctional reagents such as glutaraldehyde, toluendiisocyanate or diazotized diamine. Studies on this method are rather few, and this method is not suitable for immobilization of microbial cells in a living state.

C. Entrapping Method

So far, the method directly entrapping cells into polymer matrices has been most extensively investigated for immobilization of cells in a living state.

1. Synthetic Polymers. The following polymers are employed as the matrices for immobilization: 1) polyacrylamide, 2) polyvinylchloride, 3) photo-crosslinkable resin, and 4) polyurethane. Among these matrices, polyacrylamide gel has been extensively used for immobilization of many kinds of microbial cells. Photo-crosslinkable resin which has recently been developed, is suitable for immobilized living cell systems because the immobilization can be performed under mild conditions.

2. Natural Polymers. The natural polymers used for immobilization of cells are mainly polysaccharides such as calcium alginate, κ-carrageenan and agar. Besides polysaccharides, collagen and gelatin also have been used for the immobilization.

Since 1975, calcium alginate gel has been used for the immobilization of cells and enzymes (30). In 1977, Kierstan and Bucke found that this gel provided suitable matrix for the immobilization by entrapment of whole microbial cells, subcellular organelles, and isolated enzymes (40). Then, the gel has been extensively used for immobilization of microbial cells in a living state.

Recently, we found κ-carrageenan is a very useful matrix for immobilization of microbial cells (77,78). κ-Carrageenan, which is composed of unit structure of β-D-galactose sulfate and 3,6-anhydro-α-D-galactose, is a readily available nontoxic polysaccharide isolated from seaweed, and is widely used as a food additive.

κ-Carrageenan easily becomes a gel under the following conditions. It becomes a gel by cooling, as in the case of agar. Gelation occurs also by contacting with a solution containing one of gel-inducing reagents such as K^+, NH_4^+, Ca^{2+}, Cu^{2+}, Mg^{2+}, Fe^{3+}, amines, and water-miscible organic solvents (78). If immobilization of microbial cells is carried out with κ-carrageenan by selecting a suitable gel-inducing reagent, an immobilized preparation having high enzyme activity can be obtained. The advantage of this method is that the immobilization can be performed under mild conditions without the use of chemicals inactivating the enzyme activity. Another advantage of this method is that the various shapes of immobilized microbial cells such as bead, cube, and membrane can be easily tailor-made for particular application purposes. In addition, if immobilized microbial cells are suspended in physiological saline, the κ-carrageenan gel is rapidly dissolved and a cell suspension can be obtained. So the number of living cells can be readily counted by serial diluted drop-plate method. This is advantageous for investigating characteristics of microbial cells after immobilization, and this method is considered to be suitable for immobilized living cell systems.

III. APPLICATION OF IMMOBILIZED LIVING MICROBIAL CELLS

Immobilized living cells can be applied to various multistep enzyme reactions. In this review, we present briefly the studies on production of useful compounds such as alcohols, organic acids, amino acids, antibiotics, steroids, and enzymes by immobilized living cells.

The first application for production of useful compounds by immobilized living cell system may be the quick vinegar fermentation process with the trickle-filter developed in the beginning of the last century. This vinegar process, a carrier-binding method, had been mainly used for earlier studies on immobilized living cell; however, recently, the entrapping method is mainly used. Since we found that the yeast cells entrapped into gel grew in the gel matrix, and formed a dense cell layer near the surface of gel, the entrapping method has become extensively used for the immobilized living cell system (82). The examples for production of useful compounds by immobilized living cells are summarized in Table 1, and some of them are described in following sections.

A. *Production of Alcohols*

Over the past ten years, the increased concerns about the energy crisis originated by the petroleum shortage have urged the research for new energy substituting petroleum, and attention has been shown to ethanol production by fermentation. Since olden times, it has been well known that ethanol is produced from sugar by yeast, but the traditional processes do not give us satisfactory results in industry. Accordingly, a development of more efficient process for ethanol production is expected from the standpoints of energy and natural resources.

Among the efficient processes, continuous fermentation technique has been extensively investigated, but it has some problems on stability of process and production efficiency. On the other hand, the recently developed immobilized microbial cell system looks attractive and promising for economical ethanol production since continuous production by this system is superior to that by conventional fermentation process.

1. Ethanol. For ethanol production by immobilized living cells, carrier-binding method was first used. Afterward, entrapping method has mainly been investigated.

a. Immobilization by carrier-binding method. Navarro and Durand (61) first immobilized *Saccharomyces carlsbergensis* cells using a porous glass bead as the carrier, and compared metabolism of adsorbed cells with that of free cells. As the result, it was found that the adsorption of cells on the carrier led to accelerations of metabolism in cells and conversion into ethanol from glucose. In 1980, continuous ethanol production was carried out by using yeast cells adsorbed into pretreated inert support (26). In this experiment, 71 mg of ethanol per ml was continuously produced over 2.5 months by feeding molasses of 19.7% sugar at a residence time of 2.86 hours. This high productivity and high stability resulted from high cell population and faster growth rate of immobilized cells.

Yeast cells were also adsorbed into the ceramic Raschig rings which were coated by gelatin and sprayed with glutaraldehyde solution (72). These immobilized cells were packed into a Plexiglass tube reactor. The efficiency of a tube reactor using immobilized living cells was compared with that of the continuous stirred reactor using free cells. Both reactors were operated at various dilution rates using glucose as a carbon source. The ethanol productivity of the immobilized cell reactor was 4.2 times higher than that of the free cell reactor, because a large amount of yeast cells formed a film on the surface of the Raschig rings, and the cell densities of the immobilized cell reactor were very high. In this case, the disadvantage of the immobilized cell reactor was that yeast cells leaked out from the film formed on the Raschig rings and clogged up the reactor. However, the reactor could be regenerated by sparging with nitrogen gas and dislodging the leaked cells.

Moreover, production of ethanol from corn stover as an agricultural residue was investigated (73). The productivity in the free cell reactor was reduced with hydrolysate of corn stover, while in the immobilized cell reactor the productivity was not reduced, and was 15.9 gm/l·h which was 9 times the maximum productivity in the free cell reactor. Therefore, the performance and stability of the immobilized cell reactor were superior to those of the free cell reactor. This process using immobilized cells appears economically quite attractive.

The ability of various ion exchange resins to adsorb yeast cells was tested, and the immobilized yeast cells were used for the continuous production of ethanol (20). A column reactor was operated for over 200 hours at high dilution rate with ethanol productivity of 62.0 gm/l·h.

As stated above, a few studies on ethanol fermentation with immobilized yeast cells were carried out by carrier-binding method. This carrier-binding method has some advantages as follows: 1) immobilization can be easily carried out under mild conditions; 2) limitation of mass transfer is smaller than that of entrapping method; 3) reuse of carrier is possible; and 4) a large number of cells can be immobilized using the carrier with large surface area. However, since the binding force of cells to carriers is not strong, the cells easily leak out from carriers. Therefore, the entrapping method has become extensively used instead of carrier-binding method.

b. Immobilization by entrapping method. Kierstan and Bucke first applied entrapped yeast cells to ethanol production (40). In this report, a large amount of cells was harvested from culture broth, and entrapped homogeneously into calcium alginate gel. In this case, although the immobilized yeast cells were in a living state in gel, they were not in a growing state.

On the other hand, we found in the course of the studies on immobilized cells that microorganisms could grow in gel-matrix by feeding nutrients, and established a new immobilization method (82). That is, a small amount of yeast cells was entrapped into κ-carrageenan gel, and then the entrapped cells were grown in gel-matrix with incubation. Therefore, harvesting process of yeast cells was not necessary. Thus, immobilized living yeast cells based on the growing of cells in gel matrix are particularly designated as "immobilized growing yeast cells".

In this immobilization method, precultured broth was directly mixed with κ-carrageenan solution and made immobilized cells of bead type. These gels containing a small amount of cells (3.5 x 10^6 cells/ml-gel) were incubated in complete medium on a rotary shaker at 30°C. After 60 hours of incubation, the number of living cells in gel increased to 5.4 x 10^9 cells/ml-gel, 1000 fold that in gel before incubation. The immobilized yeast cells showed better growth than the free cells. After 60 hour incubation, the number of living cells in gel reached to 10 fold the number of free cells. As the reason of this higher growth of yeast cells in gel in comparison with free cells in conventional culture method, we consider that in gel (i) microbial cells can select suitable environmental conditions and (ii) cells may be better protected from environmental changes than free cells.

Figure 1 shows both whole gel beads and sliced beads. Before incubation, gel bead looks clear because it contains a small number of cells. After incubation, the appearance of gel bead changes similar to that of the gel homogeneously entrapping a large number of harvested cells. The slice of gel after incubation shows the formation of thin layer of condensed yeast cells near the surface of the bead, and cells are not observed at the center of the bead. Therefore, these immobilized cells forming condensed thin layer near the surface of the gel bead are expected to show efficient catalytic activities in enzyme reactions.

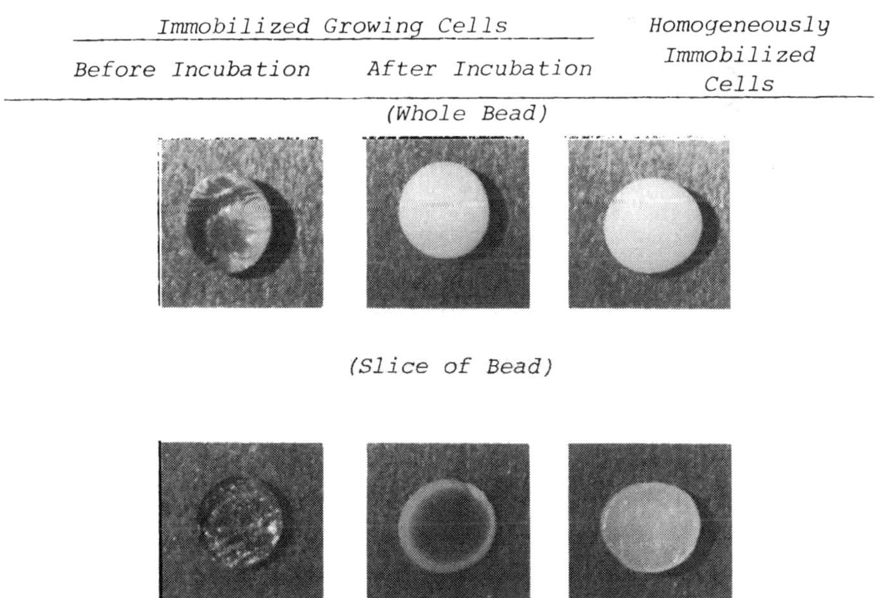

Figure 1. Photographs of immobilized yeast cells

By using the column packed with immobilized living yeast cells, ethanol production was studied. Immobilized yeast cells in layer form showed higher activity in comparison with immobilized yeast cells in homogeneous form. It is apparent that the intrinsic ethanol producing activity of immobilized yeast cells in layer form is fully utilized for ethanol production, while in the homogeneously immobilized system, the activity of immobilized cells is not fully utilized for ethanol production.

In the continuous production of ethanol, the steady state was maintained for longer than 90 days. Release of some cells from gel occurred. However, the number of cells in effluent were as low as 10^6- 10^7, while the cells in gel were maintained at the high level of 10^9- 10^{10}. Therefore, it is clear that most of the alcohol is produced by the cells in gel. Gel beads kept the shape after 90 days operation. For constant production of higher concentration of ethanol, it was necessary to increase the concentration of glucose step-wise.

As stated above, immobilized growing cell system has the advantage that stable continuous production of ethanol can be maintained with supply of nutrients under suitable conditions. In addition, this system is further improved, and the continuous production of ethanol is becoming possible even using sugar molasses of higher concentration. So we expect that this new method becomes a more advantageous industrial process compared with traditional fermentation method.

Besides our studies, several papers have been published on the ethanol production using immobilized living yeast cells as shown in Table 1. Kierstan and Bucke attempted a rapid production of ethanol from glucose with a large amount of cells immobilized into calcium alginate gel (40). However, the favorable results were not obtained because of low level of both producing activity and stability. This low productivity seems to result because cells at the central part of gel are unable to contact with glucose, while the unstable activity seems to be caused by death of yeast cells immobilized because of the absence of nutrients. From these results, it is considered to be important for the stable production of ethanol that yeast cells in gel are not only in a living state, but in a growing state.

Yeast cells entrapped into calcium alginate gel were reported to be successfully applied to continuous production of beer (87). In the production, a tower fermentor containing immobilized living cells was run for 7 months without deterioration of ethanol producing activity.

Saccharomyces cerevisiae cells immobilized into calcium alginate beads were used for a continuous production of ethanol from 10% glucose (48). About 100% conversion was obtained for 10 days at 10 hour residence time, but the ethanol producing activity gradually decreased after 10 days. To overcome this drawback, effect of nutrient supply was investigated. As the result, the activity could be restored by feeding a nutrient medium. Thus, this system was applied to ethanol production from cane molasses or malt wort (51). The effect of initial cell amounts in gel beads was examined on ethanol production from cane molasses. Ethanol producing activities of gel beads were not affected by initial cell amounts after fermentation for a few days. The final cell numbers in gel were 2 to 5 x 10^9 cells per gm of gel, which were compatible with those in our κ-carrageenan method. In packed-bed column with the biocatalyst, continuous ethanol production from cane molasses was carried out. When molasses of 25.2% sugar was employed as substrate, ethanol production rate was low and decreasing rapidly. However, when molasses was diluted to 17.5%, a nearly complete conversion to ethanol was maintained for a long period at 4.1 hour residence time.

Kluyveromyces fragilis cells entrapped into calcium alginate were applied to production of ethanol from whey (50). In column operation, 2% ethanol could be maintained at 5 to 10% lactose in whey at least for one month. Besides the carbohydrate sources stated above, pentose (14,52) and inulin sugars (54) were also utilized for production of ethanol by immobilized living cell system.

As shown in Table 1, alginate gel is mainly used for entrapping the cells. But this gel is unstable in media containing phosphate ion. Therefore alginate gel having higher resistance to this ion was prepared by coating and chemical treatment. Alginate gel entrapping cells was coated with cationic polymers such as polyethyleneimine, polypropyleneimine, and polyvinylamine (11,81). And alginate sol was preliminarily treated with a carbodiimide N-hydroxysuccinimide or periodate (81). These treatments for alginate increased a stability of immobilized living cell system.

To design the efficient reactor suitable for production of ethanol, it is necessary to minimize the unfavorable effect of CO_2 gas. At present, following reactors showing high performance are used; horizontal packed-bed reactor (70) and fluidized reactor (15). These reactors are considered to become promising for the industrial production of ethanol.

Recently, Rogers and co-workers revealed that *Zymomonas mobilis*, a kind of bacteria, has several times higher ethanol productivity than yeasts (66), and is a very promising organism for continuous ethanol production (47). For example, ethanol productivities of *Z. mobilis* cells immobilized into both calcium alginate and κ-carrageenan gels were above 50 gm/l·h at over 60 gm/l of ethanol concentration (29). But there was a decline of 30% in activity after 800 hour operation. This system would be improved in maintaining the long-term stability by the selection of optimal environmental conditions.

As stated above, immobilized *Z. mobilis* cells are more advantageous for production of ethanol at high productivity, high concentration, and high temperature compared with immobilized yeast cells. This process may be further investigated by many workers.

c. Conclusion for ethanol production. The immobilized living cell system is being further investigated to determine the conditions of optimum operation and to scale up the reaction, and it seems to be a promising technique for industrial production of ethanol.

Further, if a novel strain having resistance to higher concentrations of ethanol and/or glucose and growing well in the immobilization matrix can be obtained, the immobilized living cell system becomes a more promising method for ethanol production.

2. Other Alcohols. At present, n-butanol, isopropanol, and 2,3-butanediol are industrially produced by petrochemical synthesis. But due to the petroleum shortage, production of these alcohols by fermentation procedure will become necessary in the near future. Therefore, as in the case of ethanol production, immobilized living cell systems have also been investigated in order to increase their productivities in fermentation production as follows.

2,3-Butanediol was produced by *Enterobacter aerogenes* immobilized with κ-carrageenan (16). When immobilized cells were grown in the nutrient medium for one day, the maximum activity was obtained. Then it was kept stable for 10 days. Using these immobilized cells, formation of acetoin as a by-product was much suppressed.

Continuous production of n-butanol and isopropanol using *Clostridium butylicum* entrapped into calcium alginate was studied (44). The immobilized cell beads were placed into a conical column to avoid problems associated with gas evolution. This productivity was about 4 times higher than that of a batch fermentation using free cells. Also, spores of *Clostridium acetobutylicum* were immobilized into calcium alginate gel. Then vegetative cells were formed within the gel by incubation in a complete growth medium (31). This immobilized cell produced n-butanol from a simple substrate without growth factors as well as immobilized growing cells.

Glycerol fermentation using *S. cerevisiae* immobilized with κ-carrageenan could be shifted from ethanol fermentation, when sodium sulfite was added to the fermentation broth (12). Because the addition of sulfite caused an excess formation of $NADH_2$, which is used to reduce dihydroxyacetone phosphate to glycerol phosphate.

In the field of basic organic compounds such as alcohols described above, fermentation method will become competitive technique to petrochemical synthesis in the near future. Then, immobilized living cell system will become a promising method for alcohol production.

Table 1. Production of useful compounds using immobilized living cells

Useful Compound	Microbial Cells	Carrier for Immobilization	References
Beer	Saccharomyces carlsbergensis	Polyvinyl chloride, porous bricks	19
	Saccharomyces cerevisiae	Calcium alginate	87
Etharol	Saccharomyces cerevisiae	Calcium alginate	2,10,11,14,15,23,32, 40,45,48,51,70,81,88
		κ-Carrageenan	82,84
		Polyacrylamide	23,71
		Coated Raschig rings	72,73
		Carrier A	26-28,79
		Wood chips	56,65,67
		Ion exchange resin	20
	Saccharomyces carlsbergensis	κ-Carrageenan	83
		Porous silica	61
	Saccharomyces bayanus	κ-Carrageenan	1
	Kluyveromyces fragilis	Calcium alginate	50,54
	Pachysolen tannophilus	Calcium alginate	52
	Zymomonas mobilis	Calcium alginate	29,53
		κ-Carrageenan	1,29

Table 1. (continued)

Useful Compound	*Microbial Cells*	*Carrier for Immobilization*	*References*
	Zymomonas mobilis	Polystyren beads	4
		Borosilicate glass fiber pad	3,4
2,3-Butanediol	Enterobacter aerogenes	κ-Carrageenan	16
n-Butanol	Clostridium butyricum	Calcium alginate	44
	Clostridium acetobutyricum	Calcium alginate	31
Isopropanol	Clostridium butyricum	Calcium alginate	44
Glycerol	Saccharomyces cerevisiae	κ-Carrageenan	12
Acetic acid	Acetobacter aceti	Porous ceramic	24,25
	Acetobacter sp.	Hydrous titanium	38,39
Citric acid	Aspergillus niger	Calcium alginate	49,80
	Saccharomycopsis lipolytica	Wood chips	13
Gluconic acid	Aspergillus niger	Calcium alginate	49
Lactic acid	Lactobacillus lactis	Calcium alginate	49
	Lactobacillus delbrueckii	Calcium alginate	76
L-Glutamic acid	Brevibacterium flavum	Collagen	18
	Corynebacterium glutamicum	Polyacrylamide	74
L-Isoleucine	Serratia marcescens	κ-Carrageenan	85
Prednisolone	Arthrobacter simplex	Calcium alginate	63

Useful Compound	*Microbial Cells*	*Carrier for Immobilization*	*References*
		Polyacrylamide	64
	Corynebacterium simplex	Polyacrylamide	46
		Collagen	17
Hydrocortison	Arthrobacter globiformis	Polyacrylamide	43
	Curvularia lunata	Calcium alginate	62
		Photocrosslinkable resin	33,75
12-Ketochenodeoxy-cholic acid	Brevibacterium fuscum	κ-Carrageenan	68
Cholest-4-ene-3-one	Nocardia erythropolis	DEAE-cellulose	6,8
4-Androsten-17 on-3-(O-carboxy-methyl)-oxime	Mycobacterium phlei	Polyacrylamide	5,7
Vitamine B_{12}	Propionibacterium freudenreichii	Polyurethane	89
Ampicillin	Kluyvera citrophila	Polyacrylamide	58
Bacitracin	Bacillus sp.	Polyacrylamide	59,60
Cephalosporins	Streptomyces clavuligerus	Polyacrylamide	21
Penicillin G	Penicillium chrysogenum	Polyacrylamide	57

Table 1. (continued)

Useful Compound	Microbial Cells	Carrier for Immobilization	References
Amylase	Bacillus subtilis	Polyacrylamide	41
	Bacillus amyloliquefaciens	κ-Carrageenan	69
Cellulase	Trichoderma reesei	κ-Carrageenan	22
Protease	Streptomyces fradie	Polyacrylamide	42
Hydrogen gas	Clostridium butyricum	Agar	35,37,55
		Collagen	55
		Polyacrylamide	36,55
	Rhodospirillum rubrum	Polyacrylamide	86
		Agar	9
Methan	Methanogenic bacteria	Agar, Collagen, Polyacrylamide	34

B. Production of Organic Acids

Organic acids are widely used in food and medicine, and some of them have been produced by traditional fermentation. With development of immobilized cell systems, organic acids will be produced by immobilized living cell systems. Immobilized living cell systems are being studied on organic acid productions, as shown in Table 1.

Acetic acid, in the form of vinegar, is produced by fermentation which is performed by the oxidation of dilute aqueous alcohol. This fermentation was studied using *Acetobacter* cells adsorbed into a large-surface-area ceramic support. A pulsed flow reactor was used for raising its oxygen transfer capacity. As a result, the highest productivity was obtained in continuous production of acetic acid (24,25). Also, *Acetobacter sp.* attached on hydrous titanium chelate cellulose was used for continuous conversion of a dilute aqueous alcohol (in the form of 'charging wort') to acetic acid (in the form of vinegar) in a tower fermentor-type reactor (38,39). To obtain a high production rate by these immobilized living cells, the technical problem associated with oxygen transfer was investigated by using pure oxygen, a carrier with a greater surface area, and a reactor with good efficiency for oxygen utilization.

Citric acid is currently produced by fermentation. For increasing its productivity, *Aspergillus niger* was entrapped into calcium alginate beads and employed in an air-lift completely stirred reactor for continuous production of citric acid from sucrose (80). Further, the study on the production of gluconic acid from glucose was done using *Aspergillus niger* immobilized in a living state into calcium alginate gel with trickle-bed reactor (49).

Since L-lactic acid is produced anaerobically by fermentation, the immobilized living cell system is suitable for production of lactic acid. For example, *Lactobacillus lactis* was entrapped into calcium alginate beads for continuous production of L-lactic acid from glucose (49). This production system was investigated using a packed-bed column reactor. In this system, no major problems were encountered other than pH control in the reactor. Further, using *Lactobacillus delbrueckii*, L-lactic acid was successfully produced by immobilized living cell system without difficulty (76).

As described above, organic acids are aerobically or anaerobically produced. There are no problems in the anaerobic process, but for production of organic acids such as acetic acid, citric acid and so on, which are produced by aerobic fermentation; dissolved oxygen concentration and oxygen transfer rate remain as important problems. Accordingly, when immobilized living cell system is applied to this aerobic fermentation, these problems should be dissolved by divising both matrices for immobilization and reactor system.

C. *Production of Other Useful Compounds*

Besides the production of alcohols and organic acids, the immobilized living cell system can be applied to production of amino acids, antibiotics, steroids, enzymes and others as shown in Table 1.

1. *Amino Acids.* A few papers have been published on the production of amino acids using immobilized living cells. The formation of L-glutamic acid from glucose using immobilized whole cells of *Corynebacterium glutamicum* (a glutamic acid-producing bacteria) entrapped in polyacrylamide gel was investigated (74). In this case, it was thought that the number of living cells gradually decreased because only glucose was used as a feed medium without other nutrients. Living cells of *Brevibacterium flavum* immobilized into collagen were used for the continuous production of L-glutamic acid in a recycle reactor system (18).

We applied immobilized growing cell system to L-isoleucine production; presently L-isoleucine is being industrially produced by a conventional aerobic fermentation process. To improve this method we studied continuous L-isoleucine production using immobilized living *Serratia marcescens* cells (85). As it was necessary to supply oxygen in this system, continuous operation was carried out using fluidized 2-bed reactors in series. The numbers of *S. marcescens* cells in gel-matrix were maintained at the steady state and isoleucine producing activity of the immobilized living cells was stable for long periods. However, the concentration in reactant was still low for industrial purposes.

2. *Antibiotics.* On the production of antibiotics, whole cells of *Penicillium chrysogenum,* penicillin-producing fungi, were immobilized into polyacrylamide gel lattice, and were used for production of penicillin G from glucose and ammonium sulfate (57). Subsequently, immobilized living cell system was applied to the production of bacitracin by *Bacillus sp.* By adding the nutrient medium to the flask, the producing activity of immobilized cells increased gradually and reached to a steady-state. This activity increase apparently resulted from the growth of cells in the gel-matrix, especially at the gel surface as in the case of immobilized growing yeast cells (60). By increasing aeration rate, a bacitracin productivity in the immobilized living cell reactor increased, and became higher than that in a fermentation process (59). Immobilized living *Streptomyces clavuligerus* cells were applied to a production of cephalosporins (21). In this case, a mild new method using prepolymerized polyacrylamide partially substituted with acylhydrazide groups was employed for immobilization. This preparation was suitable for cephalosporin production, and cephalosporins were continuously produced for 96 hours with yields similar to those of free cells.

3. Steroids. Immobilized living cell system is suitable for steroid transformation requiring coenzymes. *Corynebacterium simplex* cells entrapped into polyacrylamide were first applied to production of prednisolone from cortisol (46). Subsequently, *Arthrobacter simplex* cells were entrapped into calcium alginate. In the immobilized bacterial cells, steroid Δ'-dehydrogenase activity for the production of prednisolon from cortisol could be increased by incubating immobilized cells in nutrient medium. It was revealed that this increase of the productivity was due to the rise of the number of cells in the gel-matrix by a growth in nutrients (63). A mycelium of *Curvularia lunata* entrapped into polyacrylamide gel or into calcium alginate beads in a living state was used for the conversion of cortexolone to cortisol by the action of steroid-11β-hydroxylase (62). Further, spores of *C. lunata* were entrapped with photocrosslinkable resin prepolymer and were grown into mycerial forms in nutrients containing Reichstein's Substance S as an inducer (33,75). The resulting immobilized myceria showed higher 11β-hydroxylation activity. In addition, *Arthrobacter globiformis* cells immobilized in polyacrylamide gel were used for transformation of hydrocortison in batchwise system. When this batchwise reaction was repeated in buffer, the enzymic activity of the immobilized cell system decreased due to cell lysis (43). However enzyme activity and its stability increased by the incubation of the gel in nutrient media. Besides the reports stated above, 12-ketochenodeoxycholic acid was produced from dehydrocholic acid using living cells of *Brevibacterium fuscum* immobilized with κ-carrageenan (68).

4. Enzymes. Further, the attempts to produce extracellular enzymes by immobilized living microbial cells were reported (41,42,69). Namely, production of α-amylase was attempted in a batch system using cells of *Bacillus subtilis* immobilized into polyacrylamide gel lattice (41). Amounts of α-amylase produced by the immobilized living cells under optimum conditions were about three-fold more than those produced by washed cells. *Bacillus amyloliquefaciens* cells entrapped into κ-carrageenan gel were successfully applied to continuous production of α-amylase (69). Cellulase production by *Trichoderma reesei* immobilized with κ-carrageenan was studied in continuous culture, and this had an advantage over conventional free cell fermentation (22). Similarly immobilized *Streptomyces fradiae* cells were used for protease production (42).

5. Others. Besides the studies on the production of useful compounds as described above, the studies on the production of energy sources such as hydrogen and methan gas were carried out using immobilized living cell system (9,34-37,55, 86).

Vitamin B_{12} production using immobilized living cells was studied (89). *Propionibacterium sp.* cells were entrapped with urethane prepolymer, photocrosslinkable prepolymer, κ-carrageenan, calcium alginate, and agar. Among these matrices, urethane prepolymer was the most suitable for B_{12} production by immobilized cells.

From the results of these experiments, the immobilized living cell system is expected to become a promising method not only for continuous ethanol production but also for the production of useful compounds conventionally prepared by fermentation method.

IV. CONCLUSION

As stated above, the immobilization techniques started from immobilized enzymes, have developed to immobilized whole cells and further to immobilized living cells.

This immobilized living cell system is a new technique which looks like the combined technique of both fermentation and conventional immobilized whole cell system, and has the following advantages: i) cell number used for immobilization is very small, ii) cell concentration in gel-matrix is higher than that in liquid medium, iii) productivity is higher because a dense cell layer is formed near the surface of gel beads by growing, iv) stability of productivity is higher because microbial cells are reproduced in gel during operation, and v) the dilution rate can be varied independently of the growth rate of the cells. Accordingly, immobilized living cell system is considered to be very advantageous for continuous production of useful compounds produced with multi-enzyme reaction system requiring ATP and coenzymes.

Recently the genetic engineering, as well as the enzyme engineering, has received special attention as one of the most promising techniques in biotechnology. Genetic engineering and enzyme engineering, in which immobilized biocatalysts play the important role, are not competitive technologies, and they should cooperate with each other to further progress. Fundamentally, genetic engineering can not be an independent production technology. To be an efficient production technology it should be combined with fermentation technology, enzyme engineering, and further with isolation and purification process technologies.

A novel microorganism having desired characteristics, produced by genetic engineering and used as immobilized living microbial cells, will be a very promising production technology.

Therefore, we are convinced that if the cooperation among scientists and engineers in a variety of fields related to biotechnologies such as genetic engineering, fermentation technology, enzyme engineering and separation process technology is accelerated, the biotechnology will contribute to the future welfare of human beings.

REFERENCES

1. Amin, G. and H. Verachtert, *Eur. J. Appl. Microbiol. Biotechnol. 14*, 59 (1982).
2. Apiluktivongsa, P., T. Yoshida, and H. Taguchi, *Ann. Report ICME 4*, 287 (1981).
3. Arcuri, E. J., *Biotechnol. Bioeng., 24*, 595 (1982).
4. Arcuri, E. J., R. M. Worden, and S. E. Shumate II, *Biotechnol. Lett. 2*, 499 (1980).
5. Atrat, P., C. Horhold, M. J. Buchar, and K. A. Koschtschejenko, *Z. Allgem. Mikrobiol. 20*, 239 (1980).
6. Atrat, P., E. Huller, and C. Horhold, *Z. Allgem. Mikrobiol. 20*, 79 (1980).
7. Atrat, P., E. Huller, and C. Horhold, *Eur. J. Appl. Microbiol. Biotechnol. 12*, 157 (1981).
8. Atrat, P., E. Huller, C. Horhold, M. J. Buchar, A. Y. Arinbasarova, and K. A. Koschtschejenko, *Z. Allgem. Mikrobiol. 20*, 159 (1980).
9. Bennett, M. A. and H. H. Weetall, *J. Solid-Phase Biochem. 1*, 137 (1976).
10. Birnbaum, S. and P. O. Larsson, *Appl. Biochem. Biotechnol. 7*, 55 (1982).
11. Birnbaum, S., R. Pendleton, P. O. Larsson, and K. Mosbach, *Biotechnol. Lett. 3*, 393 (1981).
12. Bisping, B. and H. J. Rehm, *Eur. J. Appl. Microbiol. Biotechnol. 14*, 136 (1982).
13. Briffaud, J. and J. M. Engasser, *Biotechnol. Bioeng. 21*, 2093 (1979).
14. Chiang, L. C., H. Y. Hsiao, M. C. Flickinger, L. F. Chen, and G. T. Tsao, *Enzyme Microb. Technol. 4*, 93 (1982).
15. Cho, G. H., C. Y. Choi, Y. D. Choe, and M. H. Han, *Biotechnol. Lett. 3*, 667 (1981).
16. Chua, J. W., A. Erarslan, S. Kinoshita, and H. Taguchi, *J. Ferment. Technol. 58*, 123 (1980).
17. Constantinides, A., *Biotechnol. Bioeng. 22*, 119 (1980).
18. Constantinides, A., D. Bhatia, and W. R. Vieth, *Biotechnol. Bioeng. 23*, 899 (1981).
19. Corrieu, G., H. Blachere, A. Ramirez, J. M. Navarro, G. Durand, B. Duteurte, and M. Moll, In "Abstracts of Papers," 5th Int'l. Ferment. Symp. (Dellweg, H., ed.), pp. 294, Berlin, (1976).
20. Daugulis, A. J., N. M. Brown, W. R. Cluett, and D. B. Dunlop, *Biotechnol. Lett. 3*, 651 (1981).
21. Freeman, A. and Y. Aharonowitz, *Biotechnol. Bioeng. 23*, 2747 (1981).
22. Frein, E. M., B. S. Montenecourt, and D. E. Eveleigh, *Biotechnol. Lett. 4*, 287 (1982).
23. Fukushima, S., *Kagaku Kogaku 44*, 770 (1980).
24. Ghommidh, C., J. M. Navarro, and G. Durand, *Biotechnol. Lett. 3*, 93 (1981).

25. Ghommidh, C., J. M. Navarro, and G. Durand, *Biotechnol. Bioeng. 24,* 605 (1982).
26. Ghose, T. K. and K. K. Bandyopadhyay, *Biotechnol. Bioeng. 22,* 1489 (1980).
27. Ghose, T. K. and K. K. Bandyopadhyay, *Biotechnol. Bioeng. 24,* 797 (1982).
28. Ghose, T. K. and R. D. Tyagi, *J. Molecular Catalysis 16,* 11 (1982).
29. Grote, W., K. J. Lee, and P. L. Rogers, *Biotechnol. Lett. 2,* 481 (1980).
30. Hackel, U., J. Klein, R. Megnet, and F. Wagner, *Eur. J. Appl. Microbiol. 1,* 291 (1975).
31. Haggstrom, L., Advances in Biotechnology, Vol. 2, 79 (1980).
32. Holcberg, I. B. and P. Margalith, *Eur. J. Appl. Microbiol. Biotechnol. 13,* 133 (1981).
33. Hoq, M. M., A. Tanaka, and S. Fukui, *Ann. Report ICME,* Vol. 4, 139 (1981).
34. Karube, I., S. Kuriyama, T. Matsunaga, and S. Suzuki, *Biotechnol. Bioeng. 22,* 847 (1980).
35. Karube, I., T. Matsunaga, T. Otsuka, H. Kayano, and S. Suzuki, *Biochim. Biophys. Acta 637,* 490 (1981).
36. Karube, I., T. Matsunaga, S. Tsuru, and S. Suzuki, *Biochim. Biophys. Acta 444,* 338 (1976).
37. Kayano, H., T. Matsunaga, I. Karube, and S. Suzuki, *Biochim. Biophys. Acta 638,* 80 (1981).
38. Kennedy, J.F., S.A. Barker, and J.D. Humphreys, *Nature 261,* 242 (1976).
39. Kennedy, J.F., J.D. Humphreys, S.A. Barker, and R.N. Greenshields, *Enzyme Microb. Technol. 2,* 209 (1980).
40. Kierstan, M. and C. Bucke, *Biotechnol. Bioeng. 19,* 387 (1977).
41. Kokubu, T., I. Karube, and S. Suzuki, *Eur. J. Appl. Microbiol. Biotechnol. 5,* 233 (1978).
42. Kokubu, T., I. Karube, and S. Suzuki, *Biotechnol. Bioeng. 23,* 29 (1981).
43. Koshcheenko, K.A., G.V. Sukhodolskaya, V.S. Tyurin, and G.K. Skryabin, *Eur. J. Appl. Microbiol. Biotechnol. 12,* 161 (1981).
44. Krouwel, P.G., W.F.M. van der Laan, and N.W.F. Kossen, *Biotechnol. Lett. 2,* 253 (1980).
45. Larsson, P.O. and K. Mosbach, *Biotechnol. Lett. 1,* 501 (1979).
46. Larsson, P.O., S. Ohlson, and K. Mosbach, *Nature 263,* 796 (1976).
47. Lee, K.J., D.E. Tribe, and P.L. Rogers, *Biotechnol. Lett. 1,* 421 (1979).

48. Linko, P., In "Food Process Engineering" Enzyme Engineering in Food Processing (Linko, P. and Larinkari, J., eds.), Vol. 2, pp. 27, Applied Science Publishers London. (1980).
49. Linko, P., In "Advances in Biotechnol" (M. Moo-Young, C.W. Robinson, and C. Vezina, eds.), Vol. 1 pp. 711, Pergamon Press, NY (1981).
50. Linko, Y.Y., H. Jalanka, and P. Linko, *Biotechnol. Lett. 3,* 263 (1981).
51. Linko, Y.Y., and P. Linko, *Biotechnol. Lett. 3,* 21 (1981).
52. Maleszka, R., I.A. Veliky, and H. Schneider, *Biotechnol. Lett. 3,* 415 (1981).
53. Margaritis, A., P.K. Bajpai, and J.B. Wallace, *Biotechnol. Lett. 3,* 613 (1981).
54. Margaritis, A. and P. Bajpai, *Biotechnol. Lett. 3,* 679 (1981).
55. Matsunaga, R., I. Karube, and S. Suzuki, *Biotechnol. Bioeng. 22,* 2607 (1980).
56. Moo-Young, M., J. Lamptey, and C.W. Robinson, *Biotechnol. Lett. 2,* 541 (1980).
57. Morikawa, Y., I. Karube, and S. Suzuki, *Biotechnol. Bioeng. 21,* 261 (1979).
58. Morikawa, Y., I. Karube, and S. Suzuki, *Eur. J. Appl. Microbiol. Biotechnol. 10,* 23 (1980).
59. Morikawa, Y., I. Karube, and S. Suzuki, *Biotechnol Bioeng. 22,* 1015 (1980).
60. Morikawa, Y., K. Ochiai, I. Karube, and S. Suzuki, *Antimicrob. Agents Chemother. 15,* 126 (1979).
61. Navarro, J.M. and G. Durand, *Eur. J. Appl. Microbiol. 4,* 234 (1977).
62. Ohlson, S., S. Flygare, P.O. Larsson, and K. Mosbach, *Eur. J. Appl. Microbiol. Biotechnol. 10,* 1 (1980).
63. Ohlson, S., P.O. Larsson, and K. Mosbach, *Eur. J. Appl. Microbiol. Biotechnol. 7,* 103 (1979).
64. Ohlson, S., P.O. Larsson, and K. Mosbach, *Biotechnol. Bioeng. 20,* 1267 (1978).
65. Robinson, C.W., M. Moo-Young, and J. Lampty, *Advances in Biotechnol.* Vol. 2, 105 (1980).
66. Rogers, P.L., K.J. Lee, and D.E. Tribe, *Biotechnol. Lett. 1,* 165 (1979).
67. Ryu, Y.W., J.M. Navarro, and G. Durand, *Eur. J. Appl. Microbiol. Biotechnol. 15,* 1 (1982).
68. Sawada, H., S. Kinoshita, T. Yoshida, and H. Taguchi, *J. Ferment. Technol. 59,* 111 (1981).
69. Shinmyo, A., H. Kimura, and H. Okada, *Eur. J. Appl. Microbiol. Biotechnol. 14,* 7 (1982).
70. Shiotani, T., and T. Yamane, *Eur. J. Appl. Microbiol. Biotechnol. 13,* 96 (1981).

71. Siess, M.H. and C. Divies, *Eur. J. Appl. Microbiol. Biotechnol. 12,* 10 (1981).
72. Sitton, O.C. and J.L. Gaddy, *Biotechnol. Bioeng. 22,* 1735 (1980).
73. Sitton, O.C., G.C. Magruder, N.L. Book, and J.L. Gaddy, *Biotechnol. Bioeng. Symp. No. 10,* 213 (1980).
74. Slowinski, W. and S.E. Charm, *Biotechnol. Bioeng. 15,* 973 (1973).
75. Sonomoto, K., M.M. Hog, A. Tanaka, and S. Fukui, *J. Ferment Technol. 59,* 465 (1981).
76. Stenroos, S.L., Y.Y. Linko, and P. Linko, *Biotechnol. Lett. 4,* 159 (1982).
77. Takata, I., T. Tosa, and I. Chibata, *J. Solid-Phase Biochem. 2,* 225 (1977).
78. Tosa, T., T. Sato, T. Mori, K. Yamamoto, I. Takata, Y. Nishida, and I. Chibata, *Biotechnol. Bioeng., 21,* 1697 (1979).
79. Tyagi, R.D. and T.K. Ghose, *Biotechnol. Bioeng. 24,* 781 (1982).
80. Vaija, J., Y.Y. Linko, and P. Linko, *Appl. Biochem. Biotechnol. 7,* 51 (1982).
81. Veliky, I.A., and R.E. Williams, *Biotechnol. Lett. 3,* 275 (1981).
82. Wada, M., J. Kato, and I. Chibata, *Eur. J. Appl. Microbiol. Biotechnol. 8,* 241 (1979).
83. Wada, M., J. Kato, and I. Chibata, *Eur. J. Appl. Microbiol. Biotechnol. 10,* 275 (1980).
84. Wada, M., J. Kato, and I. Chibata, *J. Ferment. Technol. 58,* 327 (1980).
85. Wada, M., T. Uchida, J. Kato, and I. Chibata, *Biotechnol. Bioeng. 22,* 1175 (1980).
86. Weetall, H.H., and L.O. Krampitz, *J. Solid-Phase Biochem. 5,* 115 (1980).
87. White, F.H. and A.D. Portno, *J. Inst. Brew. London 84,* 228 (1978).
88. Williams, D. and D.M. Munnecke, *Biotechnol. Bioeng. 23,* 1813 (1981).
89. Yongsmith, B., A. Tanaka, and S. Fukui, *Ann. Report ICME Vol. 3,* 263 (1980).

CHAPTER 2

BIOENERGY PRODUCTION AND POLLUTION CONTROL WITH IMMOBILIZED MICROBES

Ralph A. Messing

Corning Glass Works
Corning, New York

I. INTRODUCTION

An ever increasing population with the concomitant decrease in available land resources necessitates more efficient waste treatment processes and facilities. Previously, high population density communities disposed of their wastes in rivers, lakes, and the seas. This can no longer be tolerated, because these bodies of water have become so polluted that the wild life has become endangered and thus our food supply has diminished. Another approach to waste disposal in the past has been landfills and surface-spreading. These techniques are becoming obsolete due to the diminishing land availability and the resultant ground pollution which subsequently results in water and air pollution.

Another crisis became apparent within the same decade that our waste treatment problems surfaced. The diminishing supply of fossil energy was impressed upon the industrial nations by OPEC in the petroleum crisis of 1974. This induced the search for alternative, renewable, energy sources. One of those potential renewable energy stocks was waste. Thus, these two problems are linked.

Traditionally, the preferred treatment of biodegradable wastes in the United States has been the aerobic processes. Energy intensive procedures, vigorous agitation and the delivery of massive quantities of air or oxygen, are required for these techniques. There is no yield of an energy product in the aerobic processes. However, massive quantities of cells or biomass are evolved under these conditions. The advantage of aerobic waste treatment is that high concentrations of biodegradable wastes may be treated rapidly with only a few hours

ISBN 0-12-040306-4

of retention. These processes are readily adapted to continuous procedures. On the other hand, anaerobic waste treatment has been primarily a batch process requiring residence times of approximately 30 days. These extended treatment periods are required because methanobacter exhibit very low growth rates. As contrasted to the aerobic treatment, the anaerobic process evolves relatively low cell mass.

Although the anaerobic treatment of waste yielded an energy product, methane, the gas produced by this process contained approximately 50% carbon dioxide and corrosive quantities of hydrogen sulfide. It was necessary to remove the hydrogen sulfide and carbon dioxide by an additional scrubbing process in order to obtain the fuel value of methane. Because of the additional gas processing the cost of producing pipe line quality methane was high. Since the only advantages of anaerobic waste treatment was the production of an energy product which was rather expensive, and the very modest quantity of cell mass, the selection of aerobic treatment for waste in the past was logical. In addition, the energy costs for agitation and air delivery were very inexpensive prior to 1975; therefore, neither energy production nor energy utilization were considered to be major factors.

The anaerobic treatment of waste, although modestly fuel efficient, did not represent a feasible approach to treating large volumes of waste solutions by continuous processing until the advent of the anaerobic filter, which is essentially a vertical column containing rocks or other material for the retention of solids in the waste stream (1). An upward flow of fluid through the rocks is employed. The gas stream is separated from the fluid after exiting the column. The retention time for anaerobic treatment of waste was reduced from 30 days to less than a week and the need for agitation was eliminated with the application of the anaerobic filter. A major contribution of this technique was the film of microbes responsible for degrading the waste which was formed on the outer surface of the rocks. These immobilized microbes were retained in sufficient quantities to allow a significant reduction in processing time. With this reduced retention time and the ability to process on a continuous basis, anaerobic waste treatment becomes more competitive with the aerobic processes. Now, if one considers the increase in energy costs, it is not surprising to see the emphasis shift from aerobic to anaerobic waste treatment.

There still remain three problems that one encounters in the application of the anaerobic filter. The first of these problems is that of achieving acceptable reductions of pollutants when very low pollutant concentrations are delivered to the system. The second problem is that of processing very high concentrations of pollutants without adversely affecting the system. The final problem is that of the additional costs of scrubbing the evolved gas to produce a satisfactory fuel.

Three different modifications of the anaerobic filter appear to offer solutions to these three problems.

II. A POTENTIAL SOLUTION TO ACHIEVING VERY LOW LEVELS OF POLLUTANTS

Perhaps the most drastic departure from the original anaerobic filter is that of the anaerobic attached film expanded bed (AAFEB), Figure 1. The apparatus is a continuous upward fed recycle reactor containing porous, water-insoluble, support particles of between 5 and 200 microns in diameter for microbial immobilization. Although this reactor can be generally classified as a fluidized-bed-reactor the velocity of upward flow is adjusted at a sufficiently low rate to expand the bed only by about 20%. Two pumps are employed, one to deliver the waste material and the second to recycle a portion of the partially treated effluent from the top of the bed to the bottom of the bed. Although a variety of materials have been utilized as a microbial film support by Jewell, et al.,(2) for investigating aerobic waste treatment, many of Jewell's more recent studies have been performed with porous alumina and polyvinyl chloride plastic materials (3,4). Since the porous alumina contained pores between 100 and 400 Å the microbes could not gain access to the internal structure; therefore, the biological activity was solely due to the film formed about the particle and the free cells in the fluid.

Jewell, et al., have reported (3,4) that domestic sewage containing between 50 and 600 mg/l COD (chemical oxygen demand) delivered to the AAFEB system requires as little as 30 minutes of hydraulic retention to achieve an effluent containing less than 40 mg/l of COD. They demonstrated that the process was effective in the temperature range of 10° through 35°C, and the system was remarkably stable and resistant to shock loadings.

Although two pumps are utilized in this system, one for expanding the bed and the other to deliver the feed, the energy requirement is relatively modest since the flow required to achieve a 20% expansion of the bed does not require a major input of energy. The advantage of expanding the bed appears to be the elimination of clogging problems while accumulating significant quantities of cells on the surface of the particles. Diffusion problems are minimized by the employment of these very small particles, diameters of 20-30 microns, for the formation of the film of immobilized microbes.

Jewell, et al., (3,4) have demonstrated that this system can be utilized with a variety of wastes. They have processed dilute cow manure, glucose solutions, particulate cellulose (5), in addition to domestic sewage. The start-up process requires several weeks after which the reactor begins to produce gas with a methane content of greater than 50% by volume.

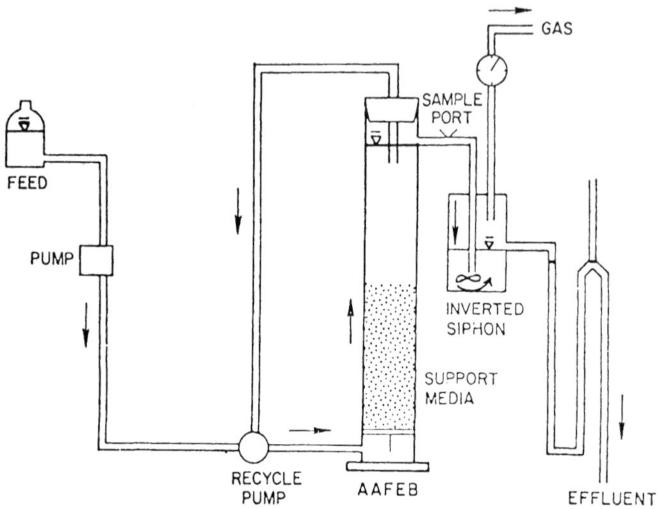

Figure 1. Anaerobic attached film expanded red apparatus (AAFEB) (reproduced from Reference (5) with permission of W. J. Jewell, J. W. Morris, and J. D. Wolszon.

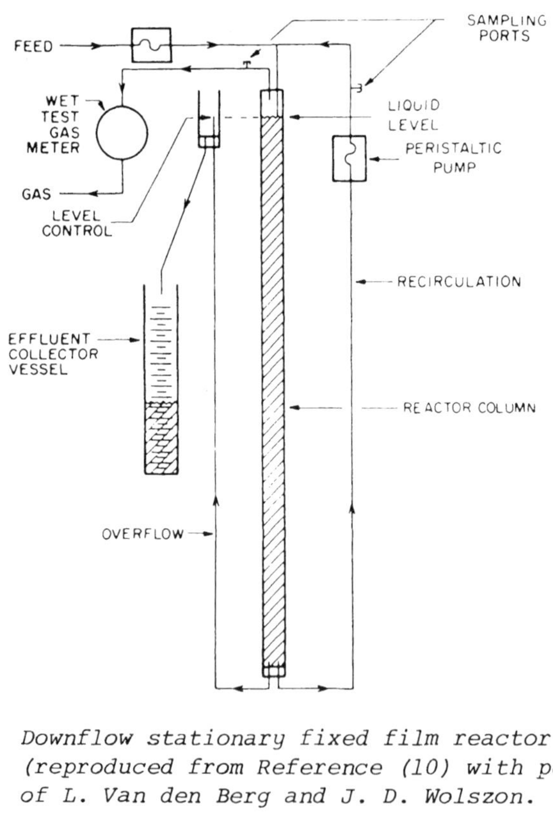

Figure 2. Downflow stationary fixed film reactor (DSFF) (reproduced from Reference (10) with permission of L. Van den Berg and J. D. Wolszon.

It is clear from the studies performed with the anaerobic attached film expanded bed equipment that this process is certainly competitive with aerobic processes with respect to retention times and achieving very low levels of COD with dilute waste streams. The energy requirements and plant size should certainly be significantly less than that required by aerobic processes. The AAFEB may prove to be even more attractive after full pilot and plant studies are completed.

III. AN AEROBIC HIGH-RATE PROCESS FOR CONCENTRATED WASTES

A modification of the anaerobic filter providing a means of processing concentrated wastes, up to 130,000 mgs of COD per liter, is that of the downflow, stationary, fixed, film (DSFF) reactor, Figure 2, developed by van den Berg, et al., (6,7,8,9,10,11). The equipment consists of a column or tank containing support material for immobilizing microbes in a fixed film configuration, an overflow level control, a pump for recycling and a pump for delivering feed. The feed solutions are delivered from the top of the reactor and the processed solution is removed from the bottom of the reactor. Recirculation is accomplished, if desired, from the bottom of the reactor to the top. The gas is removed from the top of the reactor.

Van den Berg, et al., (6,7) examined a variety of materials as supports for stationary fixed films in this reactor. Some of the materials evaluated were polyvinyl chloride sheets, potters clay, red drain tile clay, needle punched polyester and glass. The red drain tile proved to be the best support in terms of, not only, start-up, but also, in steady-state performance. The needle punched polyester proved to be the second best performer.

Although provision is made by recycling of the effluent, the DSFF reactors were able to produce methane at high rates with and without the recycling of the effluents from concentrated wastes in the range of 4,000 to 140,000 mgs COD per liter. Recycling, however, is beneficial when wastes such as pear peeling and piggery wastes, which contain easily settleable suspended solids, are employed. Recycling rates are generally 0-4 times the feed rate.

Although the DSFF system can operate from about 10°C through 60°C, no advantage could be identified by operating a thermophilic, 55°C, digestion over mesophilic conditions (8). Most of the studies have been performed at 35°C which appears to be the optimum temperature for most of the wastes investigated (11).

The National Research Council of Canada researchers (8,9, 10) report that the methane content of the gas evolved from this system is in the range of 56 to 61%; however, with heat treated sewage digested sludge liquor the gas evolved was approximately 73% methane (10).

The DSFF system has proved to be remarkably stable and resistant to shock loadings. Change-overs from one waste to another could be made with little or no loss in the rate of methane production. This was also true when changes from a low of less than 20 ppm to a high of 4,000 ppm sodium content in changing from a food processing to a chemical industry or seqage waste. When the COD was increased by 9-fold at temperatures of 35°C the reactors were back to normal within 24 to 48 hours. Reactors that were shut down for a few months required only 48 hours to recover performance after the initiation of feeding.

The outstanding attributes of this system can best be demonstrated by COD reductions with respect to hydraulic retention times. Bean blanching wastes containing 10,000 mgs COD per liter required only 0.85 days for an 86% reduction in COD. Piggery waste containing between 27,000 and 51,000 mgs COD per liter required 8 days of hydraulic retention time to reduce the COD by 70%. Rum stillage containing between 50,000 and 70,000 mgs COD per liter required 4.5 days to reduce the COD by 57%. Pear peeling wastes containing 110,000 to 140,000 mgs COD per liter required 11.8 days for a 64% reduction in COD.

The solution to processing high concentration wastes by a continuous anaerobic process which does not appear to be readily upset by shock loads may be the downflow, stationary, fixed film reactor. It is possible that this system may be capable of handling concentrated pollutants equivalent to those which may be processed by aerobic waste treatment. It is, also, possible that with additional modifications and effort, the retention times for concentrated wastes may be substantially reduced such that they will approach those of the aerobic processes. The marked advantage of this system is the positive production of a fuel, methane, while substantially reducing the pollutants.

IV. A PROCESS FOR PRODUCING HIGH BTU GAS WITHOUT SCRUBBING

A departure from the anaerobic filter is the two-stage system utilizing controlled-pore ceramics to immobilize the microbial cells, Figure 3, (12,13). This system utilizes an upward flow delivery with no recirculation. The temperature is individually controlled for each stage. The first stage, a hydrolytic-redox stage, containing predominately acid-formers is operated at a higher temperature than the second or anaerobic stage which contains predominately *Methanobacter*. The first stage is assembled vertically while the second stage is assembled horizontally. The fluid level of the second stage is maintained by a level controller which activates a gas pump to remove gas when the fluid level falls below the half full mark. A check valve is inserted in the effluent pipe to maintain pressure on the anaerobic stage.

Rather than employing a fixed, surface film on a support material, this system utilizes the three dimensional structure of controlled pore ceramics. In other words, cells are immobilized on the internal walls of the support material. Early studies of the immobilization of microbes in controlled pore ceramics (14) indicated that between 10^8 and 10^9 cells per gram of ceramics could be immobilized if the pore diameters were in the range of 1 through 5 times the major dimension of the cell. The rationale for this finding is based on Figure 4. If cells immobilize on their minor dimension on opposite walls of pores and they reproduce by fission, then at least 4 cell lengths would be required for reproduction. Now if another cell required exit another cell length is necessary, thus, 5 cell lengths would be required for continuous reproduction.

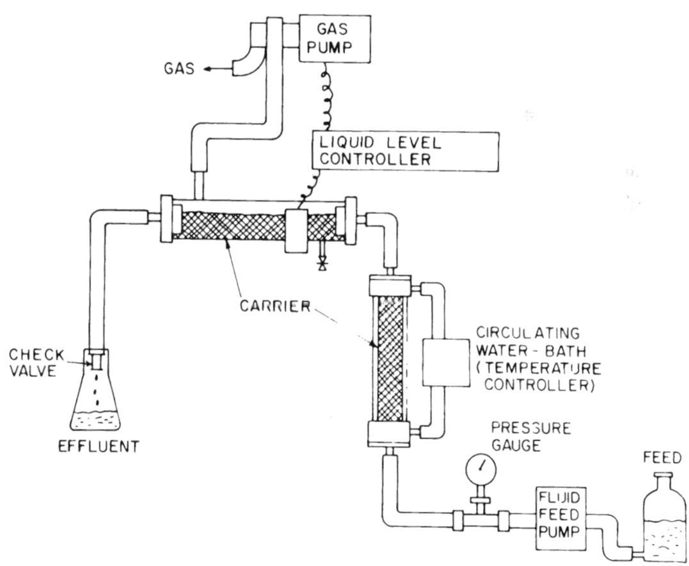

Figure 3. Two-stage waste processor (reproduced from Reference (12) with permission of John Wiley and Sons, Inc.

This hypothesis was supported by scanning electron micrographs, Figure 5, of cells immobilized on controlled pore ceramics. These cells were closely packed and were in fact immobilized on their minor dimension. A reasonable explanation for the immobilization of these cells on their minor dimension is that maximal cell surface remains available for transfer of nutrients and removal of waste products. It may be also noted in Fig. 5 that many of the cells are reproducing by fission while remaining immobilized on the carrier surface. The immobilization of microbes that reproduce by other than fission was reported by Messing, et al. (15,16).

When sewage adjusted to pH 8.6 - 8.9 containing 800 - 2600 mgs COD per liter was continuously delivered to the reactor at residence times of 2-5 hours, the COD was reduced by 70 - 88%. Approximately 45% of the total carbon delivered to the system was converted to methane. The gas delivered from the system contained greater than 90% methane with less than 5% carbon dioxide, high BTU gas.

The efficient performance of this system is due to: 1) the optimized pore dimensions for both high cell accumulation and prevention of washout; 2) the more complete conversion of carbon dioxide due to the elevation of pH to above 8 and the pressurization of the reactor by inserting a check valve in the effluent stream; and 3) the gas-fluid interface maximization for removal of methane by the horizonal assembly of the anaerobic stage and maintaining the level of about 50% fluid depth. In addition, the lower temperature of the second state contributes to the retention of the carbon dioxide in the fluid phase. Since the microbes are predominantly located in the fluid phase, in order to achieve a more complete conversion of carbon dioxide to methane, it is necessary to maintain that gas in the liquid. Thus, the three factors (high pH, pressurization, low temperature) contribute to the retention of carbon dioxide in solution for further conversion to methane.

As with the previously described systems this two-stage reactor has been found to be remarkably stable. Some of these systems have been operated continuously for periods of three years without plugging problems in the reactor and without a diminishing performance. The immobilized microbe population was not adversely affected by dramatic excursions of feed concentrations of compositions. Performance recovery times for those excursions were approximately 1 to 2 days. Performing reactors have been turned off for periods of 6 - 7 weeks and complete performance recovery was noted within 24 hours of the reinitiation of feed. One remarkable attribute of immobilization within pore structure is that the ceramic containing the immobilized *Methanobacter* can be removed from the anaerobic

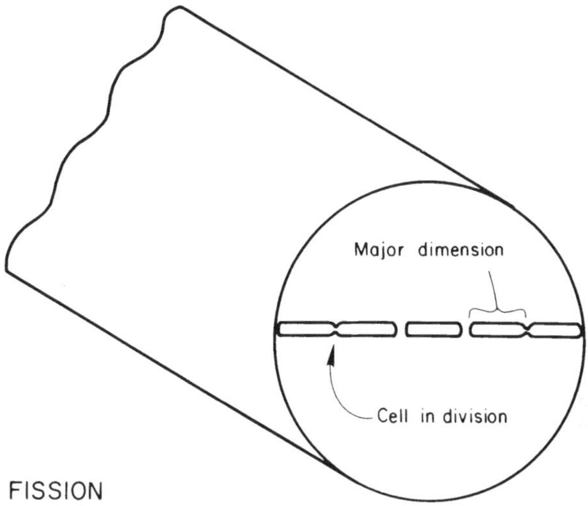

Figure 4. *Diagramatic representation of reproduction of cells within an optimized pore (reproduced from Reference (12) with permission of John Wiley and Sons, Inc.*

Figure 5. *SEM of cells from sewage immobilized on controlled-pore alumina (reproduced from Reference (12) with permission of John Wiley and Sons, Inc.*

stage and exposed to the air for several hours, replaced in the reactor and performance recovery will be noted within less than one-half day. The probable explanation for this stability is that the pore environment remains anaerobic due to the fact that there is nutrient material within the pore structure and the cells continue to metabolize even when the ceramic structure is exposed to oxygen for a limited time.

The performance of this system with the variety of feeds including manures, sulfite-pulp-waste, pumpkin waste, whey and stillage have been examined. These wastes have contained between 1000 and 60,000 mgs COD per liter. Evidence for promising performances has been collected.

This system appears to offer the promise for delivering pipe line quality gas without scrubbing while dramatically reducing pollutants.

V. CONCLUSION

Three anaerobic systems have been described for producing energy while dramatically reducing retention times. These systems certainly appear to be competitive with aerobic waste treatment with respect to retention time and waste concentration. The positive aspect of energy production certainly warrants both pilot plant and full scale evaluation of these innovative systems as future high technology waste treatment approaches. If the performance of the scaled-up versions of these units are comparable to the laboratory models, then major contributions will have been made to solving the increasing waste production from a rising population and for contributions to alternate energy.

REFERENCES

1. Young, J. C., P. L. McCarty, *J. Water Poll. Control Fed. 41, R160* (1969).
2. Jewell, W. J., R. J. Cummings, "An optimized Biological Waste Treatment Process for Oxygen Utilization,"
3. Jewell, W. J., U.S. Patent #4,284,508 (August 18, 1981).
4. Jewell, W. J., M. S. Switzenbaum, J. W. Morris, *J. Water Poll. Control Fed. 53,* 482 (1981).
5. Morris, J. W., W. J. Jewell, 36th Annual Purdue Industrial Waste Conference, Purdue University, West Lafayette, IN (May 13, 1981).
6. van den Berg, L., K. J. Kennedy, "Effect of Substrate Composition on Methane Production Rates of Downflow Stationary Fixed Film Reactors," Symp. on Energy from Biomass and Wastes VI, Inst. of Gas Tech., Lake Buena Vista, FL, (January 26, 1982). Also Issued as NRCC #19826.

7. van den Berg, L., K. J. Kennedy, *Biotech. Letters 3,* 165 (1981).
8. Kennedy, K. J., L. van den Berg, *Biotech. Letters 4,* 171 (1982).
9. Stevens, T. G., L. van den Berg, "Anaerobic Treatment of Food Processing Wastes Using a Fixed Film Reactor," 36th Annual Purdue Industrial Waste Conference, Purdue University, West Lafayette, IN (May 12-14, 1981).
10. Kennedy, K. J., L. van den Berg, "Effect of Temperature and Overloading on the Performance of Anaerobic Fixed Film Reactors," 36th Annual Purdue Industrial Waste Conference, Purdue University, West Lafayette, IN (May 12-14, 1981).
11. van den Berg, L., K. J. Kennedy, "Potential Use of Anaerobic Processes for Industrial Waste Treatment, Seminar on Anaerobic Waste Water Treatment and Energy Recovery, Pittsburgh, PA (November 3-4, 1981). Also Issued as NRCC #19727.
12. Messing, R. A., *Biotech. Bioeng., 24,* 1115 (1982).
13. Messing, R. A., U.S. Patent #4,321,141 (March 23, 1982).
14. Messing, R. A., R. A. Oppermann, *Biotech. Bioeng., 21,* 49 (1979).
15. Messing, R. A., R. A. Oppermann, F. B. Kolot, *Biotech. Bioeng. 21,* 59 (1979).
16. Messing, R. A., R. A. Oppermann, F. B. Kolot, Am. Chem. Soc. Symp. Ser. *106*, 13 (1979).

CHAPTER 3

LARGE-SCALE CELL CULTURE TECHNOLOGY

William R. Tolbert
Joseph Feder

Monsanto Company
St. Louis, Missouri

I. *INTRODUCTION*

Large-scale growth of vertebrate cells *in vitro* offers a method for the production of biomolecules not available from other sources. This has focused the attention of researchers on the development of mass culture systems which offer opportunities for large-scale growth of such cells for production of desired biomolecules. Although the recent explosion of recombinant DNA methodology offers an approach for the production of biomolecules derived from these cell types using conventional large-scale microbial fermentation, this new technology is useful only after the identification and the availability of a specific protein and its gene. The search for growth regulating molecules produced by vertebrate cells is in its early stages of development. Often these types of molecules are present in cells only in minute quantities, making their discovery, isolation and characterization difficult unless one has available processes for the culture of large quantities of the specific producing vertebrate cells. Thus even if one plans the use of recombinant methods for the final production of a given molecule, the discovery and prerequisite description of these molecules require an available method for their production by vertebrate cells in *in vitro* systems. Molecules which require post-translational processing,glycoproteins and lipoproteins presently are not amenable to biosynthesis via recombinant methods. Large-scale cell culture processes are necessary for the production of such molecules. Often biological activity is represented by several distinct molecular

ISBN 0-12-040306-4

forms, all of which may be concurrently produced in a single vertebrate culture system as in interferon production. Recombinant methods, on the other hand, allow the production of a single molecule at a time.

The purpose of this chapter is to review the literature on the vertebrate culture methods currently available and in use for the production of biologically important molecules with a particular focus on large-scale cell culture technology developed in our laboratory. An integrated approach to large-scale culture including equipment technology facility design and day-to-day operation is presented. This should be useful both for the establishment of large-scale cell culture facilities dedicated to the production of biomolecules for clinical or commercial application, as well as for the research laboratory. It is becoming more and more apparent that there are populations of regulatory biomolecules, specific for particular cells, which probably play an important role in their growth and development. Included among these are nerve growth factor, fibroblast growth factor, epithelial growth factor, angiogenic factor, the tumor transforming factors of Todaro, interferons, etc. (1-4). These undoubtedly represent only a small number of such regulator molecules yet to be discovered. Cell culture technology will play an important role in research efforts for isolation and characterization of such molecules.

II. Review of Existing Technology

The historical, commercial product for vertebrate cell culture has been vaccines produced by the growth of various animal and human viruses in cells. Most of this work has been with primary cultures, e.g., chicken embryo fibroblasts or monkey kidney cells grown in roller bottles. Almost all non-lymphoid primary cultures are anchorage dependent and require surface attachment of cells. Roller bottle techniques are highly labor intensive and increase the probability of contamination due to the many individual steps that must be accomplished (5). Use of primary cultures carries the additional hazard of inclusion of unrecognized viruses in vaccines for human injection such as avian leukosis viruses found in experimental measles vaccine and SV_{40} simian virus widely distributed in licensed and experimental polio vaccines. Traditionally primary cultures have been mandated by the U.S. Food and Drug Administration, although there has seemed to be a recent change in this attitude (6). This has stimulated research into use of both diploid and non-diploid cell lines in the production of biologicals (7). Much of this effort has been outside of the United States where restrictions have been less stringent. Larger scale, non-roller bottle systems have been used in vaccine production for foot-and-mouth disease, measles, rabies, polio, etc. to the level of several thousand liters of submerged culture (8-16). More recently, large-scale systems also have been used for the production of human lymphoblastoid interferon (17-22). Wellcome Research Laboratories have up to 2,000 liter systems for this lymphoblastoid production (23). There has been much interest in many other products that may be obtained from culture of vertebrate cells including tumor antigens, hormones, growth factors, lymphokines, various enzymes, and yet-to-be-discovered regulators of growth differentiation and function. Few of these products have progressed to the large-scale production levels of vaccines and interferon.

A. Anchorage-Independent Growth Systems

Vertebrate cells exhibit either anchorage independent or dependent growth *in vitro*. All non-lymphoid primary cultures or diploid lines and many heteroploid lines require surface attachment for proliferation while most lymphoid derived cells and some heteroploid and transformed cell lines can be adapted to single cell suspension. In addition, larger numbers of non-diploid lines can grow as aggregates in suspension (24). The design of large-scale culture systems is dictated primarily by the growth characteristics of the specific vertebrate cell. The traditional approach to scale-up of vertebrate suspension culture has been to employ systems designed for fermentation of microorganisms. These systems have been most applicable to cells which grow singly in submerged culture such as BHK, baby hamster kidney; L cells, mouse fibroblast; HeLa, human cervical carcinoma; Namalva, human Burkitt's lymphoma and others. One of the most widely used of these fermentors was developed by van Hemert at Bilthoven, The Netherlands. This "Bilthoven unit" consists of a control system and moderate-to-small fermentation vessels from one to one hundred liters and has been used both for anchorage independent and dependent cells (8,15,25-26). Modification of standard New Brunswick Scientific Company fermentors by use of bottom mounted, off-centered, magnetically driven, marine impellors has also been used (27-28). In addition to fermentors with conventional turbine or marine impellor agitators, top or bottom mounted, the vibromixer agitation system has been found particularly suited to vertebrate cell suspension systems (10,29-32). This agitation system consists of one or more discs containing conical holes attached to a vertical shaft. The shaft and discs vibrate vertically with an amplitude of 1-3mm at 60 cycles/sec. The shape of the conical holes causes vertical agitation of the suspension rather than the horizontal agitation produced by turbine blades. This system has the added advantage of an absolute static seal with vibration transmitted through a flexible diaphragm. In all of these fermentation systems, means are provided for sterilization of the culture container, temperature control in the range of 35 to 38°C, aseptic addition of inoculum and nutrient media, aseptic sampling of the cell suspension and aseptic harvest of cell suspension. In addition, systems also are provided for addition of various gases, usually 5 to 10% CO_2 in air, monitoring and control of pH, monitoring and control of dissolved oxygen, etc. Some attempts also have been made to operate such systems in a chemostat mode or to control the various parameters with a computer (8,11,33-35). In general, such attempts have been less than successful. Chemostat operation of vertebrate cell cultures maintains the cells at a relatively low concentration and washout often occurs. While many culture parameters can be monitored and partially controlled, their relationship to cell proliferation and product secretion is often quite complex. Although these conventional systems have proved quite successful for a select group of cell lines, in general they have not addressed the specific requirements for *in vitro* growth of vertebrate cells. These cells are much larger and much more fragile due to lack of a cell wall and have a much slower proliferation and metabolic rate. While microorganisms may double on the order of minutes, vertebrate cells require hours and many human cells have 30 to 100 hour doubling times. Agitation systems must provide gentle mixing of the cells at low shear rates and methods for prevention of contamination by adventurous organisms must be more stringent. In all vertebrate cell culture and especially in large-scale systems, the major problem is detection and prevention of contamination by microorganisms (10,28,36-37). Almost without exception significant levels of antibiotics have been required in large-scale systems to maintain sterility. These antibiotics not only can affect the growth of the vertebrate cells and products produced, but they also make detection of chronic, low level contaminants extremely difficult. Reduction and elimination of dependence upon antibiotics should be a primary goal for design of new systems for vertebrate cell culture.

B. Anchorage-Dependent Growth Systems

As compared to cells that grow suspended freely in a liquid medium, anchorage-dependent cells present much more difficult problems for scale-up. These cells will only grow when attached and flattened out on a solid or semi-solid substrate which is either negatively or positively charged. Glass surfaces or plastic which has been exposed to a high voltage discharge to produce a wettable charged surface have been traditionally used for growth of anchorage-dependent cells. Multiwell plates, petri dishes, flasks of varying sizes and roller bottles with surface areas up to 1,600 cm^2 have been used for growth of these cells. Maximum number of cells per vessel depends on the maximum surface density for each cell type but in general is in the range of $3x10^7$-$3x10^8$ cells in the largest vessels. Cell densities for suspension grown cells also vary for each cell type, but are in the range of 1-$4x10^9$ cells per liter of medium.

1. *Rotating Surface Systems.* Practical process scale-up of anchorage-dependent cells requires the development of systems which provide large surface area to volume ratios. The surface area necessary to provide the equivalent number of cells obtained from 100 liters of anchorage independent cell suspension is approximately 85 m^2. This corresponds to a thousand 850 cm^2 plastic roller bottles. The multiple handling of each bottle for inoculation, feeding and harvest makes scale-up problematic. Nevertheless, brute force can be used for scale-up in such a system. Benton et al., report production of 25,000 liters of mouse mammary tumor virus from cells grown in 490 cm^2 plastic roller bottles (38). Many ingenious methods have been devised to circumvent the brute force roller bottle method for scale-up of anchorage-dependent cell growth. One method to increase surface areas is to provide a roller bottle that contains a spiral plastic sheet extending the length of the bottle and spiraling from the center out to the wall. This bottle is filled so the entire spiral surface is wet with nutrient medium allowing cell growth on both surfaces while the vessel is rotated on a conventional roller deck (39). Bellco Glass, Inc., Vineland, NJ, has developed their own enhanced surface area system, the Bellco-Corbeil TM culture system. In this system a cylindrical vessel of a size similar to a large roller bottle is filled completely with small glass tubing. Each piece of glass tubing is provided with spacers so that medium and atmosphere can freely circulate around and through each of the tubes. In this system also, medium is filled to allow nutrients to reach all glass surfaces during rotation of the vessel on a conventional roller deck. Such vessels can be connected to multiple automatic filling and refeeding systems (40). A much larger version of the same idea recently was put on the market by Chemap A.G., Mannidorf, Switzerland. This system is called the GYROGEN monolayer tissue culture fermentor. It consists of a large, in-place steam sterilizable vessel which is packed with many individual glass tubes. The GYROGEN is partially filled with medium and the packed tubes rotate parallel to their long axis through and out of the medium to allow nutrients to reach all of the glass surfaces. This system comes in versions as large as 34 $meters^2$ of surface area which corresponds to four hundred 850 cm^2 roller bottles.

2. *Multiplate Systems.* Surface area has also been increased by provision of numbers of flat surfaces in various arrangements for growth of attached cells. Flat rectangular trays, 600 cm^2 of growth area each, have been stacked up to ten layers and provided with means for filling and harvesting cells. This multitray unit is sold commercially by Vangard International, Nepture, NJ. Such multitray units can also be handled in a multiple filling harvesting setup (41). Multiple flat plates perpendicular to the axis of a rotating vessel have been used by Connaught Medical Research Laboratories for production of

polio virus (42). A much larger 200-liter static multiplate propagator having a total surface area of 21.5 m^2 also has been developed (43). Another flat plate arrangement has many parallel plates contained within a rectangular box and medium is perfused through the box with a flow parallel to the long dimension of the plates. This system has been scaled up to 10 m^2 and used for production of polio virus grown in MRC-5 diploid human fibroblasts (44-45).

3. *Enhanced Surface Area to Volume Ratio Systems.* A number of attempts have been made to significantly increase surface area/volume ratio for more practical scale-up of anchorage-dependent cells. Static perfusion systems have been used for large, *in vitro* growth of anchorage-dependent cells. Glass beads or helices have been packed into large cylindrical vessels with media perfused either downward from the top or upward from the bottom allowing growth of anchorage-dependent cells on the glass surfaces (46-50). Another device, the IL410 cell culture system, has been developed by Instrument Laboratories, Inc., Lexington, MA. In this system, gas permeable membranes form a long flat tube, like wide dialysis tubing, and are rolled on to a reel similar to that used for 35 mm motion picture film. Each layer of tubing is separated by a spacer allowing air to circulate freely while cells are grown on the inside surface of the tubing. Nutrient medium is pumped continually through the length of tubing, allowing significant increases in surface area available for cell growth. Many such reels may be contained in a single CO_2 incubator with all the required equipment for monitoring and control of various growth parameters. Growth characteristics of the Vero monkey kidney cell line on such gas-permeable membranes have been described (51). A completely different method for growth of anchorage dependent cell is use of small hollow fibers having diameters of several hundred microns and contained either in cylindrical bundles or spread out in a flat arrangement. The cells grow on the exterior surface of the fibers and the medium is either perfused through the lumen of the fiber or around the exterior of the fiber with air or oxygen through the lumen (52-56). While all of the preceding systems and devices were designed for growth of anchorage dependent mammalian cells at larger scale than conveniently possible in roller bottles, they have not found widespread industrial use. The hollow fiber systems, however, may hold more significant future application.

4. *Microcarrier Systems.* The system which has had widespread industrial application and evaluation is culture of anchorage dependent cells on microcarriers first developed by van Wezel in 1967 (57). Small solid or semi-solid beads from one to several hundred microns in diameter are suspended in

nutrient medium and anchorage dependent cells, harvested from other growth systems allowed to attach to these beads. The cells can grow to cover the beads under conditions of very gently agitation in conventional submerged culture systems, spinner vessels or modified fermentors. The initial beads used by van Wezel as microcarriers were DEAE Sephadex A-50 anion exchange chromatography beads which had an exchange capacity of 3.5 meq/g A-50. Levine *et al.*, found that reduction of the charge density on the dextran beads to below 2 meq/g significantly improved growth of anchorage-dependent cells on these beads (58-59). These microcarriers called "Superbeads" are commercially available from Flow Laboratories, McLean, VA. Pharmacia Fine Chemicals (Uppsala, Sweden), the manufacturers of Sephadex chromatography materials, has developed their own microcarrier with reduced charge called "Cytodex 1" which has been extensively used (60). They also have developed two new beads called "Cytodex 2" and "Cytodex 3" which have additional properties intended to solve some of the problems associated with earlier microcarriers (61). The Cytodex 2 beads are substituted with only a very thin layer of positive charge in a shell surrounding the dextran bead and the Cytodex 3 beads are coated with a layer of denatured collagen. This latter collagen-coated bead allows subculture of certain diploid cells which cannot be removed from conventional beads, by enzymatic degradation of the collagen layer. Bio-Rad Laboratories, Inc., Richmond, CA, has developed a microcarrier bead with an acrylamide rather than a dextran base. This bead has significant advantages with respect to its physical stability (62-63). Non-spherical DEAE cellulose chromatography materials have also been used as microcarriers for some cell types (64). Treated plastic beads which do not swell or change their volume with ionic strength also have been used as microcarriers (65-67). One of these, Biosilon, is available commercially from Vangard International, Neptune, NJ. Several papers have appeared describing use and optimization of parameters for microcarrier culture of anchorage-dependent cells and application of this technology to production of vaccines and other products (37,68-75). Fifteen years after van Wezel's original paper, the microcarrier method for growth of anchorage-dependent cells is the only such system which has had significant large-scale application at an industrial level. One main advantage of the system is to allow continued use of modified fermentation vessels designed for microorganisms. Vessels in the range of hundreds of liters have been used for production of virus and vaccines and scale up an order-of-magnitude larger is possible (12-15,26). While the microcarrier growth system has many advantages, there still are basic difficulties in its application to many types of anchorage-dependent cells and there is a need for further development of specific equipment to exploit its full potential. The growth of vertebrate cells on microcarriers as opposed to growth of cells in free suspension adds significant complexity. The microcarrier system involves a two-phase growth where not only concentrations of various nutrients and cell waste products must be controlled to maintain the proper microenvironment at the cell surface, but the amount and distribution of surface area also becomes important. Inoculation of cells at too low of cell surface density can lead to poor cell growth and a significant percentage of beads without cells. Seeding at too high a cell surface density is inefficient and results in only a minimal increase in cell population. Attachment of cells to beads which may be a thousand-fold larger in volume and mass requires much more gentle agitation systems than for cells freely suspended. A major source of cell injury in the microcarrier systems is due to bead collisions. This has led to significant difficulty in maintaining the carriers in suspension without completely destroying the cells and has significantly limited the density of beads and therefore surface area available for cell growth. Consequently, more gentle agitation systems have been developed in an attempt to solve these problems. Bellco Glass, Inc., Vineland, NJ, has added a wide paddle to their conventional spinners and has decreased the minimum rpm of their magnetic stirrers. Techne Ltd., Cambridge, U.K., has developed a precessing, rigid pendulum type agitator for suspension of microcarriers in small vessels (76). For larger vessels, fluid lift pumps and wide screw-type impellors have been designed (77). In addition to agitation problems, the large amount of surface area that can be concentrated in relatively small volumes has led to rapid depletion of nutrients and build-up of toxic by-products. Continuously operating perfusion systems have been utilized to reduce this problem (8,69). The majority of large-scale applications of the microcarrier technology have been with non-diploid cell lines such as Vero, LLC-MK2 and CV-1. These cell types allow scale-up from small microcarrier cultures to larger microcarrier cultures, either by enzymatic release of cells from beads for reinoculation or by transfer of cells from confluent beads to fresh beads (5). However, with many diploid lines, particularly diploid human fibroblasts, this may not be possible. These cells, when attached to the beads, cannot be released by enzymatic treatment or scaled up by the bead transfer system (5,26,78,79). This inability to inoculate cells grown in a small microcarrier culture into a larger culture severely limits the usefulness of this technology for these cell types. Only moderate scale microcarrier systems can be grown if growth and aseptic harvesting of hundreds of roller bottles are required for inoculation. This has severely restricted production of human fibroblast interferon (HuIFN-β) which is produced by human diploid foreskin fibroblasts. Application of such new microcarriers as the collagen-coated CYTODEX 3 beads may reduce these problems.

III. CELL CULTURE AT MONSANTO

A. A New Approach to Large-Scale Culture

Because of enhanced interest in large-scale growth of mammalian cells and cell culture production of various biomolecules, Monsanto Company has developed over the last 6-10 years a broad based, new cell culture technology. This proprietary technology differs significantly in conception through practical application from the more conventional approaches to large-scale vertebrate cell culture. Systems have been developed for large-scale growth of both anchorage-dependent and independent cell lines and for products secreted into the medium, induced by special procedures, or retained within the cells. Over the past five years more than 200,000 liters of rodent and human cell suspensions have been grown and harvested without the use of antibiotics. Development of this new technology was based on the principle that new types of equipment should be designed to efficiently meet the requirements of antibiotic-free vertebrate cell culture rather than modification of conventional fermentation equipment. This required development of systems which would maintain absolute sterility to protect the cells with proliferation rates measured in hours from contaminants with proliferation rates measured in minutes and also required an extremely high confidence level for each operation to prevent loss of the expensive serum supplemented medium. The systems developed to meet these demands required that equipment be relatively simple, portable and autoclavable, and that the facility be designed with increasing levels of containment both for protection of personnel and isolation of cultures. The ability to maintain high flexibility was also extremely important as each cell type and individual cell line has its own specific requirements and idiosyncracies for optimum growth and product yield.

Figure 1 provides a floor diagram of the relatively small facility required for significant large-scale production. It consists of a kitchen area for preparation of glassware and equipment before sterilization and is connected to a nominally P3 biohazard contained area by an air shower and by a double-door vacuum autoclave. The contained area has 4° and 37°C environmental rooms; laminar flow biological safety cabinets, each within a closed cubicle; a separate media preparation area; and a 100-gallon, steam-heated kill tank, as well as standard laboratory microscopes, centrifuges, etc. Each work area, including the kitchen, has HEPA filters in supply and exhaust air ducts. All cell culture equipment is sterilized by extensive autoclaving with large vessels vented through sterile filters during cooling. These vessels are connected by flexible tubing to quick-connect couplers to allow all connections to be made within a laminar flow cabinet. The laboratory has an aggressive biosafety program with a center-wide biosafety committee consisting of a full-time biosafety coordinator, operating level managers and a representative of the Industrial Hygiene Department. Work is carried out under in-house biosafety guidelines strongly based on the 1981 *NIH CDC Proposed Biosafety Guidelines for Microbiological and Biomedical Laboratories*. All mammalian cell culture is classified at least level 2 and large scale at level 3 as recommended in discussions with Dr. W. Emmett Barkley, Director, Division of Safety, NIH, PHS, DHHS. All personnel entering a biohazard area must be completely covered (jumpsuit, face mask, gloves, hair cover) and must strictly follow safety regulations. Careful training is given for operation of cell culture equipment and development of aseptic techniques. Written procedures are followed for all steps and experienced supervisors and scientists are available at all times. All materials are autoclaved in and out of the biohazard-contained areas and all waste materials are decontaminated in an autoclave or steam-heated kill tank.

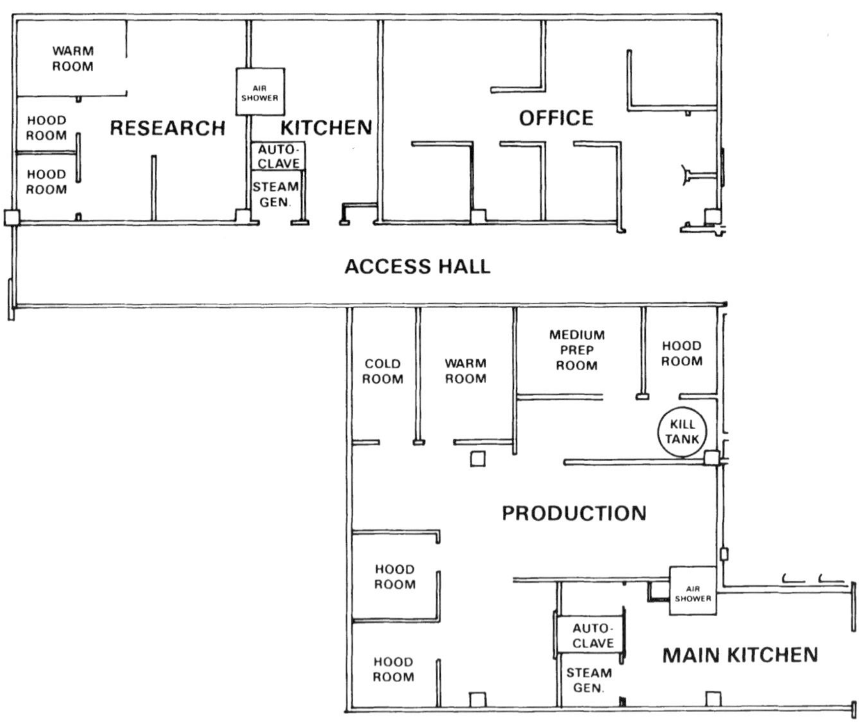

Figure 1. Diagram of large-scale vertebrate cell culture facility. Nominal P-3 contained areas are shown for production and research with HEPA filtered air in and out of each laboratory area.

In the laboratory, peristaltic pumps transfer all fluids greater than three liters through closed, sterile systems. Masterflex pumps (Cole-Palmer Instrument Company, Chicago, IL) are used for smaller volumes and Randolph pumps (The Randolph Company, Houston, TX) are used for larger volumes. Both peristaltic pumps are connected to foot-actuated switches to allow hand-free operation. These pumps are also used in place of pressure vessels for cold sterilization of medium. Unless otherwise stated, all medium used was Dulbecco's modified Eagle Medium, 4.5g/liter glucose supplemented with 5% fetal bovine serum. No antibiotics were used. Medium containing serum is processed in 100-500 liter lots through 0.45μ prefilters and sterile Pall Sealkleen 0.2 or 0.1μ pleated cartridges (Pall Corporation, Glen Cove, NY)(80). Filters are bubble tested and media samples are inoculated on blood agar or other bacterial broth and incubated at least six days prior to use. Medium samples are reinoculated after partial use of 100-liter medium holding vessels. Cell stocks are prepared in a controlled rate freezing apparatus and stored in liquid nitrogen. Representative samples are recovered and tested for viability, growth and various contaminants. Mycoplasma tests are performed on all stocks and further characterization including karyotyping, isoenzyme patterns, reverse transcriptase and electron microscope evaluations have been made by contract with the American Type Culture Collection to certify frozen stocks of particular interest. New, certified frozen cultures are frequently recovered during production to insure cell line integrity.

B. *Anchorage-Independent Growth Systems*

1. 100-Liter Suspension Culture Reactors. One of the major production units for our large-scale cell culture production is a 100-liter vibromixer agitated suspension growth vessel (30,32). It was designed for operation without antibiotics under the following criteria: (a) vessels should be simple in construction with absolute static seals, (b) vessels should be autoclavable so that both inside and outside of vessels, particularly at connectors and sampling ports, conditions were reached and sufficiently maintained to kill bacterial spores, (c) vessels should be portable to allow all connections for transfer of medium, inoculation and harvest to be made inside a laminar flow hood, (d) systems developed for in-place sampling of vessels must maintain absolute sterility. The 100-liter vessels were constructed from 30-gallon sanitary processed 316 stainless steel drums with removable heads sealed by Neoprene gaskets and bolt locking rings. All attachments to the drums were made of 316 stainless steel TIG welded through the removable head (Figure 2). A double septum port has been developed for this system which allows multiple aseptic sampling

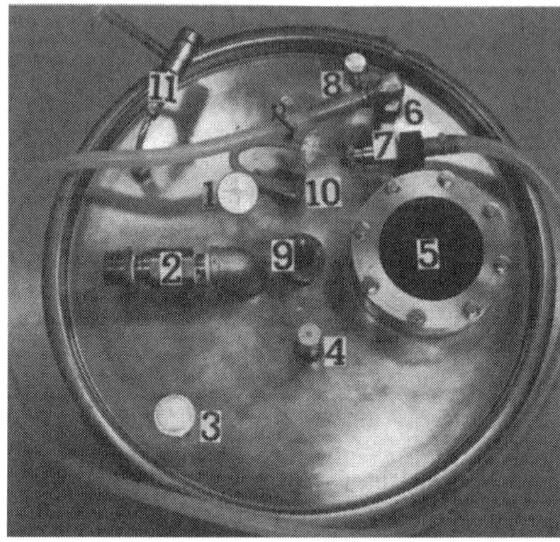

Figure 2. Modifications to head of 30-gal. 316 stainless steel drum are shown: 1) gas vent, 2) check valve for entering steam during autoclaving, 3) overlay gas port, 4) pop-up check valve for pressure release, 5) sight glass, 6) liquid port, 7) flexible coupler, 8) double septum sample port, 9) agitator shaft, 10) oxygen port, 11) sterile air-shielded sample port. Modifications 1-8 are required for medium storage vessels and 9-11 are added for growth vessels. Reprinted from Tolbert et al., (32).

in non-sterile environments (81). The vessels are autoclaved at 124°C for 40-90 min. For the 100-liter reactor or growth vessel, agitation is supplied by an E-1 vibromixer (Chemap, A. G., Mannidorf, Switzerland)(Figure 3). A rubber diaphragm maintains an absolute static seal for this agitation system. Means to slowly sparge 100% O_2 is provided either through the hollow shaft or by an auxiliary tube. This sparge is at a rate between 0.03 and 2 ml/min through scintered glass. A 5-10% CO_2 in air mixture is passed over the liquid surface. An additional larger sampling system (82-83) is used to allow sampling of aggregates (24,84), as well as single cell suspensions. These reactors are inoculated with cell suspension and filled with medium using the Randolph peristaltic pump with all connections made with a laminar flow hood. The vessels then are transferred to a 37°C warm room for growth. Cells are harvested in a Western States STM1000 basket centrifuge (The Western States Machine Company, Hamilton, OH). Suspended cell-free media may be further processed for isolation of secreted products. All

Figure 3. *100-Liter vibromixer agitated culture reactor is shown in warm room during growth stage (a) and during harvest with Western States STM1000 centrifuge (b). Agitator disks and other internal parts are shown in (c). Reprinted from Tolbert et al.,(32)*

Table 1. Continuous cell lines produced in 100-liter system

Line	Suspension Type[a]	Growth Interval (Days)	Cell Yield[b] (ml/liter)
Human Embryonal Rhabdomyosarcoma	Aggregate	5	3.1
Human Hepatoma	Aggregate	6	2.0
SV40 Transformed Human Fetal Lung	Aggregate	7	0.8
Human Osteosarcoma	Aggregate	7	0.6
Human Melanoma	Aggregate	7	0.5
Human Colon Adenocarcinoma	Aggregate	4	2.1
Human Medulloblastoma	Aggregate	6	0.8
Rat Carcinosarcoma	Single	3	2.3[c]

[a] *Cells grew as small aggregate clumps or as single cells in suspension.*

[b] *Representative yields determined from volume of wet packed cells from at least 200 liters of culture.*

[c] *Corresponds to ca. $1.5x10^6$ cells/ml for this line.*

Reprinted from Tolbert et al.,(32).

waste is pumped into a steam-heated kill tank. These 100-liter vessels have been extensively used over the past few years in production of large volumes of suspension culture. Table 1 lists several continuous mammalian cell lines and representative cell yields obtained in this system. A growth curve of the human hepatoma line is presented in Figure 4 with concurrent values of several other parameters. Cell volume density measurements were made by use of conical hematocrit tubes as previously described (85).

While the utility of this batch culture system for sustained large-scale production has been demonstrated over several years of operation, it could also be of significant advantage to relatively small laboratories. General cell culture facilities that include a large autoclave, laminar flow biological safety cabinet and a warm room could have the added capability of occasional large-scale production at relatively low cost.

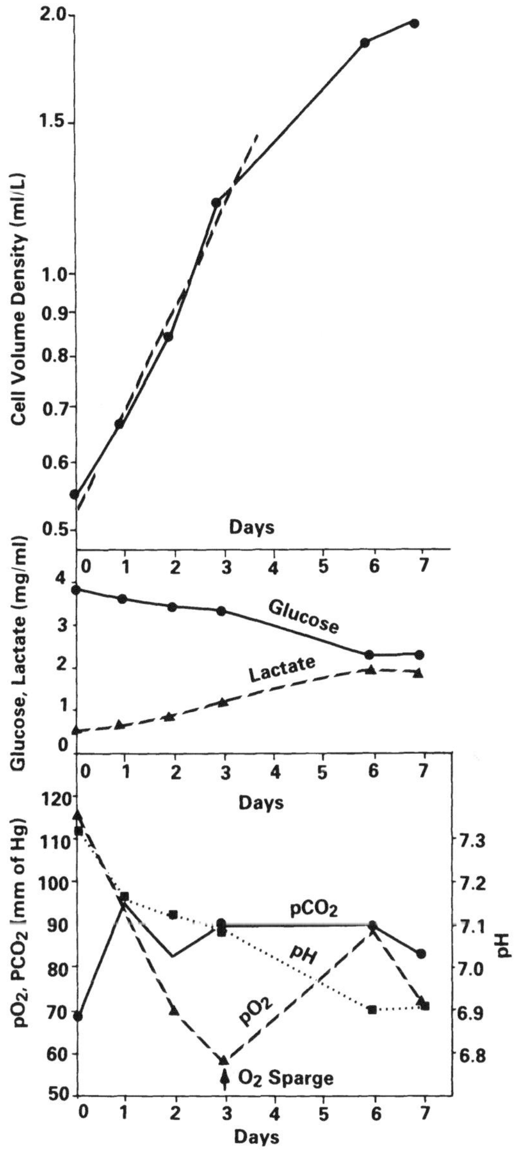

Figure 4. *Human-hepatoma culture in 100-liter reactor. The upper logrithmic growth curve of ml packed cells per liter of suspension is shown with population doubling time of 62 hrs (dashed line). The lower curves indicate other monitored culture parameters. Values for pH, pCO_2 and pO_2 were determined on withdrawn samples in a blood-gas analyzer. A sparge of 0.2 ml/min 100% O_2 was required after 3 days. Reprinted from Tolbert et al., (32).*

2. Anchorage-Independent Cells - Perfusion Culture Systems. Conventional cell culture systems expose in vitro growing cells to a constantly changing medium environment which contrasts with the near homeostatic condition maintained in vivo. Cells are inoculated into media rather rich in all nutrients and devoid of metabolic products. With time the nutrient concentration drops significantly together with elevation in metabolic products. The use of perfusion systems to achieve a steady-state environment has long been known to provide significant advantage over static growth methods (86). Application of such systems to vertebrate cell suspensions was demonstrated by Himmelfarb and his co-workers to produce much higher cell densities than that obtained by conventional techniques (87). We have developed a vertebrate cell suspension system which combines the advantage of medium perfusion and high cell density growth with the ability for extended culture at large scale and chemostat-like operation (88-90). This system consists of a suspension vessel containing a rapidly rotating vertically disposed porcelain filter for removal of expended medium. The filter is connected to a separate rotating seal and bearing assembly above the surface of the culture to prevent grinding of cells at the junction of rotating and stationary surfaces. This becomes even more important during extended growth of fragile vertebrate cells at high density. The added constraint of antibiotic-free development and use of this system ensures high levels of sterility maintenance both in the equipment and procedures designed for its use. High density perfusion growth of vertebrate cells is particularly useful when desired products are released by the cells directly into the medium. Harvest of cells is also facilitated by higher densities and volume reduction as compared to equivalent conventional methods. An unexpected and important advantage found with the perfusion system was significant increase in the cell yield per volume media/sera. The cylindrical 1.0μ porcelain filters were obtained from Selas Flotronics, Huntington Valley, PA, and were rotated at from 200-300 rpm by bottom attached stirrers. To maintain sterility, all components of the seal and filter assembly were contained within the culture vessel and only static penetrations were made through Neoprene stoppers (Figure 5). Shear forces at the boundary between the rapidly rotating filter surface and the more slowly moving liquid minimize clogging of the filter pores during operation. More than 30 liters of expended medium for the 4-liter reactor and 250 liters for the 40-liter reactor may be withdrawn through these rotating filters. Reservoirs for fresh and expended medium are connected to the perfusion reactor through peristaltic pumps. An additional pump added 0.5 M $NaHCO_3$ as required to maintain pH above 7.0. This pump was activated by a pH controller and an autoclavable

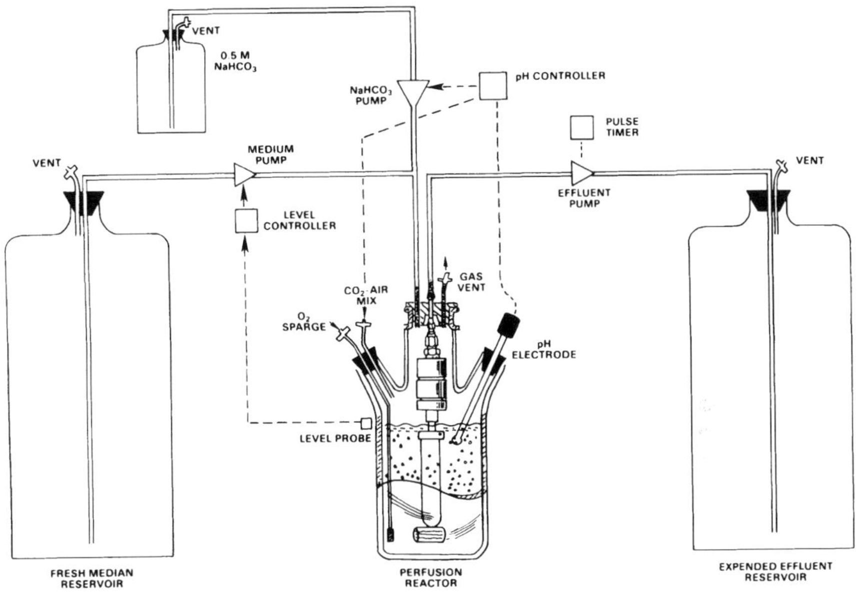

Figure 5. High density rat carcinosarcoma cell suspension is grown in the central reactor with agitation by 200 to 300 rpm rotation of the vertical filter. Cell-free expended medium is withdrawn through this filter to the effluent reservoir at a rate set by the peristaltic pump/timer combination. The capacitive level control system maintains the reactor volume by initiating fresh medium flow to replace the effluent. pH is monitored by an autoclavable INGOLD(R) electrode and controlled both by varying the CO_2 concentration of the overlay gas and by addition of 0.5 M $NaHCO_3$ when required. Dissolved oxygen concentration is maintained by a very low volume 100% O_2 sparge through sintered glass. All gases enter vessels through 0.2μ membrane filters. The double septum sampling port and bottom magnetic drive are not shown. Reprinted from Tolbert, et al., (88).

INGOLD(R) electrode penetrating the reactor. During growth of a rat carcinosarcoma cell line expended medium was withdrawn at increasing rates through the rotating filter to roughly maintain a 10-20 ml/hr/10^9 cells flow. A capacitive level control system attached to the outside of the reactor maintained the level of the cell suspension by controlling the fresh medium pumped to replace withdrawn expended medium. A 5-10% CO_2 in air mixture was passed over the suspension surface at 0.5

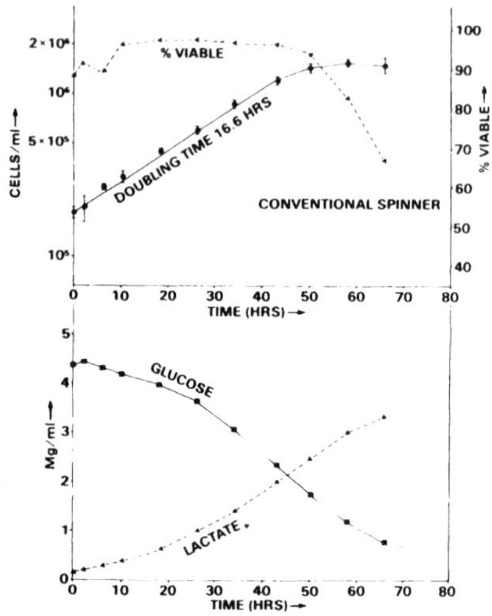

Figure 6. Rat carcinosarcoma suspension was cultured in a conventional Bellco spinner. Cell density determined by hemocytometer ± standard deviation of replicate counts. Viability is measured by trypan blue dye exclusion. Cells were pelleted by centrifugation for 10 min at 300 x g and supernatant solutions frozen for later glucose and lactic acid determination. Reprinted from Tolbert et al., (88).

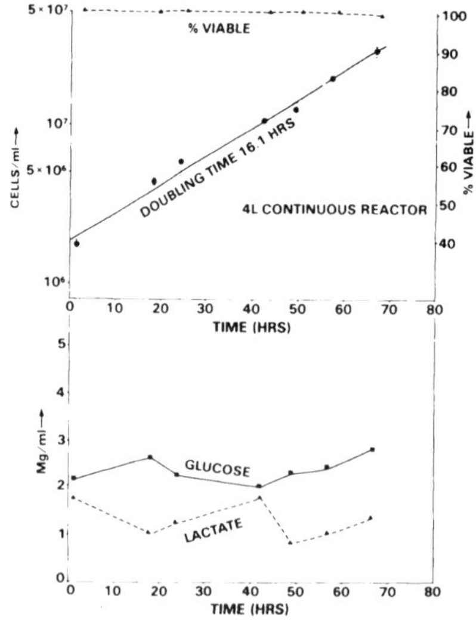

Figure 7. Rat carcinosarcoma suspension was cultured in the 4-liter rotating filter perfusion reactor as described in the text. Cell density, viability, glucose and lactic acid determinations were made as described under Figure 6. Reprinted from Tolbert et al., (88).

to 2 liters/min and a very low sparge of 100% oxygen (0.01 to 2 ml/min) was used as required to maintain dissolved oxygen levels. All connections to the reactor were made by flexible tubing terminating in SWAGELOK quick-connects. The flexible tubing allowed aseptic connections to be made within a laminar flow hood. Temperature control was supplied by locating the reactor and associated equipment in a 37°C warm room. Samples were removed from the reactor by a syringe and 21-gauge needle through the double-septum sampling port (81). Cell concentration was determined by replicate hemocytometer count and percent viability calculated from trypan blue dye exclusion. Dissolved oxygen and CO_2 were measured on withdrawn samples in a blood gas analyzer. After removing cells by centrifugation, glucose and lactic acid analyses were made. Figure 6 shows a typical growth curve for the rat tumor cell suspension in a conventional Bellco spinner. The cells grew with a doubling time of 16.6 hrs to a maximum density of 1.x 10^6 cells/ml. The cell viability in this non-perfused system markedly decreased as the cells reached their maximum density. In this static system, nutrients were depleted and metabolic products built up as illustrated by the glucose and lactic acid curves. Results with the 4-liter perfusion system are shown in Figure 7. Essentially the same doubling time of 16.1 hrs was seen to a maximum cell density of 30.8 x 10^6 cells/ml with no decrease in viability. The continuous perfusion of fresh medium at a rate proportional to the cell density roughly stabilized both nutrient and by-product concentration. The total medium usage for this run was 31.3 liters with a cell yield of 3.9 x 10^9 cells/ml. This represents a 2.6-fold increase in the cell yield over the conventional system. Table 2 shows the results of several 4 and 40-liter runs with this system.

Table 2. Results with rotating filter perfusion system

Reactor Size, L	*Inoculation Density x 10^6 cells/ml*	*Maximum Density x 10^6 cells/ml*	*Doubling Time hr*	*Total Cells x 10^9*	*Total Medium L*	*Cell Yield 10^9 cells/L*
4	*0.912*	*11.1*	*18.3*	*44.4*	*14.7*	*3.0*
4	*0.366*	*9.66*	*18.1*	*38.6*	*15.1*	*2.6*
4	*1.91*	*21.5*	*15.0*	*86*	*25.0*	*3.4*
4	*1.77*	*30.8*	*16.1*	*123*	*31.3*	*3.9*
40	*0.76*	*13.6*	*16.5*	*545*	*175*	*3.1*
40[a]	*0.55*	*9.02*	*16.7*	*868*	*360*	*2.41*

[a] *Continuous harvest*

Reprinted from Tolbert et al. (88)

The last entry in Table 2 describes use of a 40-liter perfusion reactor in a chemostat mode (Figure 8). The reactor was operated as above until a cell density of 9×10^6 cells/ml was obtained. At that time the cell suspension was continually pumped from the reactor into an agitated holding vessel maintained at 4°C. Fresh medium replenished the reactor for the combination of suspension removed and filter withdrawn expended medium. The harvested cells were collected daily and processed. The total cell production for the 40-liter reactor operated in the chemostat mode was 8.68×10^{11} cells. This is equivalent to a conventional suspension culture of 600-800 liters. A total of 1.7×10^{12} cells was harvested in perfusion runs using a total of 621 liters of medium. The 100-liter batch system produced 13.4×10^{12} of these same cells using 14,000 liters of medium. The medium usage efficiency was 2.7-fold higher in the perfused system.

Development of perfusion systems has continued with design of a perfusion chemostat that incorporates several improvements over the system described above (91). In the latter system, the cells were prevented from clogging the 1μ filter pores by tangential shear forces between the rapidly rotating cylindrical filter surface and the more slowly moving liquid. The rotating filter separated cells from the expended medium and supplied agitation for the perfusion reactor. Eventual clogging of the filter by subcellular debris limited the extent and duration of cultures. In the new system a stationary filter surrounded by a rotating concentric agitator was substituted for the rotating filter (92). This greatly simplified the system by removing the need for a rotating seal assembly making all liquid seals static. Clogging of the filter pores still was prevented by tangential shear caused by the concentric agitator induced motion of the suspension around the stationary cylindrical filter. This system also was designed so that cell suspension was not exposed to junctions between rotating and stationary surfaces to prevent cell damage at high cell densities. A second improvement was to remove the filter entirely from the growth vessel and locate it in a separate satellite vessel. This satellite vessel was connected in a recycle circuit to the main growth vessel with suspension pumped back and forth to maintain equivalent component concentrations. Expended medium was withdrawn from the filter in the satellite vessel and replaced with fresh nutrient medium in the growth vessel. The chemostat mode of operation was implemented by pumping both cells and expended medium from the growth vessel to a separate holding vessel maintained at 4°C. The perfusion system provided the additional nutrient addition necessary for high density cell growth. The third improvement to the earlier system involved use of a novel, extremely gentle agitation system utilizing large, slowly rotating, flexible sheets or "sails" (93).

These "sail" agitators operated at as low as 10 rpm to provide the required mixing with greatly reduced stress to the high density fragile vertebrate cells and were also employed in a different arrangement as the concentric rotating agitator surrounding the stationary filter. The gentle sail agitation system consisted of multiple flexible sheets made from monofilament nylon cloth available under the trademark NITEX(R) (Tetko, Inc., Elmsford, NY). These relatively large, flexible sheets were attached at the bottom and top to allow them to billow like sails as liquid spilled from their trailing edges during rotation of the agitator. This system supplied the required agitation energy over a very large surface area so that shear forces seen by individual cells and cell aggregates were much less than conventional systems. Such gentle agitation was particularly important at high cell densities to maintain the integrity of the fragile vertebrate cells. Four large flexible sheets were used in the growth vessel and in the satellite filter vessel the concentric agitator consisted of nine more narrow flexible sheets running the length of the stationary porcelain filter. Magnetically driven rotation of this agitator at about 50-75 rpm was sufficient to prevent filter clogging during removal of several hundred liters of expended medium. The satellite vessel was designed for easy replacement, if necessary, without requiring termination of the run.

In addition to these improvements of the perfusion chemostat, an aseptic connection system was devised (82,83). Operation of the perfusion chemostat or other large systems, particularly in the absence of antibiotics, was facilitated by the ability to make aseptic connections in non-sterile environment. This aseptic connection system consisted of commercially available quick-connects such as SWAGELOK QC series connectors surrounded by a cylindrical jacket. As shown in Figures 9 and 10, this jacket has a side port for injection of sterile air which flows around the connector and is forced out through a narrow gap between the end of the connector and the flanged end of the jacket creating a conical barrier of sterile air between the external environment and the ends of each connector.

This perfusion chemostat system has been used for the growth of a variety of cell lines including a human hepatoma cell line. Data for both 12 and 40-liter runs with this system are shown in Table 3, producing 253 and 780 ml of packed cells respectively. In addition the 12-liter run was operated at 2.5% bovine calf serum instead of the normal 5% BCS level implying much higher overall efficiency. The growth curve for this run with additional monitored parameters is shown in Figure 11.

PERFUSION CHEMOSTAT SYSTEM

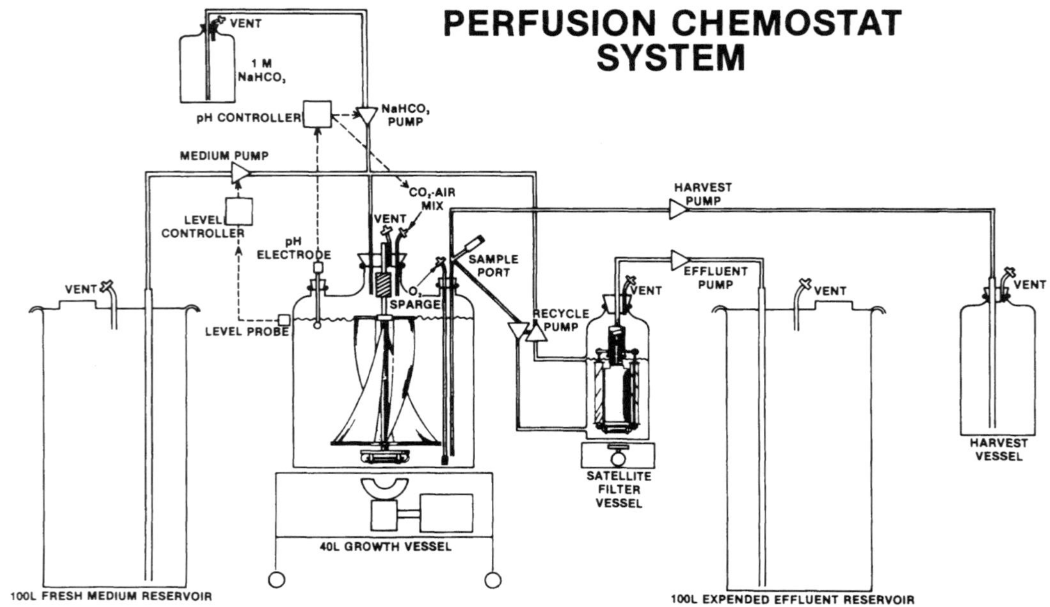

Figure 8. *A diagram of the entire perfusion chemostat system is shown. Cell-free expended medium was removed from the interior of the porcelain filter at a rate set on the effluent pump to provide the required perfusion of fresh medium. Cell suspension was recycled to and from the satellite vessel by a double-headed pump at a rate sufficient to maintain similar component concentrations in the two vessels. While operated as a chemostat, a harvest pump removed cell suspension directly to a harvest vessel contained in an adjacent 4°C cold room. The resulting reduction of the liquid level in the growth vessel was compensated by an exterior capacitive level probe which actuated the medium pump to supply fresh nutrient medium. An autoclavable pH electrode penetrating the growth vessel allowed control of the pH by varying the CO_2 in the mixture of overlay gas and by the addition of sodium bicarbonate when necessary. A sterile air-shielded sample port and a means for low rate oxygen sparge through a sintered glass dispersion tube were also provided. Except for the harvest vessel, the entire system was contained with a 37°C warm room.*

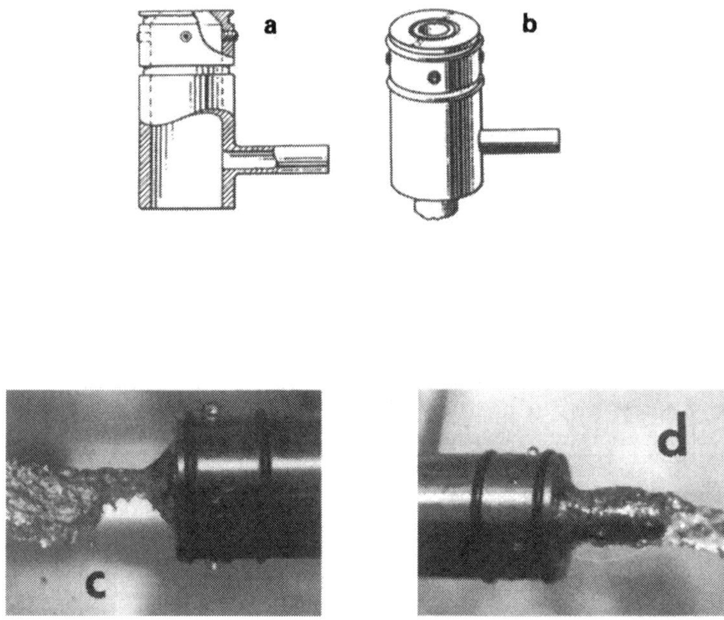

Figure 9. *The cylindrical jacket for the air-shielded connectors is shown (a) in partial cross-section illustrating the 45° internal flange which forms the conical air barrier and (b) with a female quick-connect inserted and O-rings in place. Both female (c) and male (d) shielded connectors are shown with water in place of sterile air to illustrate formation of the protective cone over the ends of the couplers. Reprinted from Tolbert et al. (83).*

Table 3. Perfusion Chemostat

Description	12-L System	40-L System
Total Volume	12L	44L
Total Medium Perfused	203L (2.5% BCS)	709L (5% BCS)
Maximum Cell Volume Density	12.0 ml/L	13.3 ml/L
Average Density During Harvest	8.7 ml/L	10.5 ml/L
Total Cell Volume	253 ml	780 ml

Figure 10. Use of plastic covers made from 50 ml disposable centrifuge tubes protects the sterility of the connectors before (a) and after (b) connections have been made. Reprinted from Tolbert et al.(83)

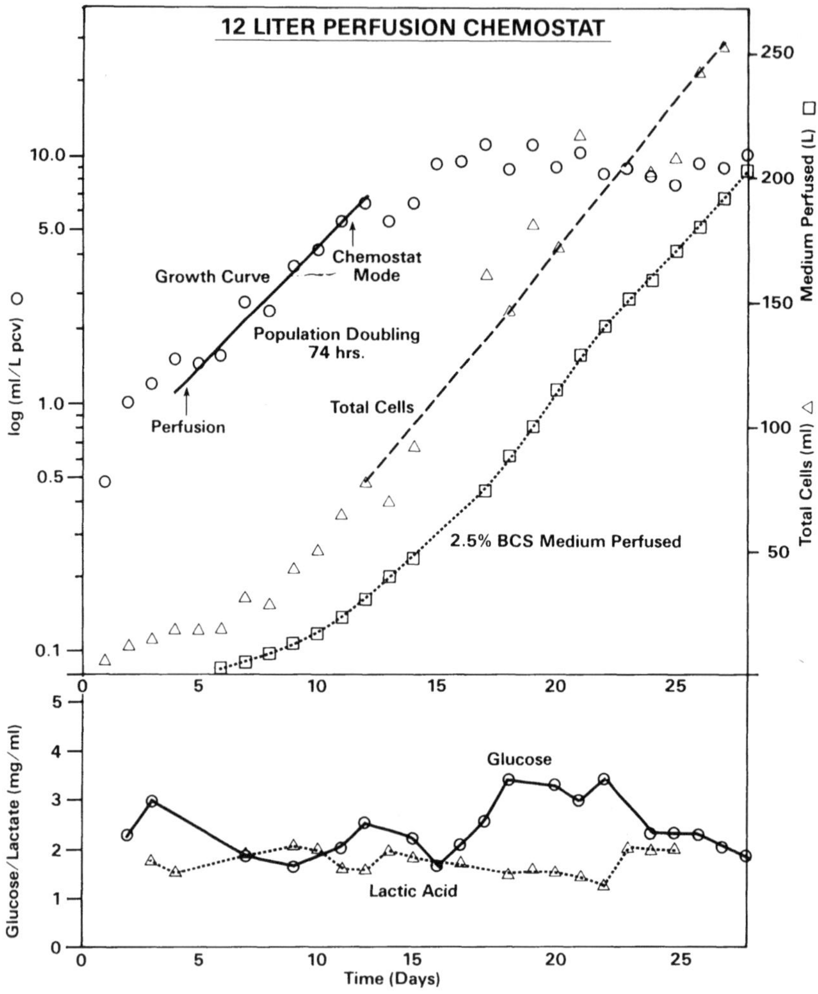

Figure 11. The growth curve for human heptoma cells in the 12-L perfusion chemostat system is shown. Initial growth was in 5% BCS with perfusion of 2.5% BCS medium started at day 6. Chemostat operation started at day 10.5. Other parameters shown include total cells produced both within the reactor and cumulatively harvested, total medium perfused, glucose and lactic acid values (see Figure 6).

C. *Anchorage-Dependent Systems*

While suspension culture, either single cell or aggregate, is an efficient method for large-scale vertebrate cell production, not all cell lines can be adapted to this growth mode. In particular, the vast majority of primary and finite diploid cultures is inflexibly anchorage-dependent. Exceptions are cultures of lymphoid origin which may be grown with the previously described suspension culture technology. Conventional culture of anchorage-dependent cells involves the highly labor-intensive use of glass or plastic flasks and roller bottles as has been described. Two systems have been developed for efficient growth and product secretion for anchorage-dependent cells.

1. *Hollow Fiber System*. Studies carried out concurrently in our laboratory (94) and elsewhere (52) demonstrated that cells could be grown in a hollow fiber system. Our system provides a large surface area to volume ratio as well as excellent aeration through the lumen of the fiber. A constant medium perfusion is also provided to cells grown on the outer surface of the fiber. Evaluation studies using a cartridge configuration led to the development of a flat bed, hollow fiber culture system that maximized the utilization of the large fiber surface and minimized the problems inherent in the long flow path associated with the cartridge (95,96). In particular, development of nutrient and metabolic product gradients were greatly reduced (Figure 12). The reactor core consists of a shallow bed of hollow fibers sandwiched between two stainless steel microporous filter plates through which medium flow is directed, normal to the plane of the fiber bed. Reactors with both 930 and 9,300 cm^2 of fiber surface have been successfully constructed and operated. A variety of cells has been grown in these reactors including SV3T3 cells, baby hamster kidney cells, Vero cells and Rhesus monkey kidney cells. Plasminogen activator (urokinase), migration inhibition factor (MIF) and tumor angiogenesis factors were produced in this system. The hollow fiber reactor offers a system that not only allows the culture of large quantities of cells, but provides a method for their maintenance for long-time periods with minimal stress. This system, therefore, best represents a method for long-term cellular production of a specific biomolecule in which the cell integrity must be maintained and the product is found in the exocellular culture medium such as urokinase or hormones (53, 94-96).

2. *Microcarrier System*. As described above, the microcarrier system has been the most effective method for large-scale growth of anchorage-dependent cells. Nevertheless, difficulties have been encountered in attempts to scale up this

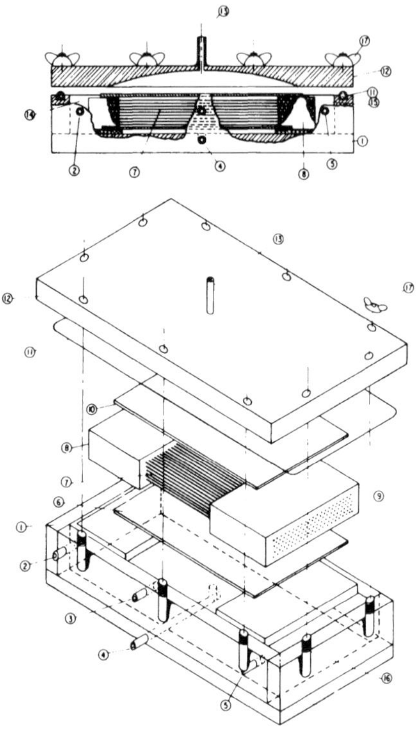

Figure 12. Schematic diagram of flat-bed hollow fiber cell culture reactor. 1) lower reactor housing, 2) air/CO_2 inlet, 3) cell inoculation port, 4) fresh medium inlet, 5) air/CO_2 outlet, 6) lower stainless steel microporous filter, 7) hollow fiber bed, 8) polypropylene hollow fiber potting, 9) open ends of hollow fibers, 10) upper stainless steel housing, 13) spent medium outlet, 14) prereactor chamber, 15) post-reactor chamber, 16) bolt, 17) winged nut. Reprinted from Ku et al., (53).

system, particularly with human diploid cells (5,26,78,79). These problems included sensitivity to antibiotics and inability to scale up from smaller to larger microcarrier cultures, either by enzymatic cell release or by bead-to-bead transfer. A novel perfusion system has been developed in our laboratory which allows efficient scale up of microcarrier cell culture (97,98). The large surface area, flexible sheet sail agitator described above for the perfusion chemostat allowed successful operation with much higher microcarrier concentrations than previously possible (Figure 13). Agitation of these reactors was in the range of 8-15 rpm and concentrations as high as 12 g/liter of BioCarriers (Bio-Rad Laboratories, Richmond, CA) were successfully used. These carriers were sterilized by autoclaving at

124°C for 60 min in pH 5 buffer containing 0.05 M NaCl and 0.05 M 2(N-Morpholino) ethane sulfonic acid (MES). During autoclaving the depth of settled beads was less than two inches. After cooling the beads were aseptically transferred to the microcarrier reactor and washed twice with phosphate buffered saline and once with serum-containing media. At this bead concentration approximately one-fourth of the operating volume of the reactor consisted of settled beads. The 4-liter reactor containing between 30-50 grams of beads was inoculated with human diploid foreskin fibroblasts grown in fifteen to twenty 690 cm^2 glass roller bottles. Cells attached rapidly to the beads and spread out with 60-80% of the beads containing one to three cells with a volume density of between 1 and 3 x 10^5 cells/ml. Constant agitation was maintained during inoculation, cell attachment and throughout cell growth. Cell numbers were determined by modification of the nuclei counting procedure reported by Sanford et al., (97,99). After the cell density had reached approximately 1 x 10^6 cells/ml and many of the beads were partially to totally covered with cells, perfusion was initiated. The major difference between this system and the perfusion chemostat previously described is the addition of a settling bottle in the recycle line between the growth reactor and the satellite filter vessel. During operation of the recycle pump, the suspension with cells and microcarriers was continually removed from the culture reactor into the settling chamber. Here a relatively dense slurry of cells and microcarriers was allowed to settle and aggregate during this temporary residence outside of the vessel's agitated volume. Essentially cell and microcarrier-free medium continued in the recycle circuit to the satellite filter vessel where portions were removed to the effluent reservoir as programmed and the remainder returned to the culture vessel. During this operation fresh medium was added to the growth vessel to replace expended medium removed through the filter. During the settling procedure beads and cells at high density were brought into close contact while being continually perfused with medium from the growth vessel. This close contact between the essentially confluent microcarriers stimulated bridging of cells from one carrier to another and eventual formation of large aggregates of microcarriers as shown in Figure 14. The continued perfusion of media through the settling bottle during the formation of aggregates prevented the limiting environment that would occur if microcarriers and cells were allowed to settle to the bottom of the reactor by periodically stopping agitation. This continued maintenance of optimum cell environment becomes even more important as the density of cells increases and the requirements for nutrients and dissolved oxygen become more critical. The formation of aggregates continues with inclusion on non-cell carrying beads and formation of a cocoon-like structure with

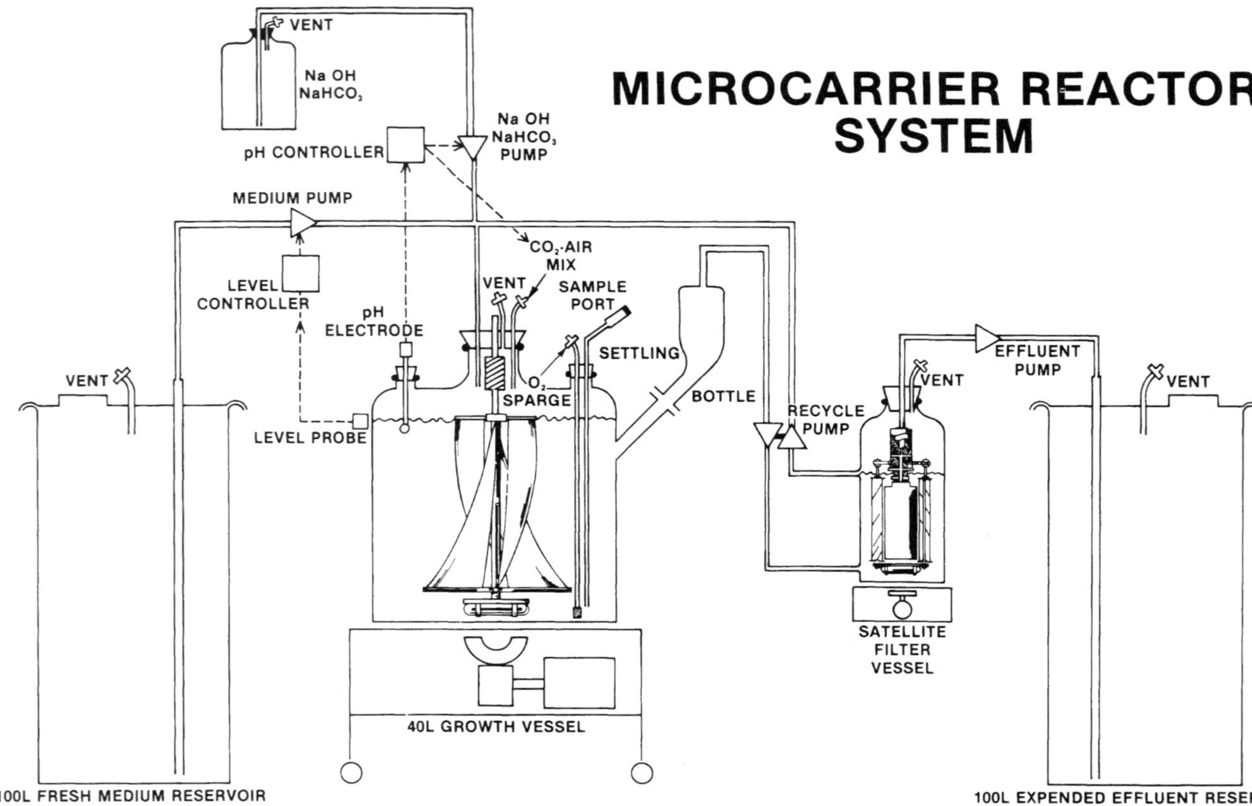

Figure 13. A diagram of the microcarrier reactor system is shown. It differs from that shown in Figure 8 by the addition of a settling bottle in the recycle circuit between growth and satellite filter vessels. This sequestering of the cells and microcarriers outside of the agitated volume of the growth vessel enhances formation of cell-bead aggregates and allow essentially cell-carrier free medium to be pumped to the filter.

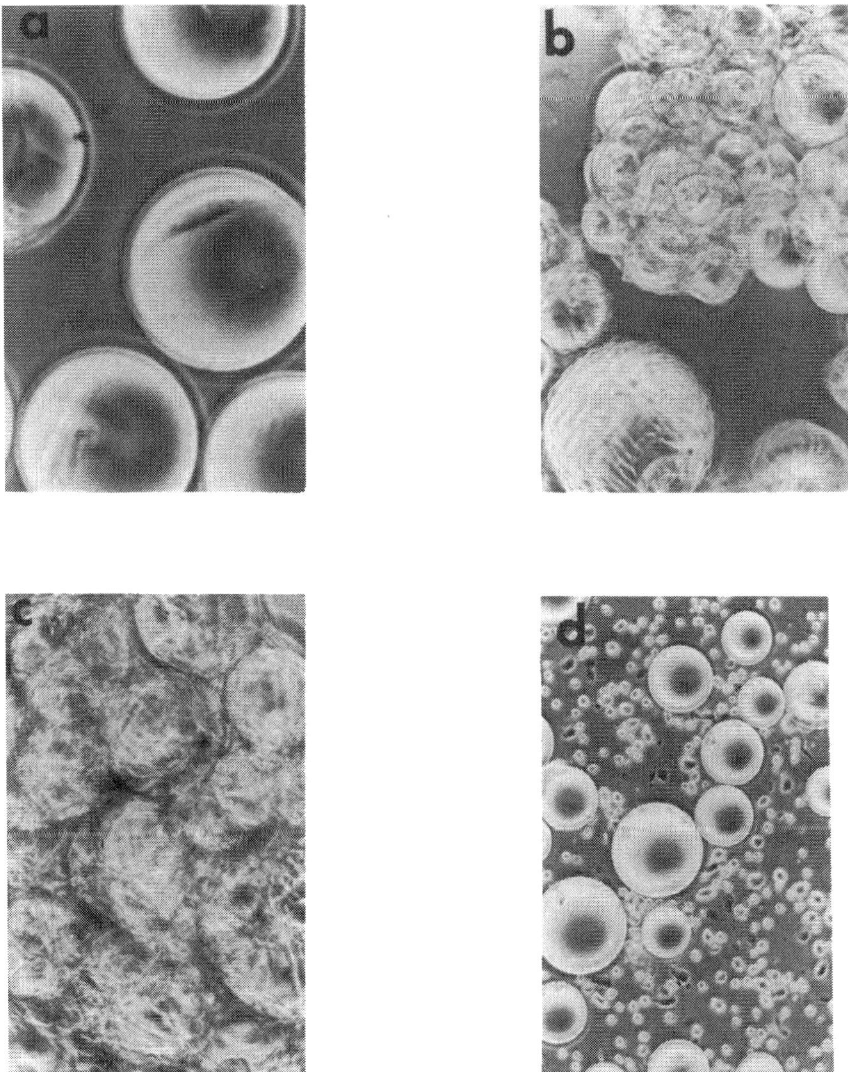

Figure 14. Photomicrographs of human diploid foreskin fibroblasts are shown on Bio-Carriers: a) one day after inoculation of the 4-L reactor, 240X, b) after 15 days growth, 80X, c) higher magnification of aggregate, 160X, and d) after short trypsin treatment, 80X. (see text)

growing not only on, but also between and around all of the beads in the aggregate. This allowed increase in cell number significantly beyond that projected for the nominal bead surface area. An unexpected advantage was obtained when the micro-carriers and cells were in this highly aggregated condition. In this situation, the attachment of individual cells to the bead surface seemed to be reduced and replaced by cell-cell contacts so that short enzymatic treatment completely released the cells from the beads in a highly viable condition. After maximum growth of the cells, reaching a cell concentration of approximately 10^7 cells/ml, medium was removed and the bead aggregates washed three times with PBS containing 0.02% EDTA and then incubated with 25 rpm agitation for 10 min at 37°C with 0.5 grams/liter porcine trypsin and 0.2 grams/liter EDTA in Hank's balanced salt solution. Note the volume of the bead was included in these concentration calculations. When this was done at any of the growth stages before formation of the large cell-bead aggregates, the cells were very poorly removed from the beads as has been reported by others (78-79). But after formation of the cell-bead aggregates, this treatment procedure completely removed all of the cells from the beads as shown in Figure 14. With this procedure it was possible to produce an inoculum in a 4-liter microcarrier reactor of 4 x 10^{10} human diploid foreskin cells equivalent to one thousand three hundred 690 cm^2 roller bottles. This inoculum then was used for a 44-liter reactor containing 400 grams of biocarriers with an actual surface area of 188 m^2 as shown in Table 4. A total of 3.4 x 10^{11} cells was produced or the equivalent of eleven thousand 690 cm^2 roller bottles. Figure 15 presents growth

Table 4. Microcarrier Reactor: Production of Normal Human Cells

	Surface Area (m^2)	*Total Cells Produced*	*Equivalent Roller Bottles*	*Medium Efficiency*
1 Roller Bottle	*0.069*	*$3x10^7$*	*1*	*1*
4-Liter Reactor (Average of 5 Runs)[a]	*21.6*	*$4.0x10^{10}$*	*1,300*	*3.9*
44-Liter Reactor	*188*	*$3.4x10^{11}$*	*11,000*	*4.3*

[a] *Two runs made with 5% bovine calf serum in place of 10% fetal bovine serum used elsewhere.*

curves from the 44-liter reactor and the 4-liter run used for its inoculation. Also shown are various other parameters that were monitored during operation of this perfusion system. In all of these runs, media usage was approximately four-fold less than would have been required to produce the same number of cells in roller bottles. Human diploid foreskin cells grown in

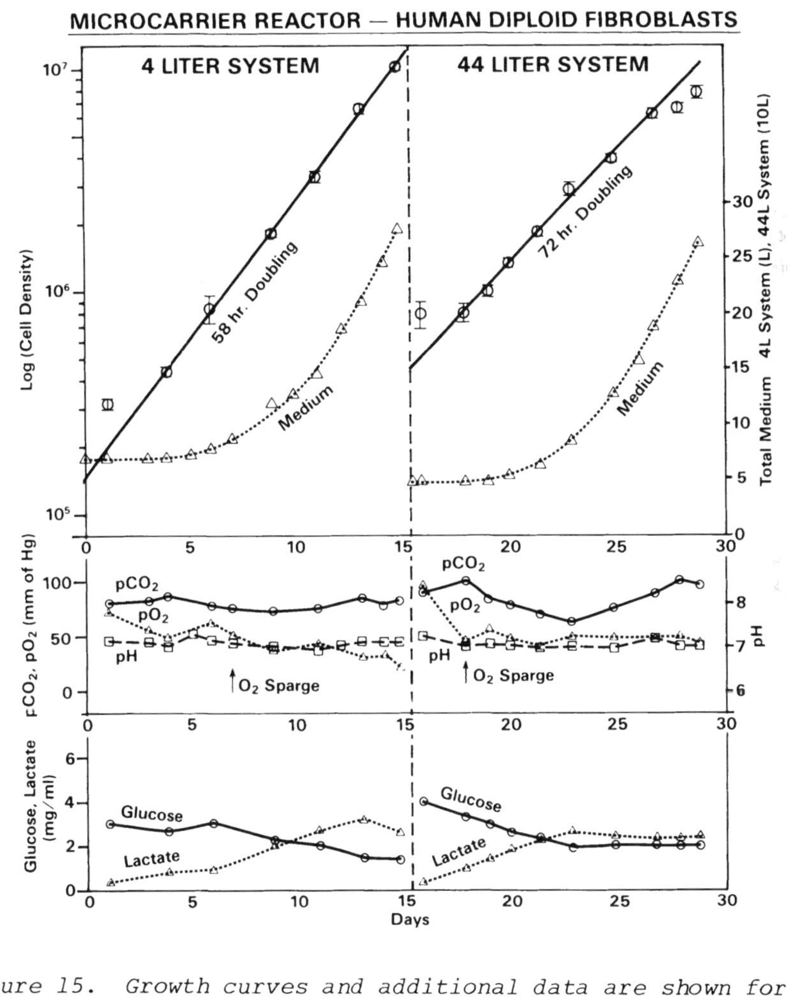

Figure 15. Growth curves and additional data are shown for 4 and 44-liter microcarrier reactor runs. The 4-liter reactor was used to inoculate the 44-liter reactor at day 15.4. See text for further description.

this system were used to produce a normal angiogenesis factor (100) and plasminogen activator. Aliquots of the cell-bead aggregates were also shown to be useful in the production of human fibroblast interferon by the superinduction method. This system then seems to solve many of the problems previously reported for use of microcarriers at large scale. It allows efficient scale-up from smaller to larger microcarrier cultures even for the sensitive, human diploid fibroblasts and provides for cell densities 5-10 times higher than earlier reported with significantly higher efficiency in utilization of expensive animal sera.

IV. CONCLUSIONS

The need exists and will increase for the development of efficient, effective methods to mass produce a wide variety of specific vertebrate cells and cell products. As knowledge of growth, differentiation and cell control at the molecular level advances, the requirement for such large-scale culture technology will be magnified. The primary criterion for this technology is that it must have the flexibility to meet the spectrum of varying growth requirements expressed by individual vertebrate cell types and provide a means for production and isolation of biomolecules of importance and interest. Present technology that has demonstrated at industrial-scale involves modification of existing fermentation systems designed for microorganisms. Such systems require the use of antibiotics and are limited to growth of the few lymphoid or dedifferentiated cell types that will grow as single cells in suspension. Addition of the micro-carrier concept has enhanced the variety of cells that is applicable to these systems. Problems still exist, however, with efficient use of this technology for growth and scale up of more sensitive and important types such as human diploid cells.

We have described a variety of mass culture systems which can be operated antibiotic-free and which can be selected to meet the specific requirements for each cell/product combination. A set of basic principles was established for the design and development of these culture systems which include:

a. The condition that all cell culture will be performed in the absence of antibiotics. This imposed upon the entire system a design both in equipment, facility, and personnel training which is quite stringent. It obviates the major problem of low level contaminants and introduction of unknown/uncontrollable factors affecting the cells and resultant products. While there may be specific instances where limited use of antibiotics for defined purposes would be acceptable, they should be excluded in the general plant design.

b. The second concept is that of a modular design. Modular design implies that the size of reactors will be limited to semi-large scale such as a 100-liter reactor. This allows the best conditions for maintaining sterility of the equipment before use and simplifies plant design in not requiring steam sterilization in-place. It also allows for flexibility and is an insurance policy for an approach which has excluded the use of antibiotics. Any scale is possible by duplication of individual modules.

c. The facility design is also of paramount importance. It should provide P-3 level containment both for protection of personnel and the environment and to potentiate sterility maintenance. The facility should also be modular with isolated areas for laminar flow hoods and dedication of specific areas to specific cells or products. Separate controlled areas should be maintained for glassware and equipment preparation/sterilization, media preparation, inoculation, growth, and harvest operations.

d. Appropriate training and supervision of personnel and dogmatic compliance with defined safety regulations and operating procedures are necessary for successful results. When a single microorganism can destroy large amounts of materials and labor, each operation and procedure must be perfect. Use of laminar flow hoods or air-shielded quick-connects and other devices to maintain sterility greatly simplifies these procedures.

In addition to these principles, growth of a specific cell type should be matched to the system which can best meet its individual needs and which can most efficiently provide for product harvest and isolation. A batch or semi-batch system as described with the 100-liter vessels can be used with cells that grow well in either single or aggregate cell suspension. It may often be chosen for those cells that are more resistant to shear forces or cells that release large amounts of subcellular particles that would clog filter systems. Both cells and expended medium can be harvested from this system by continuous centrifugation. More efficient growth of suspension cells can be obtained with the continuous perfusion and chemostat systems. These permit much more gentle agitation and allow optimization of steady-state growth conditions with reduction of metabolic waste materials. This results in much greater cell densities with increased medium/serum use efficiency. Products that are secreted into the medium can be continually recovered from the filter separated, expended medium and cells can be continually harvested in the chemostat mode. While more fragile cell types are applicable to this system, characteristics of excessive shedding of subcellular debris or the tendency to form very large aggregates which is enhanced at higher densities are contraindicative factors. In the decision between the batch and continuous systems, the requirement for more highly trained personnel for the latter method should also be considered. No one system will be best for all applications. For anchorage-dependent cells, two methods were described: hollow fiber and microcarrier perfusion reactors. The hollow fiber system is most useful for cells that will be maintained for long time periods, continually secreting a desired product into the medium. It is difficult, if not impossible, to recover the individual cells from this system and short-term procedures such as superinduction of diploid human fibroblasts for interferon would not be appropriate. While total surface area is limited as compared to large microcarrier systems, cells have been maintained in hollow fiber reactors approaching tissue densities for much longer periods than possible with other anchorage-dependent systems. The perfusion microcarrier system described here has demonstrated that scale up of even diploid human cells is possible to areas of hundreds of m^2 and densities as high as 10^{10} cells/liter. In this system both perfused medium and cells can be recovered and possibilities for production of secreted or intercellular materials are unlimited. Culture characteristics can also be optimized and controlled to meet a wide spectrum of cellular needs. Finally, the development of new and less expensive media and growth factors for vertebrate cells, while not reviewed in this chapter, will have a significant impact upon future utilization of this technology. Research in this area is accelerating and new developments soon may make biomolecular production by large-scale vertebrate culture much more economically feasible.

In summary, we have presented a review of present large-scale cell culture technology and have described development in our laboratory of significant advances in a new, multi-faceted approach to this area. We visualize the present state as only the beginning of a field that will have significant impact on future advances in molecular biology and medicine.

REFERENCES

1. Todaro, G.J., De Larco, J.E., Fryling, C., Johnson, P.A. and Sporn, M. B., *J. Supramol. Struct. Cell Biochem. 15,* 287 (1981).
2. Twardzik, D. R., Ranchalis, J. E., and Todaro, G. J., *Cander Res. 42,* 590 (1982).
3. Sherwin, S. A., Minna, J. D., Gazdar, A. F. and Todaro, G.J., *Cancer Res. 41,* 3538 (1981).

4. Gospodarowicz, D., Greenburg, G., Bialecki, H., and Zetter, B.R., *In Vitro 14,* 85 (1978).
5. Grespi, C.L. and Thilly, W.G., *Biotech. Bioeng. 23,* 983 (1981).
6. Petriccianni, J.C., Hoppes, H.E., and Chapple, P.J., *Advances in Experimental Medicine and Biology, 118,* Plenum Press, NY, (1979).
7. van Wezel, A.L., *Dev. Biol. Stand. 47,* 7 (1981).
8. van Hemert, P., Kilburn, D.G., and van Wezel, A.L., *Biotech. Bioeng. 11,* 875 (1969).
9. Jensen, M.H. and Soerensen, F.O., *Dev. Biol. Stand. 35,* 45 (1976).
10. Girard, H.C., in "Cell Culture and Its Applications (R.T. Acton and J.D. Lynn, Eds.), p. 111, Academic Press, NY (1977).
11. Spier, R.E., *Dev. Biol. Stand. 46,* 159 (1980).
12. Meignier, B., *Dev. Biol. Stand. 42,* 141 (1979).
13. van Wezel, A.L. and van Steenis, G., *Dev. Biol. Stand.40,* 69 (1978).
14. Meignier, B., Mougeot, H., and Favre, H., *Dev. Biol. Stand. 46,* 249 (1980).
15. Montagnan, B.J., Fanget, B., and Nicolas, A.J., *Dev. Biol. Stand. 47,* 55 (1981).
16. Klein, F., Jones, W.I., Jr., Mahlandt, B.G., and Lincoln, R.E., *Appl. Microbiol. 21,* 265 (1971).
17. Bridgen, P.J., Anfinsen, C.B., Corley, L., Bose, S., Zoon, K.C., Ruegg, U.T., and Buckler, C.E., *J. Biol. Chem. 252,* 6585 (1977).
18. Zoon, K.C. and Buckler, C.E., *Tx. Rep. Biol. Med. 35,* 145 (1977).
19. Reuveny, S., Biano, T., Rosenberg, H., Traub, A., and Mizrahi, A., *Dev. Biol. Stand. 46,* 281 (1980).
20. Klein, F., Ricketts, R.T., Jones, W.I., DeAvnon, I.A., Temple. M.J., Zoon, K.C., and Bridge, P.J., *Antimicrob. Agents Chemother. 15,* 420 (1979).
21. White, R.J., Klein, F., Chan , J.A., and Stroshane, R.M., *Ann. Rep. Ferm. Proc., 4,* 209 (1980).
22. White, R.J. and Klein, F., *Cancer Treat. Rev. 7,* 245 (1980).
23. Beale, A.J., *Dev. Biol. Stand. 47,* 19 (1981).
24. Tolbert, W.R., Hitt, M.M., and Feder, J., *In Vitro 16,* 486 (1980).
25. van Hemert, P., *Biotech. Bioeng. 6,* 381 (1964).
26. van Wezel, A.L. and van der Velden-deGroot, C.A.M., *Process Biochem. 6,* March (1978).
27. Lynn, J.D. and Acton, R.T., *Biotech. Bioeng. 17,* 659 (1975).
28. Zwerner, R.K., Cox, R.M., Lynn, J.D., and Acton, R.T., *Biotechnol. Bioeng. 23,* 2717 (1981).

29. Ulrich, K. and Moore, G.E., *Biotech. Bioeng. 7,* 507 (1965).
30. Moore, G.E., Hasenpusch, P., Gerner, R.E., and Burns, A.A., *Biotech. Bioeng. 10,* 625 (1968).
31. Fontanges, R., Beschaux, P., and Beaudry, Y., *Biotech. Bioeng. 13,* 457 (1971).
32. Tolbert, W.R., Schoenfeld, R.A., Lewis, C., and Feder, J., *Biotech. Bioeng. 24,* 1671 (1982).
33. Tovey, M.G., Mathison, G.E., and Pirt, S.J., *J. Gen. Virol. 20,* 29 (1973).
34. Nyiri, L.K., in "Cell Culture Applications" (R.T. Acton and J.D. Lynn, Eds.), p. 161, Academic Press, NY (1977).
35. Peraino, C. and Eisler, W.J., Jr., in "Tissue Culture Methods and Applications" (P.F. Kruse, Jr. and M.K. Patterson, Jr., Eds.), p. 351, Academic Press, NY (1973).
36. McGarrity, G.J., in "Methods in Enzymology, Vol. 58," (W.B. Jakoby, Ed.), p. 18, Academic Press, NY (1979).
37. McLimans, W.F., in "Methods in Enzymology, Vol. 58" (W.B. Jakoby, Ed.), p. 194, Academic Press, NY (1979).
38. Benton, C.V., Harshman, J.S., Robinson, O.R., Jr., Shibley, G.P., *Appl. Environ. Microbiol. 37,* 148 (1979).
39. House, W., in "Tissue Culture Methods and Applications" (P.F. Kruse, Jr. and M.K. Patterson, Jr., Eds.), p. 338, Academic Press, NY (1973).
40. Corbeil, M., Trudel, M., and Payment, P., *J. Clin. Microb. 10,* 91 (1979).
41. Skoda, R., Pakos, V., Hormann, A., Spath, O., and Johensson, A., *Dev. Biol. Stand. 42,* 121 (1979).
42. MacMorine, H.G., Laurence, G.D., Parisius, W., and Cucakovich, N.B., *Dev. Biol. Stand. 37,* 13 (1976).
43. Reiss, R.E. and Schleicher, J.B., *Biotech. Bioeng. 10,* 601 (1968).
44. Mann, G.F. and de Mucha, M.J., Bull. Pan. Am. Health *Organ 10,* 205 (1976).
45. Mann, G.F. and de Mucha, J., *Dev. Biol. Stand. 37,* 255 (1976).
46. McCoy, T.A., Whittle, W., and Conway, E., *Proc. Soc. Exp. Biol. 109,* 235 (1962).
47. Burbidge, C., *Dev. Biol. Stand. 46,* 169 (1980).
48. Robinson, J.H., Butlin, P.M., and Imrie, R.C., *Dev. Biol. Stand. 46,* 173 (1980).
49. Whiteside, J.P., Whiting, B.R., and Spier, R.E., *Dev. Biol. Stand. 46,* 187 (1980).
50. Mann, G.F., Herrero, L., and Allison, L.M.C., *Dev. Biol. Stand. 46,* 289 (1980).
51. Jensen, M.D., Wallach, D.F.H., and Lin, P.S., *Exper. Cell Res. 84,* 271 (1974).
52. Knazek, R.A., Gullino, P.M., Kohler, P.O., and Dedrick, R.L., *Science 178,* 65 (1972).

53. Ku, K., Kuo, M.J., Delente, J., Wildi, B.S., and Feder, J., *Biotech. Bioeng. 23,* 79 (1981).
54. David, G.S., Riesfeld, R.A., and Chino, T.N., *J. Natl. Cancer Inst. 60,* 303 (1978).
55. Rutzky, L.P., Tomita, J.T., Calenoff, M.A., and Kahan, B.D., *J. Natl. Cancer Inst. 63,* 893 (1979).
56. Knazek, R.A., Gullino, P.M., and Frankel, D.S., U.S. Patent No. 4,184,922 (1980).
57. van Wezel, A.L., *Nature 216,* 64 (1967).
58. Levine, D.W., Wong, J.S., Wang, D.I.C., and Thilly, W.G., *Somatic Cell Genetics 3,* 149 (1977).
59. van Wezel, A.L., *Dev. Biol. Stand. 37,* 143 (1976).
60. Hirtenstein, M., Clark, J., Lindgren, G., and Vretblad, P., *Dev. Biol. Stand. 46,* 109 (1980).
61. Gebb, C., Clark, J.M., Hirtenstein, M.D., Lindgren, G., Linkskog, U., Lundgren, B., and Vretblad, P., *Dev. Biol. Stand.* in press (1982).
62. Monthony, J.F., Schwartz, N.D., Hollis, D.F., and Polastri, G.D., U.S. Patent No. 4,237,218 (1980).
63. Mitchell, K.J. and Wray, W., *Exp. Cell Res. 123,* 452 (1979).
64. Reuveny, S., Bino, T., Rosenberg, H., and Mizrahi, A., *Dev. Biol. Stand. 46,* 137 (1980).
65. Miller, R.E., Kuo, M.J., and Lewis, C., Jr., U.S. Patent No. 4,266,032 (1981).
66. Kuo, M.J., Lewis, C., Jr., Martin, R.A., Miller, R.E., Schoenfeld, R.A., Schuck, J.M., and Wildi, B.S., *In Vitro 17,* 901 (1981).
67. Johansson, A. and Nielsen, V., *Dev. Biol. Stand. 46,* 125 (1980).
68. Horng, C.B. and McLimans, W., *Biotech. Bioeng. 17,* 713 (1975).
69. McLimans, W.F., Kwasniewski, B., Robinson, F.O., Chu, T.M., Sufrin, G., and Gailani, S., *Cancer Treat. Rep. 61,* 161 (1977).
70. Girard, D.J., Loeb, D.H., Thilly, W.G., Wang, D.I.C., and Levine, D.W., *Biotech. Bioeng. 21,* 433 (1979).
71. Levine, D.W., Wang, D.I.C., and Thilly, W.G., *Biotech. Bioeng. 21,* 821 (1979).
72. Clark, J., Hirstenstein, H., and Gebb, C., *Dev. Biol. Stand. 46,* 117 (1980).
73. Crespi, C.L., Imamura, T., Leong, P.M., Fleischaker, R.J., Brunengraber, H., Thilly, W.G., and Girard, D.J., *Biotech. Bioeng., 23,* 2673 (1981).
74. Prestidge, R.L., Sandlin, G.M., Koopman, W.J., and Bennett, J.C., *J. Immunol. Methods 46,* 197 (1981).
75. Mered, B., Albrecht, P., Hopps, H.E., Petricciana, J.C., and Salk, J., *Dev. Biol. Stand. 47,* 41 (1981).

76. deBruyne, N.A. and Morgan, B.J., *Amer. Lab. 52,* June (1981).
77. Clark, J.M. and Hirtenstein, M.D., *Ann. N.Y. Acad. Sci. 369,* 33 (1981).
78. Lewis, D.H. and Volkers, S.A.S., *Dev. Biol. Stand. 42,* 147 (1979).
79. Edy, V.G., 2nd Intnl. Cell Culture Congress, Birmingham, AL (1981).
80. Sinclair, J., *Dev. Biol. Stand. 46,* 39 (1980).
81. Tolbert, W.R. and Feder, J., U.S. Patent No. 4,187,149 (1980).
82. Tolbert, W.R. and Feder, J., U.S. Patent No. 4,253,684 (1981).
83. Tolbert, W.R. and Feder, J., *Biotech. Bioeng. 24,* 1885 (1982).
84. Tolbert, W.R. and Feder, J., U.S. Patent No. 4,059,485 (1977).
85. Tolbert, W.R., Hitt, M.M., and Feder, J., *Anal. Biochem. 106,* 109 (1980).
86. Patterson, M.K., Jr., *TCA Manual 1,* 243 (1975).
87. Himmelfarb, P., Thayer, P.S., and Martin, H.E., *Science 164,* 555 (1969).
88. Tolbert, W.R., Feder, J., and Kimes, R.C., *In Vitro 17,* 885 (1981).
89. Tolbert, W.R., Feder, J., and Kimes, R.C., U.S. Patent No. 4,166,768 (1979).
90. Tolbert, W.R., Feder, J., and Kimes, R.C., U.S. Patent No. 4,184,916 (1980).
91. Tolbert, W.R., Lewis, C., Jr., Hudson, G.N., and Feder, J., in preparation.
92. Tolbert, W.R., Feder, J., and Kimes, R.C., U.S. Patent No. 4,178,209 (1979).
93. Tolbert, W.R., Hitt, M.M., and Feder, J., U.S. Patent No. 4,289,854 (1981).
94. Delente, J.J., U.S. Patent No. 3,998,396 (1976).
95. Feder, J., Ku, K., and Kuo, M.J., U.S. Patent No. 4,087,327 (1978).
96. Feder, J., Kuo, K., and Kuo, M.J., U.S. Patent No. 4,201,845 (1980).
97. Tolbert, W.R., Hitt, M.M., Feder, J., Kimes, R.C., U.S. Patent No. 4,335,215 (1982).
98. Tolbert, W.R., Hitt, M.M., Feder, J., in preparation.
99. Sanford, K.K., Earle, W.R., Evans, V.S., Waltz, J.K., and Shannon, J.E., *J. Natl. Cancer Inst. 11,* 773 (1951).
100. Tolbert, W.R., Hitt, M.M., and Feder, J., U.S. Patent No. 4,273,871 (1981).

CHAPTER 4

MIXING IN FERMENTATION PROCESSES

James Y. Oldshue

Mixing Equipment Company
Rochester, New York

Mixing in the Fermentation Industry involves a large number of different kinds of operations. There are many kinds of laboratory mixing requirements, mixing involved in seed tanks, and the agitation involved in the typical large size fermentation tanks in the production plant.

In addition, there are many auxiliary tanks, some of which are involved in changing pH in the harvesting and recovery steps, and of course the many types of mixing operations that are carried out auxiliary to the main fermentation system.

The fermentations themselves have many different characteristics, from the traditional mycelial fermentation to the many viscous type processes, as well as the low viscosity bacterial type.

This review will relate to the general use of mixing in all these operations and lay down some of the basic principles involved that apply to a wide variety of fermentation characteristics.

PUMPING CAPACITY, FLUID SHEAR RATES, MACRO SCALE AND MICRO SCALE

All that the energy that is applied to a fluid mixer eventually appears as heat. The only way that energy can be converted to heat is through the mechanism of viscous shear. In viscous materials, this can happen uniformly through the entire mixing tank. In low viscosity, turbulent flow, large scale turbulent eddies transfer energy between smaller scale eddies. The transformation to heat takes place at the micro scale level where the scale is small enough for viscous shear effects to occur.

ISBN 0-12-040306-4

Many standard references (1,2) describe in detail the mechanisms of flow and fluid shear but we will just summarize some of these concepts here, and show their application in later sections.

Fig. 1 shows a typical velocity from a radial flow impeller blade as measured by a high frequency response probe, such as a hot wire velocity meter, hot film velocity meter, or a laser doppler velocity meter. This is a general case where turbulence is present. The analysis is made by calculating the average velocity at a point as well as the fluctuating velocity component at that point.

By using the average velocity, and plotting velocity profiles at selected planes in the mixing tank, the shear rate, Fig. 2, is the velocity gradient, dv/dy and has the units of reciprocal seconds.

In viscous fluids, these shear rates operate throughout the tank at different magnitudes and scales. In turbulent flow however, the shear rate between the average velocities operates primarily on particles on the order of several hundred microns. The fluctuating velocity shear rates deal with micro scale processes that have particle sizes or cluster sizes of two hundred microns of less.

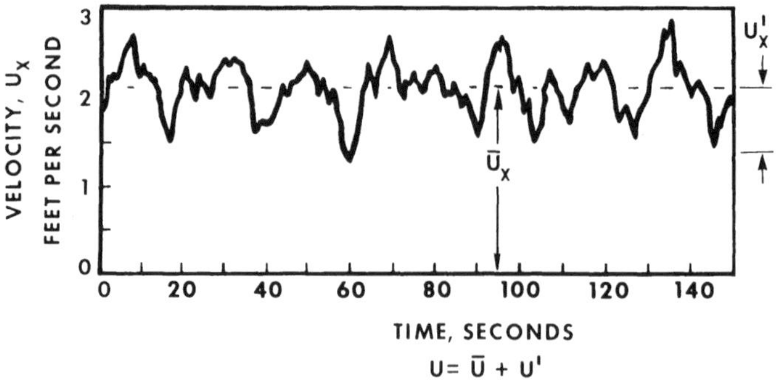

Figure 1. Typical velocity vs. time curve made by high frequency response velocity instrument in mixing tank.

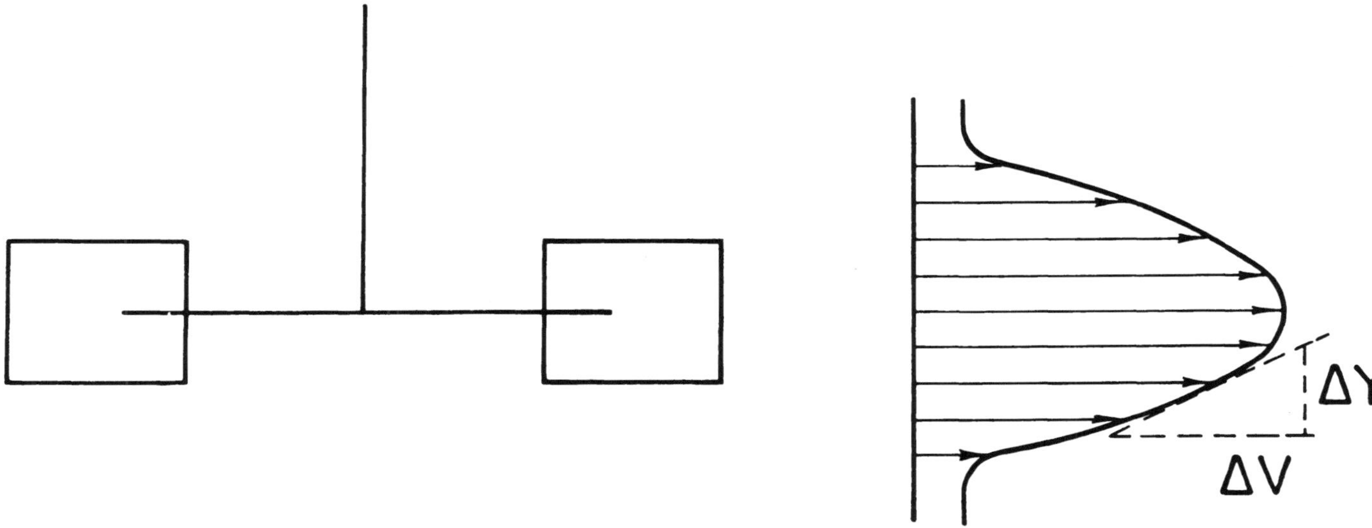

$$\textbf{SHEAR RATE} = \frac{\Delta V}{\Delta Y}$$

Figure 2. Velocity profile leaving the blade or radial flat blade turbine using average velocities at the measuring point to illustrate the calculation of shear rate.

Large-scale bubbles respond to the macro scale shear rates, and Fig. 3 shows how these shear rates can change on scale-up. There is typically a maximum shear rate around the impeller that increases with impeller tip speed, while there is also an average shear rate around the impeller that decreases with impeller speed. Fig. 3 shows that these two diverge on scale-up, tending to give a greater variety of shear rates in large tanks than in small tanks.

In addition, we can define an average in-tank shear rate, from a process standpoint, which is normally a factor of 10 or so less than the average shear rate around the impeller. In the remote parts of the tank there are also minimum shear rates that can be defined, which also decrease on scale-up.

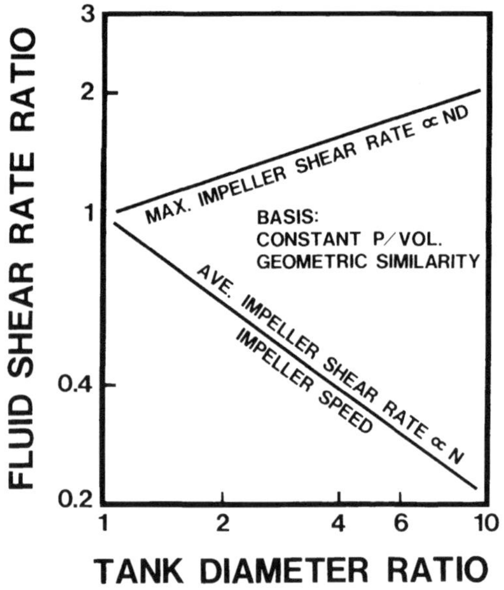

Figure 3. Typical illustration of increase in maximum impeller macroscale shear rate with scale-up and a decrease in the average impeller zone macroscale shear rate on scale-up.

Micro scale shear rates on the other hand are more related to the energy dissipation at a point, thus we would have to know more about the energy dissipation around the impeller and other parts of the tank to predict the distribution of micro scale shear rates.

It turns out that macro scale shear rates are highly sensitive to scale and geometry, and the use of non-geometric scale-up and scale-down techniques, as reported in other references (7) are powerful tools to control selected parameters in the macro scale shear rate range. On the other hand, micro scale relationships, which are often involved in chemical reactions, are more sensitive to energy input, and are not as sensitive to scale and geometry of the equipment.

In all cases, it must be recognized that shear stress is the product of viscosity and shear rate, shear stress = μ (Shear Rate), thus at any shear rate, the shear stress produced is directly proportional to the viscosity at that shear rate.

Within any given tank size, we have the ability to change the pumping capacity of the impeller, Q, and the average shear rate, (AIZSR), around the impeller by the relationship, $Q/AIZSR \propto D^{8/3}$, where Q is impeller pumping capacity, AIZSR is the average impeller-zone shear rate, and D is the impeller diameter.

SOME RECENT WORK ON VELOCITIES

A laser doppler velocity meter is of very great value in measuring the velocities and energy dissipation in transparent liquids. Fig. 4, shows the schematic of the laser doppler equipment, while Figs. 5, 6, 7, 8, and 9 show the typical traces of flow patterns studied by means of a laser doppler velocity meter.

With the ease of measuring these velocity profiles, better characterization of impeller flow pattern has been possible, and impellers with high pumping capacity and consequently low shear rates have been developed, to extend that end of the impeller spectrum.

Fig. 5 shows an axial pitched-blade turbine 2D off bottom illustrating the radial component. Fig. 6 shows an airfoil type axial flow impeller at the same 2D elevation showing greater axial flow for that impeller, with almost zero radial component.

Figs. 7 and 8 show the same impellers about 1D off bottom.

Fig. 9 shows a radial flow impeller with blades below a full disc covering the top of the blades.

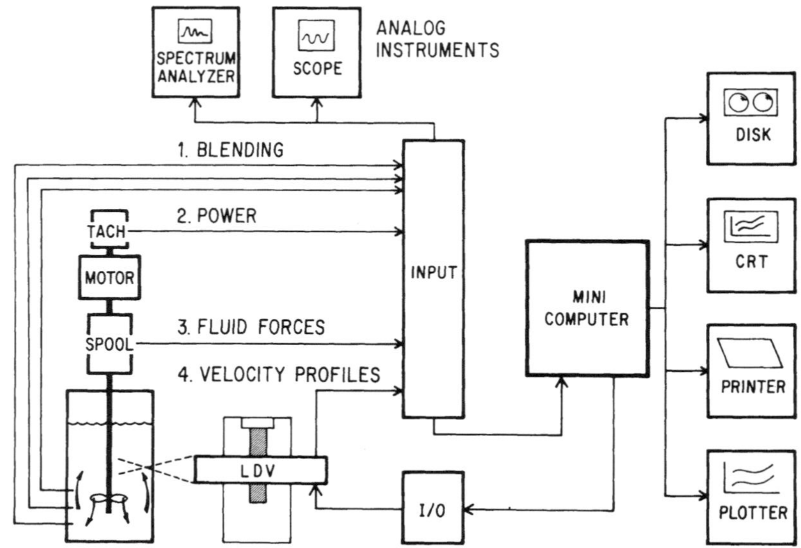

FIGURE 5. INTEGRATED LABORATORY.

Figure 4. Schematic of laser doppler velocity measuring equipment.

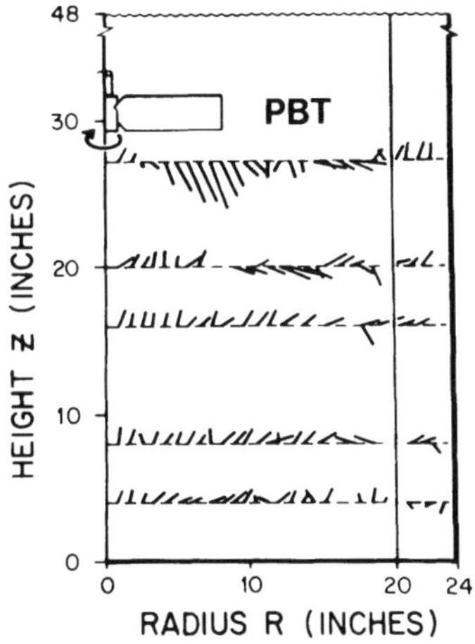

Figure 5. Typical velocity scan produced by laser equipment for pitched blade turbine mounted two impeller diameters above the tank bottom.

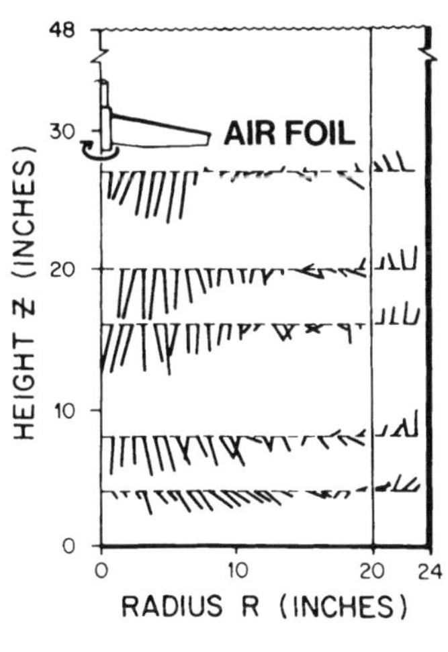

Figure 6. Typical laser scan velocity profile of air foil axial flow impeller two diameters above tank bottom.

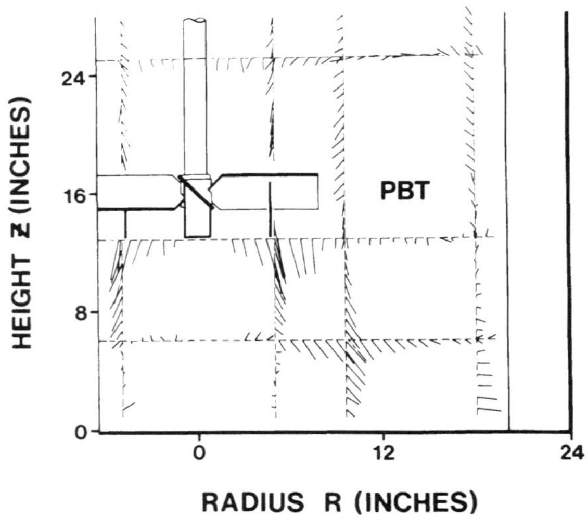

Figure 7. Typical laser scan for pitched blade turbine mounted one impeller diameter above tank bottom.

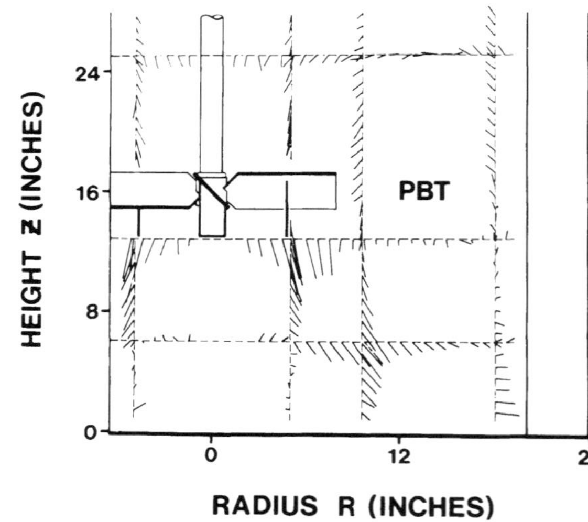

Figure 8. Typical laser scan velocity profile for air foil axial flow impeller located one diameter above tank bottom.

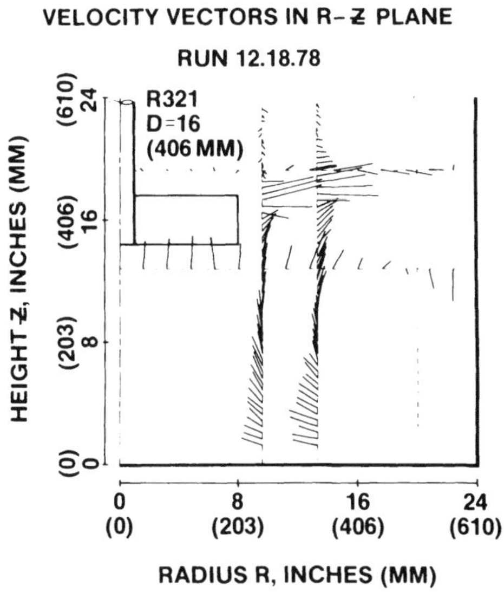

Figure 9. Typical velocity profile from laser doppler velocity meter for radial flow impeller having full diameter disc on top and radial blade extending beneath the disc.

GAS-LIQUID MASS TRANSFER FERMENTATION STEP

The traditional flat blade turbine impeller operated on a central mixing shaft with gas introduced at the bottom of the tank is still the mainstay of most production-scale fermenters and many pilot plant systems. These impellers provide good mass transfer characteristics throughout the fermentation batch. They are essentially high shear rate, low pumping capacity impellers which are generally indicated for mass transfer applications. However, the D/T ratio in many fermenters is around 0.35 to 0.45, which increases the pumping capacity vs. the shear rate of these impellers over what would be optimum for purely gas-liquid mass transfer alone.

Typical curves correlating the mass transfer coefficient, K_Ga, which is used for scale-up rather than mass transfer rate, are given in Fig. 10. In order to calculate the mass transfer coefficient, a suitable gas-liquid mass transfer driving force is needed. Fig. 11 shows a typical pilot fermenter. At Z/T = 1, (where Z = batch depth and T = tank diameter) and at high power levels, it is possible to have a well-mixed gas phase in which the exit gas composition is suitable to use for calculating the mass transfer driving force.

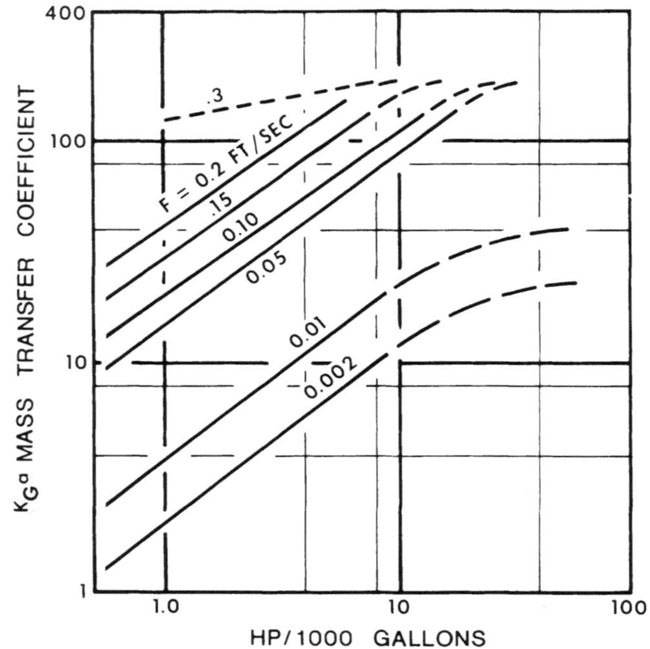

Figure 10. Typical correlation of mass transfer coefficient, K_Ga, vs. impeller power and constant linear superficial velocity.

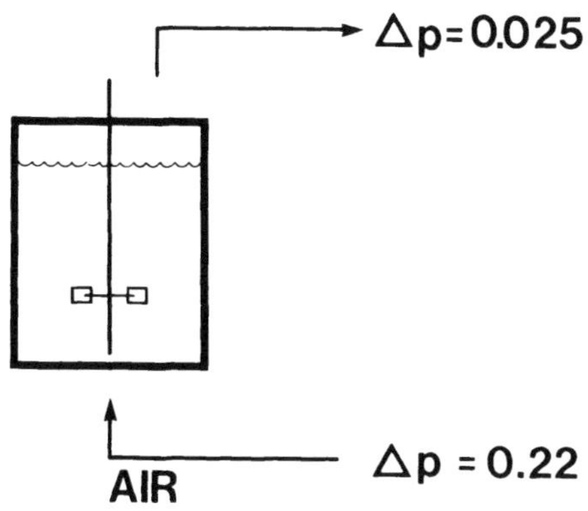

Figure 11. Schematic of gas-liquid mass transfer driving force for pilot scale tank.

In all cases, a P* value is needed, which is the equilibrium partial pressure corresponding to the dissolved oxygen level in the tank. Depending upon the fermentation, this may or may not be readily available. If it is not available, then P* must be estimated in terms of percentage of saturation. Even if there is some error in this estimate, if it is used both in the pilot plant and the full scale calculation, the error involved is largely negated by being used in calculations.

A large fermenter, Fig. 12, shows that it is very unlikely that the gas phase will be mixed from top to bottom. It has been my experience that a log mean driving force between the bottom and the top of the tank is most suitable.

There has been much interest in the use of air agitation only; either in terms of sparging air into the tank and with no internal fittings, or in using a draft tube arrangement and introducing the air in the draft tube area.

The general comparison which was presented in a summary paper by Robinson and Moo-Young (3) indicates that at any gas rate, adding a mixing impeller to the system increases the mass transfer rate over that for using air alone. When high mass transfer rates per unit volume of the fermenter are required, an impeller and gas system is more suitable than air alone. However, for medium or low uptake rates, it is possible to use air sparged fermenters, and there are

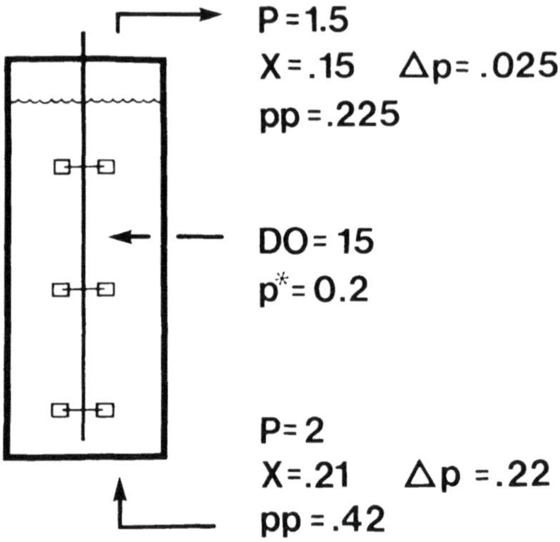

Figure 12. Typical gas-liquid concentration driving force measurements for large scale plant fermenter.

some examples of those in operation around the world. However, air mixing is not particularly conducive to effective blending, solid suspension and/or heat transfer, so that using impeller-type systems is necessary most of the time.

In a typical batch fermentation, Fig. 13, viscosity increases with time, and several investigators (4) have shown that there is a reduction in mass transfer coefficient, K_Ga, with increase in viscosity as the run progresses.

Fig. 14 illustrates the situation that can happen on fermentation with two different full-scale mixers. Most industrial fermenters today are still single-speed. Let's assume that in Fig. 14, Mixer B is installed. It has a high mass transfer rate at the beginning of the run, but as the viscosity increases, the mass transfer ability of the mixer is decreased where it no longer can meet the uptake rate of the organism. Productivity of the fermenter will be less than maximum since the required process of mass transfer during the later stages is not available.

Another possibility is to install a larger mixer, Mixer A, Fig. 14, which will have the mass transferability at the higher viscosity of the fermentation. It will meet the full mass transfer uptake requirement of the organism and allow full productivity to be established.

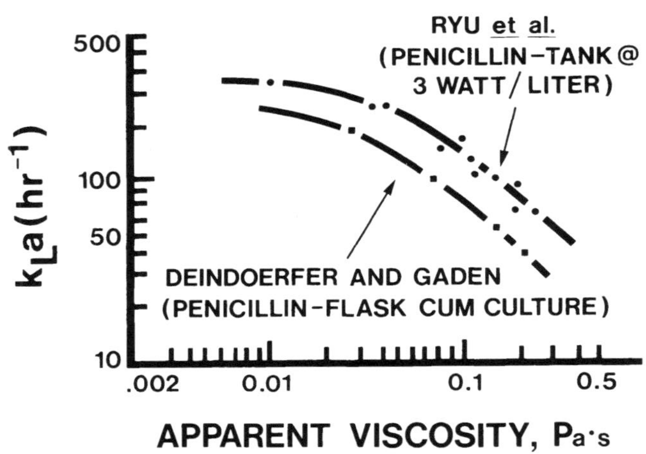

Figure 13. Data from Ryu, et al., on decrease in K_Ga with increase in viscosity of fermentation broth.

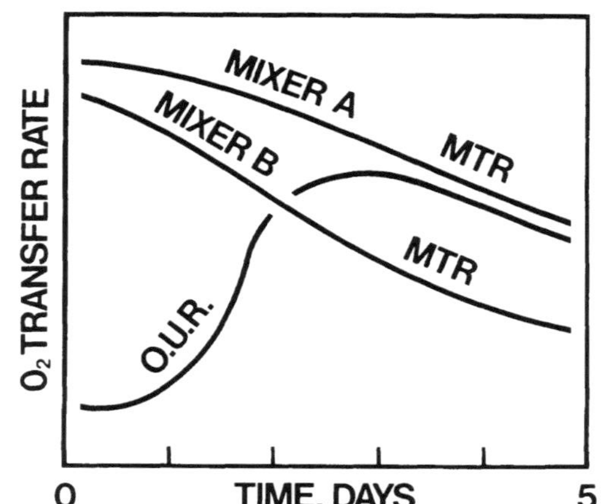

Figure 14. Oxygen mass transfer rate for two difference mixers, A and B. Mixer A satisfies oxygen uptake requirement during the entire fermentation cycle. Mixer B satisfies it initially, but is deficient in latter part of fermentation cycle.

This is further illustrated in Fig. 15 by Mixer I. However, it has such a high mass transfer ability during the first few days of fermentation, that the organisms may be overstimulated. There is some concern that this may cause an initial rapid increase in yield, or may result in lower total yield, Fig. 15 (Yield I), compared to Mixer II. Mixer II has a two-speed motor or a dual power number impeller such that the power can be reduced during the first part of the fermentation. This will maintain the mass transfer rate at just what is needed, and will not overstimulate the organisms with high dissolved oxygen levels during this first period. This may give a higher ultimate yield, Yield II, Fig. 15. There is some indication that the ability to continuously monitor dissolved oxygen level and adjust the mixer speed accordingly (with a complete, variable speed drive) will result in maximizing yield for fermentation. It would also be helpful in minimizing horsepower input. Pilot plant equipment is usually equipped with this kind of monitoring equipment and variable speed capability, but this has not yet been used on any significant number of full-scale fermenters.

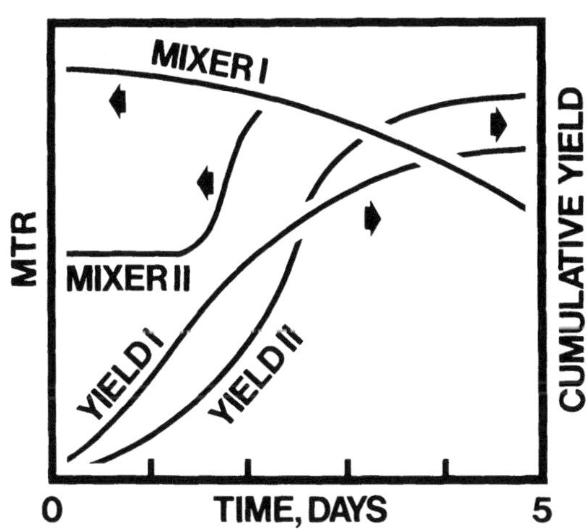

Figure 15. The constant speed mixer, Mixer I, which satisfies the oxygen requirement during the entire fermentation cycle, but may overstimulate organism to give yield I curve. Also illustrated is Mixer II, which is a 2-speed unit, which reduces oxygen mass transfer in the first part of the cycle, but is sufficient for requirement in the more viscous, latter portion of the cycle, giving yield II curve.

SCALE-UP

Scale-up is a primary concern of fermentation investigators. In order to understand the characteristics of scale-up relationships, a few general remarks are in order about the fluid mixing capability and characteristics of small and large tanks. Quite typically, a large tank will have a longer blend time, a higher maximum macro scale shear rate around the impeller, a lower average macro scale shear rate around the impeller, and a lower average tank shear rate. In addition, if equal volumes of air per unit volume of liquid are used, the large tank will have a higher linear superficial gas velocity, which is obtained by dividing the air flow rate by the cross-sectional area of the empty tank. This higher superficial gas velocity will raise the power per unit volume due to air expansion in the tank. This will tend to give more of an air-controlled flow pattern in full scale rather than a mixer-controlled flow pattern.

These high superficial velocities will have an effect on the power drawn by the impeller, and on foaming characteristics in the tank, and will actually tend to increase the blend time over what it was in the pilot tank. This means that the large-scale fermenter mixing parameters are from the small side and some people wonder why they perform as similarly to small scale equipment as they usually do.

Assuming that the pilot plant unit has been run with a conventional flat blade disc impeller for which the blade width is 1/5 of the impeller diameter and the blade length is 1/4 of the impeller diameter, drastic changes in impeller or system geometry must be made to a full scale unit if pumping capacity is to be increased and shear rates are to be decreased. This may require the use of more impellers, their blade width ratios may be increased from 1/5D to possibly 1/4D or 1/3D, and the D/T ratio increased in the full-scale plant.

All of these changes will probably still not produce the same blend time, turnover time or shear rates in the full-scale fermenter. This means that estimates must be made for the length of time a cluster of fermentation cells is away from a high level mass transfer zone, particularly if the tank is equipped with baffles and helical transfer coils. This is especially important when particles reside in remote parts of the tank, away from the main mass transfer field.

It is difficult to use a standard geometry in the pilot plant and be able to change that geometry sufficiently in the full scale to maintain most of the mixing parameters in the desired range. Therefore, we must think seriously about using a modified geometry in the pilot plant so that scale-up to more conventional type impeller combinations will be possible in full scale.

This means that the impeller blade height should be reduced to probably 1/15D or 1/20D, with the important correlary requirement that the blade height shall never be less than two or three times the diameter of the largest gas bubble that is of interest in the mass transfer step.

It also means that the D/T ratio in the pilot plant ought to be more like 0.2 or 0.25 so that a more typical 0.35 to 0.45 D/T can be used in the plant with effectiveness.

A typical pilot plant run gives a plot of $K_G a$ vs. gas rate with a lower rate and power level in the pilot plant as shown in Fig. 16. The box on the right-hand side of the figure shows the typical pilot plant area, and since the superficial gas velocity is increased in the plant scale-up, it is possible to move plant unit design to the left box of the figure. This indicates a much lower power per unit

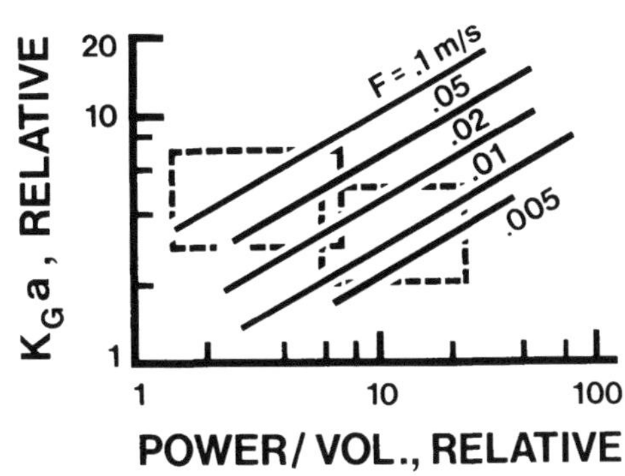

Figure 16. Typical data of $K_G a$ vs. power and constant gas rate parameters illustrating typical pilot plant area at right and full scale design area at left.

volume is possible. However, the blend time will be increased nevertheless in the scale-up. Since the liquid-solid mass transfer rate is affected by the mixer power level, it is not usually possible to reduce the power level very much from what it was in the pilot plant, even though the plant unit has a much higher linear superficial gas velocity.

The net effect is that mass transfer ability of a large size fermenter is usually more favorable to the process than it was in the pilot plant.

The geometry of the full-scale plant tank should be considered from many different aspects. The traditional, tall, thin tank came from the fact that there is a maximum diameter allowed for tanks to be transmitted over highways. So to obtain the higher volumes required for large scale production, the tanks had to be tall and relatively small diameter. This results in a good combination from many aspects of gas-liquid mass transfer. However, it does have the effect of giving increased superficial gas velocities over that which a lower profile geometry, with a Z/T ratio of 1 or 2, would yield.

Actually, various lower profile tanks with a Z/T ratio of 1 to 1.5 can be considered, and the use of horizontal cylindrical tanks can also be part of the geometry study. With the right type of mass transfer data from the pilot plant, a correlation will be available so that full-scale design can be developed for a wide variety of tank shapes. Thus, an overall evaluation can be made.

CARBON DIOXIDE DESORPTION

Desorption of CO_2 is thought to be an important part of the fermentation process. However, many of the things that improve gas absorption, such as increased liquid depth and head pressure in the tank, inhibit the desorption of CO_2.

CO_2 desorption should be measured on both the pilot scale and full scale runs. Fig. 17 shows a typical result.

BLENDING IN FULL SCALE

It is possible to estimate the mass transfer coefficient at various places within the mixing tank. Obviously, they are much higher near the impeller zone than in the rest of the tank. The uptake rate of the organism can be used to estimate the reduction in DO around the particle based on its circulation time in the system. If these calculations show that an appreciable fraction of the particles are deficient in dissolved oxygen, it may be that irreversible damage has occurred, and which may be one cause for yields not being as high as they should be.

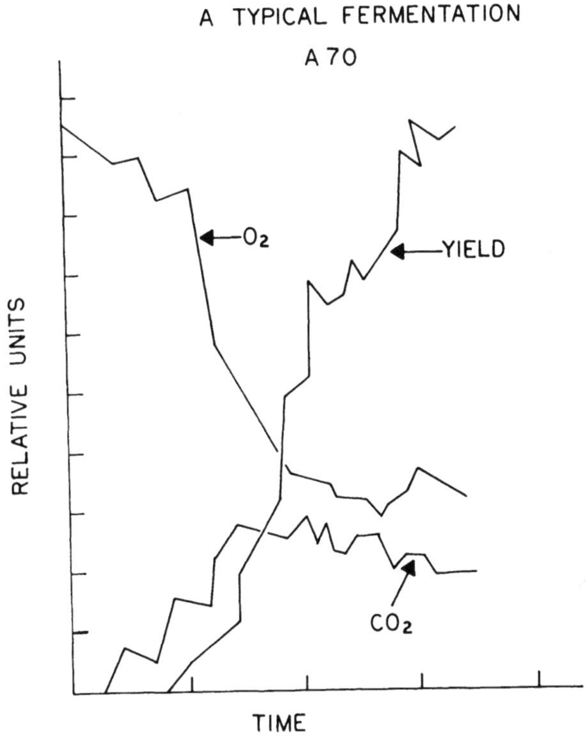

Figure 17. Typical measurements of dissolved oxygen, dissolved CO_2, and fermentation yield from pilot or production size fermenter.

Circulation time in large tanks has been shown by Middleton (8) to be much longer than would be calculated by the ratio of pumping capacity to the volume of the tank. In fact, the particles do not follow a general overall flow pattern in the tank but assume smaller circulation patterns within the main pattern. The flow pattern in a 1,000 gallon tank with a Z/T = 1.0 and a single impeller can be equivalent to 2 to 10 tanks in series as far as the particle velocity in the system is observed.

The presence of gas bubbles also increases the blend time in an homogeneous liquid. An increase in superficial gas velocity similarly increases the blend time.

MIXER COSTS

As the viscosity of the fermentation broth increases, the mols of oxygen that can be transferred for a given power level decreases. For example, to give a relative viewpoint, at 10/grams/liter of solids, let's say that the ability of the mixer to transfer oxygen will be ten mols of oxygen per megajoule. At 20/grams/liter solid, the mixer can transfer 6.4 mols of oxygen per megajoule and at a concentration of 40/grams/liter, the mixer can transfer 4 mols of oxygen per megajoule. The productivity of a fermenter is increased dramatically with the increased cell concentration, but it is true only at a higher power cost. However, the net total cost of production will be less with the higher cell concentration (5).

CHEMICAL REACTIONS

The micro scale velocity fluctuations shown in Fig. 1 can be calculated as a Root Mean Square, RMS, value and plotted as in Fig. 18. The ratio of fluctuations to average velocity are shown in Fig. 19. Around the impeller the fluctuations are about 50%. This means that a reagent added to a reactive solution in the tank at the impeller will experience an instantaneous concentration change of plus or minus 50%. If the reaction speed was such that it could detect this concentration fluctuation, there would definitely be a difference in the reaction rate. If there were a series of competitive, consecutive reactions, there might well be a difference in the selectivity of the products produced by that sort of situation. Bourne (6) has been studying the principles involved when mixers make a difference in the selectivity of chemical reaction, and further reference to

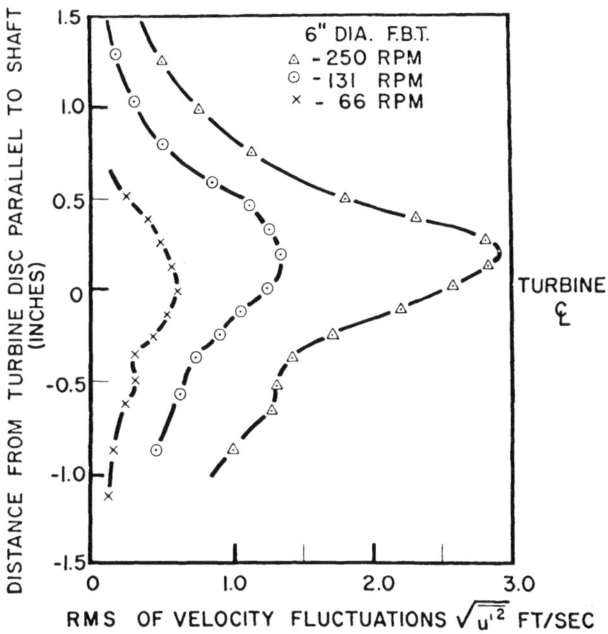

Figure 18. Plot of the Root Mean Square of velocity fluctuations in turbulent fluid at three different impeller speeds at various distances above and below the impeller center line for radial flat blade disc turbine.

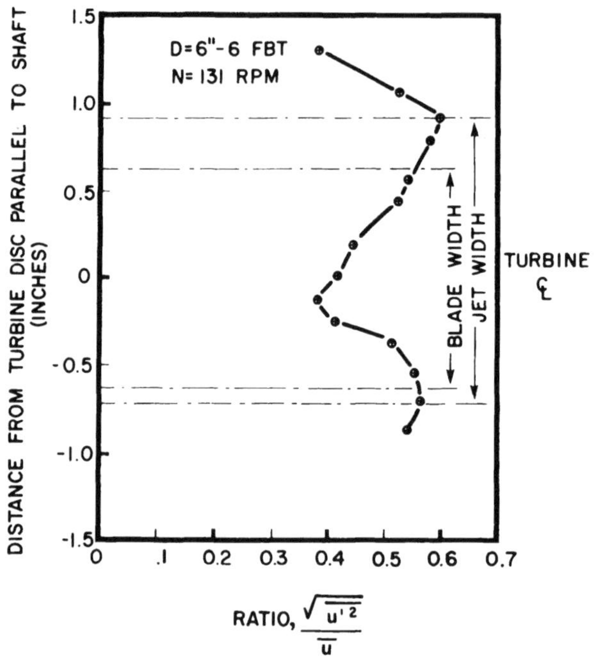

Figure 19. The ratio of the Root Mean Square velocity fluctuation divided by the average velocity at a point for a radial flow flat balde turbine.

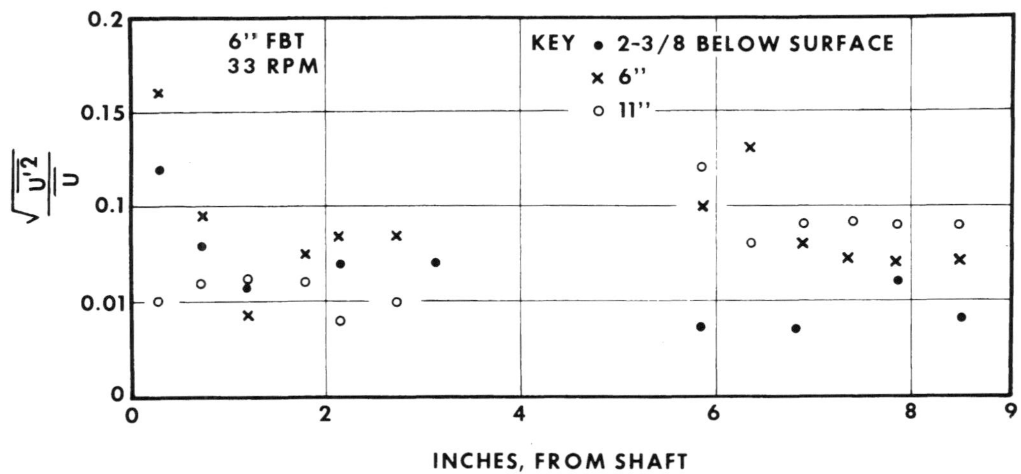

Figure 20. Ratio of Root Mean Square velocity fluctuation to the average velocity of the point for various distances from the shaft at various depths in the tank.

his work is recommended. Paul (9) also showed that the Root Mean Square, RMS, value at the point of introduction did correlate with the observed selectivity of a competitive, consecutive reaction system.

In the more remote parts of the batch, near the surface, the RMS value is only 5% to 10% of the mean, Fig. 20. Thus, for chemical reaction, experiments on the effects of feed introduction-points and other variables are important. Tests should also be conducted to determine the best way to introduce the gas, i.e., by open pipe, multiple pipes or sparge rings. The results of these experiments will provide a good indication of the best methods to be used for a given process.

NOMENCLATURE

N	Impeller speed
D	Impeller diameter
Z	Liquid level
T	Tank diameter
$K_G a$	Gas-liquid mass transfer coefficient, based on gas phase
$K_L a$	Gas-liquid mass transfer coefficient, based on liquid phase
D.O.	Dissolved oxygen
$\frac{dv}{dy}$	Shear rate
RMS	Root Mean Square
Q	Impeller pumping capacity
H	Impeller velocity head
OUR	Oxygen Uptake Rate
MJ	Megajoule

REFERENCES

1. Oldshue, J.Y. "Let's Understand Mixing," *Chemtech 9*, 554 (1981).
2. Oldshue, J.Y. "Spectrum of Fluid Shear Rates in a Mixing Vessel," *Chemeca '70*, Australia (Butterworth) (1970).
3. Robinson, C.W., Moo-Young, M., and Bello, A. "Mixing and Mass Transfer Characteristics of Air Life Fermentors," *Achema, June 11*, Frankfort, Germany (1982).
4. Ryu, D.Y., and Humphrey, A.E. "A Reassessment of Oxygen Transfer Rates in Antibiotics Fermentations," *J. of Ferment. Tech. (Japan) 50*(6), 424 (1971).
5. Ryu, D.Y. and Oldshue, J.Y. "A Reassessment of Mixing Cost in Fermentation Processes," *Biotech. Bioeng. 19*, 621 (1977).
6. Bourne, J. and Rohani, S. "An Evaluation of a 2-Environment Model of Micromixing," Proceedings 4th European Conference on Mixing, Paper J-2, September 15-17 (1982).
7. Oldshue, J.Y. "Fermentation Mixing Scale-Up Techniques," *Biotech. Bioeng. 8*, 3 (1966).
8. Middleton, J.A. Proceedings 3rd European Conference on Mixing, Vol. 1, Paper A-2, 15, York, England (1979).
9. Paul, E.L. and Treybal, R.E. *AIChE J. 17*, 718 (1971).

CHAPTER 5

AERATION

Young H. Lee
Sydney Luk

Drexel University
Philadelphia, Pennsylvania

I. INTRODUCTION

In aerobic fermentation processes and waste water treatment, oxygen transfer from gas to liquid is one of the major criterion for reactor design and scale-up. This is because oxygen is extremely sparingly soluble in water, and, therefore, considerable energy input is required for supplying adequate amount of oxygen to growing microorganisms. Aeration is one of the many gas absorption processes, and it is interesting to note that nearly 25% of all the articles published in 1980 in various chemical engineering journals were concerned with gas-liquid topic (1). Such an active research effort is certainly helpful in advancing our knowledge on aeration. This review is in continuation of the two previous ones (2,3) in this series, and covers mainly the progress in research since 1979.

Many excellent reviews appeared in the past few years regarding various aspects of aeration. Moo-Young and Blanch (4) discussed the design aspect of biochemical reactors based on fundamental mass transfer principles involved. Andrew (5) reviewed the current state of the art of aeration from the viewpoint of large-scale operation. Finn (6) classified fermentation into two general categories, namely, the two phase system (yeast and bacteria culture) and the three phase system (non-Newtonian, mycelial culture), and stressed need for further

ISBN 0-12-040306-4

investigation on the three phase system. Shah, et al. (7), gave an extensive review on design parameters estimations in bubble columns. The mass transfer aspects of stirred tank reactors were given by Charpentier (8) and Joshi, et al. (9). Gas liquid reactors were reviewed by Van Landeghem (10), and the recent advances in general aspects of gas absorption were summarized by Charpentier (1). A new book on gas-liquid reaction has been published (11).

The format of this review is similar to the previous ones: fundamental aspects of mass transfer are presented followed by advances in experimental methods. Specific aeration devices are then reviewed. Oxygen transfer in mycelial pellets (liquid-solid mass transfer) is excluded in this report.

II. ESTIMATION OF PARAMETERS

In biochemical reactors, oxygen is transferred from gas bubbles to suspended microbial cells via several mass transfer resistances: gas and liquid films around the bubble, liquid bulk, and the liquid film around the cell. Among these, the liquid film resistance around the bubble is controlling the mass transfer (2). Therefore, the oxygen transfer rate, OTR, is expressed as:

$$\text{OTR} = k_L a(c^* - c_L) \tag{1}$$

where k_L, a, c^* and c_L are liquid phase mass transfer coefficient, interfacial area per unit volume, saturation concentration of dissolved oxygen at bubble interface, and bulk concentration, respectively.

Eq. (1) shows that the oxygen transfer rate depends on k_L, 'a', and the concentration driving force. Therefore, the oxygen transfer rate can be enhanced by agitation (thus increasing k_L), by bubbling air into liquid (thus creating large interfacial area), and by increasing the gas phase pressure (thus increasing the driving force). The design and scale-up of biochemical reactors generally require a relationship between volumetric mass transfer coefficient, k_La, and the physical and the operational parameters of the system. Although information on k_La is sufficient for the design and scale-up, knowing the dependence of the individual parameters, k_L and 'a', on liquid properties and operating conditions, will lead us to a better design and operation of reactors.

A. *Liquid Phase Mass Transfer Coefficient*

1. *k_L from Gas Absorption Models.* An expression for k_L can be obtained from the solution of a diffusion equation set up for the region close to gas-liquid interface. Since only diffusivity and a parameter or parameters representing fluid mechanics of the system are involved, it is expected that k_L be a function of the two. In fact, the research has been in the direction of elucidating the exact nature of the dependency of k_L on diffusivity and fluid mechanics.

The classical models are Whitman's film theory (12), Higbie's penetration theory (13), and Danckwerts' surface renewal theory (14). Whitman assumed a steady diffusion through a stagnant liquid film with a thickness, δ, and derived an expression for k_L as:

$$k_L = D/\delta \tag{2}$$

where D is the diffusivity of the gas in the liquid. Higbie, assuming an unsteady diffusion into a liquid element at gas-liquid interface during an exposure time, t_e, derived:

$$k_L = 2(D/\pi t_e)^{.5} \tag{3}$$

Danckwerts modified Higbie's model by considering a statistical distribution of the exposure time to obtain,

$$k_L = (Ds)^{.5} \tag{4}$$

where s is the fraction of liquid surface renewed per second.

The film model is still widely used in modeling absorption and reaction mainly because it is simple and gives reasonably accurate prediction. The film model is quite convenient in obtaining an estimation of the concentration boundary layer thickness at the interface. Actual experiment (15) shows that the prediction is indeed reasonably good. Higbie's model gives better prediction for mass transfer, especially, in turbulent flow involving unsteady mass transfer such as in bubbles moving fast in liquid. Danckwerts' model is unique in that the effect of fluid mechanics is lumped in a single parameter, s, which has statistical significance.

More recent models (16,17,18) included measurable fluid mechanic parameters such as the length and the velocity scales in the formulation of convective diffusion equations, and the resulting k_L has the following form (18):

$$k_L = K(DV/L)^{.5} \tag{5}$$

where L and V are the length and the velocity scales (either macro- or microscales) of fluid elements, respectively, and K is an adjustable constant (approximately between .5 to 1.2) depending on the type of flow. For example, for turbulent bubble flow, Theofanous, et al. (18), suggest K=0.63, with L and V being microscales. Note that the microscales can be calculated solely from the rate of energy dissipation per unit volume, and liquid kinematic viscosity (19).

More recently, Henstock and Hanratty (20), Petty and Wood (21), and Ruckenstein (22) suggested using two rather than one length scales for describing k_L. Petty and Wood incorporated velocity fluctuations normal to the interface in their model in addition to the usual hydrodynamic time scale (corresponding to L/V in Eq. (5)). The velocity component normal and approaching to gas-liquid interface is most effective for mass transfer (23) and this was shown experimentally by Lee and Luk recently (15).

All models except that of the film theory (Eqs. (3) to (5)) predict that k_L is proportional to $D^{.5}$. This has been confirmed experimentally for clean liquid surfaces that are renewable by liquid flow. For non-slip rigid spheres, k_L is proportional to $D^{2/3}$ according to boundary layer theory (24).

2. k_L for Gas Bubbles. In biochemical reactors, aeration is achieved either by sparging air (or oxygen) into the broth (bubble columns of various configurations) or by a combination of sparging and agitation (stirred tanks). The bubbles play an important role in aeration.

Bubbles can be classified in terms of surface mobility. When bubbles are sufficiently small (less than 300 μm diameter), trace amount of surface active impurities existing in all liquids contaminate bubble surface, and make the bubble virtually rigid. For larger bubbles moving at high velocity, the effect of impurities becomes insignificant and the bubble surface moves freely. In addition, there are bubbles belonging to the intermediate category.

Blanch (3) and, more recently, Moo-Young and Blanch (4) gave a summary of correlations for k_L for bubbles under widely varying conditions. For both the laminar and the turbulent flow conditions, the correlations are of the following type:

$$Sh = A\ Re^{m} Sc^{n} \qquad (6)$$

where

$$Sh = k_L d/D,\ Re = vd/\nu,\ \text{and}\ Sc = \nu/D \qquad (7)$$

where d, v and ν are bubble diameter, velocity, and kinematic viscosity, respectively. For rigid surface bubbles, n = 1/3 and m is either 1/3 for creeping flow or 1/2 otherwise. For mobile surface bubbles, m = n = 1/2. The constant A is close to 1 except the mobile surface bubbles in creeping flow (A = 0.65). For non-Newtonian liquid, A becomes a function of power law index as shown by Blanch (3). More recently, Kawase and Ulbrecht (25) presented a correlation that takes into account of surfactant effect in power-law liquid.

The velocity, v, in Re of Eq. (7) has to be the slip velocity, i.e., the velocity difference between liquid and bubble. If the bubble rises straight through a stagnant liquid, v is equal to the free rise velocity, U, which can be readily calculated from the bubble diameter and liquid properties (7). When both the liquid and the bubbles are moving, it is rather difficult to estimate the slip velocity. For bubble swarms, the equivalent slip velocity can be calculated as follows (7):

$$v = u/H \tag{8}$$

where u and H are superficial gas velocity and gas holdup, respectively. The slip velocity approaches the terminal rise velocity of a single bubble at low gas holdup. Note that, for bubbles, the Higbie's exposure time is equivalent to:

$$t_e = d/U \tag{9}$$

The dependence of k_L on power input per unit volume, (P/V), can be estimated by considering a limiting case. When the liquid is highly turbulent such that isotropic turbulence prevails, energy is dissipated mainly by microscale eddies. According to Kolmogoroff's theory, the length and the velocity scales of micro eddies are expressed by (18):

$$\ell = (\nu^3/e)^{.25} \tag{10}$$

$$v = (\nu e)^{.25} \tag{11}$$

where ν and e are kinematic viscosity and energy dissipation per unit mass, respectively. When ℓ and v are substituted in Eq. (5) for L and V, respectively:

$$k_L = K\,D^{.5}(e/\nu)^{.25} \tag{12}$$

Since $e = (P/V)/\rho$, where ρ is liquid density, k_L is proportional to $(P/V)^{.25}$. In stirred tanks of different geometry, Levec and Pavkos (24) showed that the experimentally obtained k_L is proportional to $(P/V)^{.28}$. For highly turbulent bubble columns, Eq. (12) can be used to estimate k_L.

Since k_L depends on bubble diameter and the local hydrodynamics, it would be appropriate to discuss bubble size distribution and the types of flow involved in actual fermentation. Andrew classified the bubble flows in biochemical reactors in two categories from the standpoint of mass transfer: free bubble rise type (bubble columns, tower fermenters, gas-lift fermenters, and virtually all commercial scale stirred tanks), and high turbulence type (confined mainly to laboratory, small-scale agitated vessels). The large scale stirred tanks were classified as the free bubble rise category because the mass transfer as a whole is little affected by the turbulence condition existing close to the impeller. This indicates that the correlations developed for single bubbles can be quite useful. For bubble columns, Akita and Yoshida (26) gave the following expression for k_L:

$$k_L = 0.5D^{.5}d^{.5}(\rho/\sigma)^{3/8}g^{5/8} \qquad (13)$$

where d, ρ, σ, and g are bubble diameter, liquid density, surface tension, and gravitational acceleration, respectively. When Eq. (13) is compared with Eq. (5) by letting L = d, the V in Eq. (5) becomes proportional to d^2. In bubble rise velocity correlations such as the Stokes equation and the Hadamard-Rybczynski equation (4), we find that it is indeed the case.

Biochemical reactors are in general operated in bubble flow regime (7). This flow regime is characterized by almost uniformly sized bubbles, and generally occurs when the superficial gas velocity is approximately less than 5 cm/s and the rise velocity of bubbles is between 18 and 30 cm/s. The mean bubble size changes widely depending on the liquid composition, superficial gas velocity, and turbulence. Small amount of salts and/or alcohols added to water decrease the bubble size considerably. The increase in microbial cell concentration has the same effect (27).

The average bubble diameter ranges between 0.5 to 3 mm in non-viscous, low density, single cell cultures. However, at relatively high concentrations of cells (> 5 g/l) and alcohols (such as ethanol), the average bubble diameter tends to be on the lower end (0.5 to 1.2 mm) (27). Bubble size data in actual viscous fermentation broth are very scarce. Work has been done mainly with simulated viscous media employing various concentrations of glycerol, CMC (carboxymethyl cellulose) (28), and PEG (polyethylene glycol) (29). In viscous media, bubbles exist in two distinctive categories: very small ones with diameter less than 0.5 mm, and the large ones with diameter ranging from 0.5 to 10 mm. The small bubbles stay almost indefinitely in the medium and, thus, contribute very little to oxygen transfer (6). However, as will be discussed later (see Measuring Techniques section III), the small bubbles may influence k_La estimation.

In free rise of a single bubble in aqueous media, the rise velocity ranges between 4 cm/s (for d of .3 mm) to 25 cm/s (30): the velocity first increases with d_B of up to 2 mm and, then, levels off. In single cell cultures containing relatively low concentrations of extracellular long chain molecules, viscosities are close to that of water. However, due to the cells, proteins, and surface active materials existing in the medium, most of the bubbles become rigid except the larger ones (d > 3 mm), and the rise velocity is considerably reduced (5).

The values of k_L as a function of bubble diameter are plotted in Fig. 1. With tap water (A in Fig. 1), the experimental values of k_L for single bubbles have a maximum at around d of 2 mm (30). Addition of small amounts of alcohols

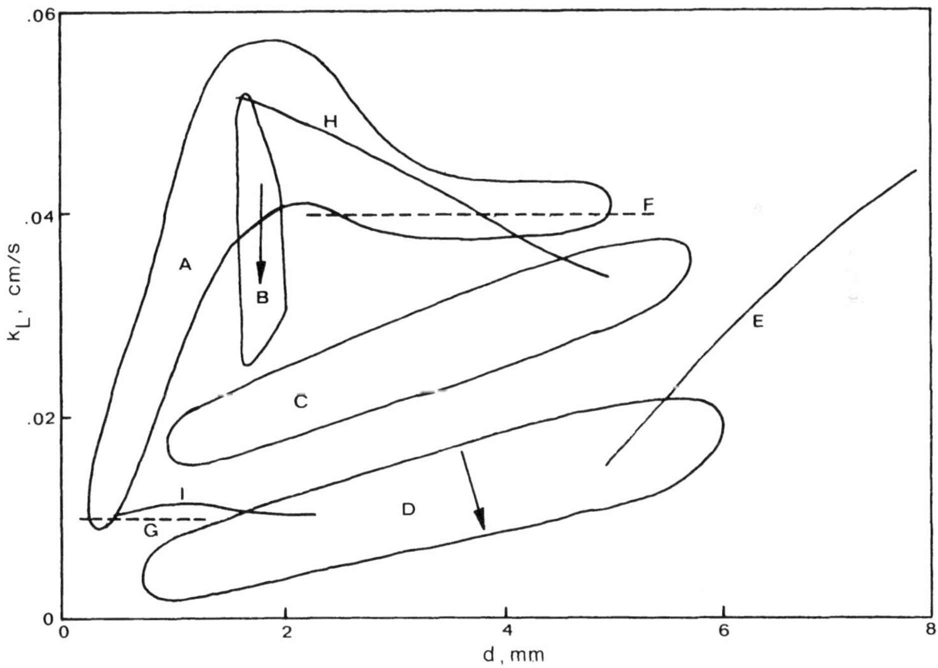

Figure 1. Effect of bubble diameter on k_L. A: in water (30); B: with increasing amount of alcohol (31); C: bubble column, water (32); D: bubble column, glycerol (32); E: 1% CMC (28); F: Calderbank's large bubble correlation; G: Calderbank's small bubble correlation; H: Higbie's model; I: Froessling's equation.

(B in Fig. 1) reduces k_L appreciably (31). In bubble columns (C in Fig. 1), k_L is always lower than the corresponding values of single bubbles (32). An increase in viscosity lowers the k_L considerably (D in Fig. 1). As the bubble size increases, Higbie's model predicts experimental data reasonably well for both the single bubble and the bubble swarms. Since the bubble column data (C in Fig. 1) are for clean water, it is expected that, in non-viscous media, the k_L values lie between 0.05 to 0.02 cm/s. Since the average bubble size is much smaller in actual fermenter broth (0.5 to 2 mm (27)), the rigid sphere model such as the Froessling's equation (Eq. (6) with $A = 0.6$, $m = .5$ and $n = 1/3$) seems to be more appropriate than others in predicting k_L. Generalization cannot be made with the non-Newtonian broth due to the lack of sufficient data. However, since it appears that the average bubble size is much larger (disregarding the non-moving, small ones), mobile surface model seems to be more appropriate for the prediction.

3. Effects of Microbial Cells. The concentration of microbial cells in aerobic fermentation ranges between 1 to 50 g/l. The respiration rates of single cell microorganisms such as yeast and bacteria are on the order of 100 to 600 mg O_2/(gm cell-hr) (6). In bacteria, oxygen is consumed mainly at the cell membrane where most of the respiratory enzymes are located, whereas, in yeast, oxygen has to diffuse through the cell membrane to mitochondria where the respiration occurs. The microbial cells also have a tendency to adsorb at interfaces. It is thus expected that the cells may influence the interfacial transport mechanism of oxygen. As summarized earlier (2), there can be three different effects: absorption enhancement by reaction, and a decrease in k_L due to the alteration of oxygen diffusivity near the interface, and the change in hydrodynamics near the interfacial zone.

The oxygen absorption enhancement due to reaction by microbial cells has been of considerable interest in recent literature, and Blanch (3) summed it up by employing 'diffusion time' and 'reaction time' approach of Astarita (33): the absorption enhancement is significant only in cases when the diffusion time is on the same order as the reaction time such as in low-turbulence surface aeration, whereas, in bubble aeration, the absorption enhancement is negligible because the reaction time (3 to 500 sec) is, in general, considerably longer than the diffusion time (.005 to .04 sec).

There have been quite a few investigators proposing an alternative mechanism of oxygen transfer at the gas-liquid interface in the presence of microbial cells. Phillips (34, 35) has been advocating adsorption mechanism working in parallel with absorption. As an evidence, he indicated the observed dependency of k_L on pressure. Recent studies on adsorption of oxygen in aqueous slurry of activated carbon (36,37) suggest a possibility of such a mechanism. Direct absorption of gaseous oxygen by the cells has been suggested by several investigators (2). More recently, Sobotka, et al. (38), showed that the oxygen consumption via the direct route was significant in their yeast fermentation and that the effect tends to be more pronounced in dense culture exhibiting strong tendency of cell floatation.

Sobotka, et al. (38), pointed out that the estimation of k_La based on dissolved oxygen measurement alone cannot reveal such a mechanism; rather, a total oxygen balance based on both the gaseous and the dissolved oxygen measurements were necessary. With respect to experimental aspect, the comments made earlier by Donaldson and Palmer (39) are especially worth noting: "The measurement of dissolved oxygen in the bulk liquid phase will indicate only enhancement mechanisms which increase the rate of appearance of dissolved oxygen. Enhancement mechanisms which irreversibly remove oxygen from solution in the mass transfer boundary layer will actually decrease the rate of appearance of dissolved oxygen, and such experiments might be interpreted as depressing the absorption rate rather than enhancing it."

Lozowski, et al. (40), calculated relative effects of the diffusivity change and the absorption enhancement due to respiration. If the decrease in diffusivity is taken into account, the absorption enhancement is always greater than 1, although the effect decreases with increase in k_L.

It appears that the oxygen transfer mechanism in the presence of respiring microbial cells needs to be clarified further. More careful experimental designs are necessary for differentiating different effects. If the direct absorption route is indeed significant for certain cultures, it will have a significant implication in biochemical reactor design.

4. *Factors Affecting k_L.* The mass transfer coefficient, k_L, depends on diffusivity and hydrodynamics as shown in Eq. (5) and (6). Kinematic viscosity does not appear explicitly in the expression for k_L except for the case when $m = 1/3$ and $n = 1/2$ in Eq. (6) (rigid sphere, intermediate Reynolds number). Even then, the effect is very small since k_L is proportional to $\nu^{-1/6}$. Other than the bulk liquid properties such as the diffusivity and the viscosity, k_L is also affected by any phenomenon that affects gas-liquid interfacial zone. The effects of surfactants, electrolytes and antifoaming agents fall into the latter category.

The effect of diffusivity on k_L is significant because k_L is proportional to either $D^{1/2}$ (for free surface bubbles) or $D^{2/3}$ (for rigid surface bubbles). Also, diffusivity data are necessary to use correlations on mass transfer. Oxygen diffusivity data in water were compiled by St-Denis and Fell (41) and they showed that the oxygen diffusivity in water follows Stokes-Einstein equation reasonably well:

$$D = K\ (T/\mu) \tag{14}$$

where T and μ are absolute temperature and liquid viscosity, respectively, and $K = 6.92 \times 10^{-10}$ gm-cm/sec^2-$^{\circ}$K. Eq. (14) shows that the diffusivity increases with increase in temperature: for example, by raising the temperature from 25 to 57°C, the diffusivity can be doubled. The temperature effect has significance in fermentation of thermophilic organisms.

Note that Eq. (14) applies only for non-viscous liquids (4). For viscous liquids, Hikita, et al. (42), showed that the following equation gives a better fit for their data in sucrose solution and Yagi and Yoshida's data on glycerol solution (43):

$$D/D_w = (\mu/\mu_w)^{-2/3} \tag{15}$$

where the subscript w stands for water. A two-fold increase in medium viscosity will result in 37% decrease in diffusivity. Note that Eq. (15) is based on O_2 or CO_2 data in sucrose and glycerol solution. Recent diffusivity data by Brignole and Echarte (44) fit Eq. (15) reasonably well.

Recently, Sovova (45) proposed an empirical correlation based on diffusivity data on water and organic acids, which predicts:

$$D \propto \mu^b \tag{16}$$

where b is -1.15 for water, -1.15 for organic liquids with approximately spherical molecules, and -0.5 for liquids with linear molecules, respectively, in the range of μ = 0.3 to 5 cp.

Oxygen diffusivity data on non-Newtonian liquids are not extensive. However, available data (43,46,47) show that, in general, the diffusivity decreases with increase in polymer concentration but the extent of the decrease depends on the type of polymer. Yagi and Yoshida (43) reported D/D_w of close to 1 up to 2 wt. % of CMC, whereas, Reuss, et al. (48), reported D/D_w of approximately 0.5 in 0.5 wt. % of xanthan gum solution.

When the concentration of microbial cells in the fermentation broth is high, the oxygen diffusivity is likely to be affected. Actual data are not available. However, for proteins, it was shown that the diffusivity decreases with increasing concentrations (49). The diffusivity in protein or polymer solution can be predicted by knowing the rigid sphere diameter and the solution property (50,51). The microbial cell may be treated as a protein and the diffusivity can be estimated using the prediction equations.

Fermentation broth normally contains some electrolytes. Gubbins, et al. (52), presented a method for estimating diffusivities in electrolyte solutions. In general, the diffusivity decreases linearly with increase in ionic concentration.

The oxygen diffusivity in fermentation broth may differ from that of pure water due to a combination of different effects. It is expected that the diffusivity changes with time in batch fermentations. This effect will be pronounced in fermentations involving viscous broth.

The effect of viscosity on k_L is mainly due to the diffusivity change as described above. Therefore, if correct diffusivity and the proper length and velocity scales that are responsible for mass transfer are used for calculating k_L, there is no need to consider the effect of viscosity separately (except in the case when viscosity appears explicitly in k_L expression as illustrated in the beginning of this section). Confusion arises when k_L is expressed as a function of less-well defined parameters. For example, in interpreting Calderbank's data (53), Baykara and Ulbrecht (54) stated: "Much less is known about the influence of viscosity upon the true mass transfer coefficient, k_L....k_L in 3% CMC is about 1/3 of that in water." Calderbank's data show that it is the decrease in terminal velocity (approximately 1/3 that of water) that causes decrease in k_L. Therefore, as long as we keep track of the basic parameters (D, V, and L in Eq. (5)), the viscosity effect will be self-evident.

For small, rigid sphere bubbles in creeping flow (when $m = n = 1/3$ in Eq. (6)), the terminal velocity is inversely proportional to viscosity (55). Therefore, k_L is inversely proportional to the square root of viscosity as long as the diffusivity is not much affected by viscosity (this is the case for some, but not all, non-Newtonian liquids). Andrew (5) suggested such a relationship for non-Newtonian fermentation broth. However, it is necessary to check the validity of the assumptions involved before attempting to use the relationship. It is shown that the oxygen diffusivity is reduced significantly in dilute xanthan gum solution (48), which shows non-Newtonian behavior.

The surface active materials such as surfactants, proteins, and some microbial cells affect mass transfer via two different mechanisms (2): surface film resistance and hydrodynamic effect. Bubble rise velocity is significantly reduced in both Newtonian and non-Newtonian liquids in the presence of surface active materials due to existence of surface tension gradient at the bubble interface. Kawase, et al. (25), recently summarized the previous efforts on modeling terminal velocity reduction due to surfactants. For small bubbles whose surface is laden with surfactants and microbial cells, the mass transfer resistance due to the adsorbed species may become significant. According to Schuegerl, et al. (27), surface resistance effect can be offset by adding particular salts such as LiCl or organic compounds such as urea.

B. Gas-Liquid Interfacial Area

The gas-liquid interfacial area, 'a', is expressed as:

$$a = 6H/d_s \tag{17}$$

where H is gas holdup (volume gas per volume reactor) and d_s is Sauter mean bubble diameter defined by:

$$d_s = \Sigma m_j d_j^{\,3}/\Sigma m_j d_j^{\,2} \tag{18}$$

where m_j is the number of bubbles of diameter d_j. Note that 'a' defined in Eq. (17) is based on combined gas and liquid volume. The value of 'a' based on liquid volume alone can be obtained by dividing Eq. (17) by (1-H). If the bubble size in the reactor has a Gaussian distribution, and k_L increases linearly with bubble diameter (this is usually the case with fermentation medium: see Fig. 1), then the product of 'a' defined by Eq. (17) and k_L corresponding to d_S give an accurate estimate of the actual k_La of the reactor. The error involved in such an estimate does not exceed 2% to 3% (56).

Note that the Sauter mean bubble diameter, d_S, is different from the arithmetic mean bubble diameter, $\bar{d}$. According to Ueyama, et al. (57), the following approximation can be used for conversion:

$$d_s = 1.5\,\bar{d} \tag{19}$$

Bubble diameter and gas holdup are affected by liquid properties, and gas and liquid flow patterns caused by power input via gas expansion (bubble column) or impeller (stirred tank). Excellent reviews on bubble dynamics are available from Shah, et al. (7), and Moo-Young and Blanch (4).

1. Bubble Size. In general, the bubble size and its distribution differ between the region where the bubble is generated and the rest of the reactors (3). Close to the sparger, gas flow rate, liquid property, and, in the case of stirred tank, impeller power input affect bubble diameter. In regions far from the sparger, liquid property and the local hydrodynamics control the bubble size: a balance between the rates of bubble coalescence and breakup determines the bubble size distribution.

In sparged stirred tanks, and bubble columns employing efficient spargers such as liquid jet spargers, the Sauter mean bubble diameter is correlated by the following form (7):

$$d_s = CH^n\ (\mu_g/\mu_L)^{.25}\sigma^{.6}(P/V)^{-.4}\rho^{-.2} \tag{20}$$

where C and n are constants that depend on the type of stirrer and the type of liquid phase, and σ, ρ, μ are surface tension, density, and viscosity, respectively. Note that the group $\sigma^{.6}(P/V)^{-.4}\rho^{-.2}$ in Eq. (20) comes from theoretical consideration: the term results from the assumption of constant Weber number in Kolmogoroff's isotropic turbulence field (4). The Weber number represents the ratio between the dynamic pressure force, τ, acting on bubbles and the surface tension force:

$$We = \tau/\ (\sigma/d)$$

The values of C and n are given by Calderbank (53): in electrolyte solutions, C = 2.25, n = 0.4; and in alcohol solutions, C = 1.90, n = .65. The actual dependence of bubble diameter on power input can be estimated when the gas holdup in Eq. (20) is expressed in terms of (P/V).

For bubble columns with conventional spargers (single or multiple hole solid spargers), Shah, et al. (7), recommended Akita and Yoshida's correlation (32):

$$d_s = 26g^{-.56}D_c^{-.3}\rho^{-.74}\mu^{.24}\sigma^{.5}u^{-.12} \tag{21}$$

where D_C, ρ, μ, σ, and u are column diameter, liquid density, viscosity, surface tension, and superficial gas velocity, respectively. Although the correlation is based on columns with diameters of less than 30 cm, bubble diameter becomes independent of the column diameter beyond D_C = 30 cm. Eq. (21) is applicable for u of up to 7 cm/s, which is generally in the range of those employed in biochemical reactors.

Eq. (20) and (21) show that the properties of liquid, especially, the surface tension, has a strong influence on Sauter mean diameter. Operational parameter, i.e., the superficial gas velocity, has only a slight effect in the direction of decreasing d_S. However, it is shown that the d_S increases mildly with gas superficial velocity in non-viscous, simulated fermentation media (58) and in non-Newtonian media (28). Especially, Schuegerl, et al. (58), showed that 1% methanol in water causes d_s to increase markedly with superficial gas velocity, whereas, 1% ethanol suppresses such tendency. Schuegerl, et al. (58), reported that the empirical correlation of Koide, et al. (59), fits their data well for water and 1% methanol solution.

Literature survey shows that the type of bubble generating device has a pronounced effect on bubble diameter and its distribution especially at low superficial gas velocities ($u < 5$ cm/s). Two-phase jet spargers (60) and rubber spargers (61) are capable of generating very fine, stable bubbles, and thus increase interfacial area and gas holdup compared with the conventional single- or multi-hole spargers. In two-phase jet spargers, the gas is sprayed in very fine bubbles via nozzle through which gas and liquid are injected at high speed. Correlations are available for predicting bubble diameters from single orifice (3,62) and for two-phase jet spargers (60). If bubble coalescence is not very significant, the bubble diameter from the sparger is more or less maintained in the vessel, especially in bubble columns. In such cases, the bubble diameter correlations for spargers have significance.

For a given liquid and fixed operational parameters, bubbles dispersed in liquid show approximately a Gaussian distribution. Experimental bubble size distributions are presented for dilute alcohol solutions (58), for viscous CMC solutions (29), and for glycerol solutions (28). Buchholz, et al. (63), compared bubble size distributions obtained by different experimental methods and found out that, although the shapes of the distribution function differ somewhat between different experimental methods, when averaged, the resulting distribution was close to the Gaussian. In very viscous media, there are two distinctive classes of bubbles as described earlier, and only the large ones contribute to mass transfer.

The standard deviations of bubble size distributions give an indication of the degree of scatter. Akita and Yoshida (32) reported the normalized standard deviation of 0.2 to 0.6 for their bubble columns, and observed that the deviation increased with increase in superficial gas velocity.

The effect of electrolytes and surfactants for decreasing the rate of coalescence and thus making the bubble size more uniform compared with coalescing system is summarized by Shah, et al. (7). For electrolytes, the increase in surface tension, and, for surfactants, the dipole arrangement of molecules at the interface play major roles in hindering coalescence.

The fermentation medium usually contains low concentrations of salt and sugar, and the mean bubble diameter is somewhat lower than that in distilled water. As the cell concentration increases, the mean bubble size decreases progressively. Schuegerl, et al. (27), showed that when the cell (yeast culture) concentration was in the range of 5 to 45 gm/L, the mean bubble diameter remained constant at approximately 1 mm.

2. *Gas Holdup*. The gas holdup, H, depends on superficial gas velocity, power input, and the physical properties of the medium. Theoretical development on prediction of holdup has been rather slow, and the prediction relies mainly on empirical correlations. Shah, et al. (7), recently gave an extensive review on gas holdup correlations on bubble columns.

For both stirred tanks and various bubble columns, the following correlation appears to hold (64):

$$H \propto (P/V)^m u^n \qquad (22)$$

where m, and n are empirical constants that depend on liquid properties and the type of gas-liquid contacting device. For stirred tanks in air-water system, m = .53 and n = .47, whereas m = .23 and n = .77 for non-coalescing electrolyte solutions (4). For bubble columns, n varies from 0.7 to 1.2 in bubbly flow regime (7). Although simple in its form, Eq. (22) is not convenient to assess the effects of liquid properties.

For bubble columns, the most widely used gas holdup correlation is that of Akita and Yoshida (26):

$$H/(1-H)^4 = K\, u\mu^{-1/6}\, \sigma^{-1/8}\, g^{-7/24}\, \rho^{21/72} \qquad (23)$$

where K = 0.2 for pure liquids, and 0.25 for electrolyte solutions. Note that Eq. (22) is based on data on water, aqueous glycol, and methanol solution, with u of up to 0.4 cm/s. Hikita, et al. (65), improved the correlation somewhat by taking into consideration of gas property. It is interesting to note that Eq. (23) is almost identical to the semi-theoretical correlation given by Merseman (66):

$$H/(1-H)^4 = K'\, u\mu^{-1/6}\, \sigma^{-1/8}\, g^{-7/24}\, \rho^{26/72} \qquad (24)$$

where K' is a constant.

Gas holdup data for non-Newtonian liquid (mainly dilute aqueous CMC solution) are available from Schuegerl (29) and Franz, et al. (28). It appears that bubbly flow regime is achievable only at low gas flow rates, and the bubbly and the slug flow co-exist even at low gas flow rates (7).

In stirred tanks, gas holdup was recently studied by Sridhar and Potter (67), Hassan and Robinson (68) and Einsele and Finn (69). Recently, Hughmark (70) suggested the following correlation for the gas holdup:

$$H = 0.74\ Q^{.5}\ (N/V) D_I^{\ 3}\ D_i^{\ -.5}\ d_s^{\ 1/4}\ \ g^{-.5}\ \sigma^{-.25} \tag{25}$$

where Q, N, V, D_I, D_i, and σ are gas flow rate, impeller speed, liquid volume, impeller diameter, impeller blade width and surface tension, respectively. Taking d_s as 2.5 mm, Hughmark reported that Eq. (25) fits 77 data sets with 8% deviation. Note that Eq. (25) was obtained based on data from relatively small tanks (less than about 50 cm vessel diameter), air-water system, gas flow rate of 1 to 36 VVM, and H of .02 to .17.

In stirred tank fermenters of low viscosity culture, liquid pumping action by the impeller increases the gas holdup by draggin down the gas bubbles (5): bubbles risen to the surface foam layer are immediately dragged back by the axial liquid flow. Andrew (5) presented an expression for the critical power input above which bubble drag-down occurs. Below this critical power input, the gas holdup remains more or less constant (in this range, the stirred tank is similar to a bubble column). However, once the power input exceeds the critical value, the gas holdup increases with power input. The critical (P/V) is lower in non-coalescing system.

In various ionic solutions, the gas holdup increases linearly with ionic strength (due to hindrance of bubble coalescence), but levels off when the ionic strength exceeds 1.0 gm-ion/L (65). The effect of increased pressure, small suspended particles, and column geometry do not appear to affect the gas holdup (7). Considerable scatter in H among different investigators is attributed to different sparger design and the liquid compositions.

Schuegerl, et al. (27), did extensive measurements on the gas holdup in actual fermentation media. The broth containing corn steep liquor showed a marked increase in the gas holdup compared with tap water, especially, at high superficial gas velocities (beyond 4 cm/s). They attributed this phenomena to the decreased bubble diameter and the coalescence depression due to acids (lactic and amino acids) existing in the broth. Addition of antifoaming agent decreased the gas holdup considerably (to the level close to tap water). They also showed that much higher gas holdup was achieved with injector nozzle compared with a perforated plate sparger.

3. *Interfacial Area.* There are basically two different methods for determining gas-liquid interfacial area: chemical and physical. In the chemical method, the interfacial area, a, is estimated directly from absorption experiments, while,

in the physical method, Sauter mean bubble diameter and gas holdup are first measured and then Eq. (17) is used to calculate 'a'. Recent reviews on interfacial area correlations are given by Shah, et al. (7), for bubble columns, and by Joshi, et al. (9), for stirred tanks.

Shah, et al. (7), recommended Akita and Yoshida's correlation (32) for conservative estimates of interfacial areas at relatively low gas velocities:

$$a = (.3/D_c)(gD_c^{\,2}\rho/\sigma)^{.5} \quad (gD_c^{\,3}/\nu^2)^{.1}H^{1.13} \tag{26}$$

where D_c, ρ, σ, and ν are column diameter, liquid density, surface tension, and kinematic viscosity, respectively. Note that the correlation is based on coalescing liquid (water, 0 - 100% glycerol), and for gas holdup of 0.04 to 0.2. If the condition of isotropic turbulence prevails in the system such as when using a two-phase nozzle sparger, Shah, et al. (7), recommended the correlation of the following type:

$$a = K(P/V)^{.4}H^n \tag{27}$$

where (P/V) is energy dissipation per unit volume, and K and n are constant.

Interfacial area correlations for non-Newtonian liquids are scarce. For dilute CMC solutions (0.5% to 1.6% wt) in a bubble column exhibiting slug flow, Schumpe and Deckwer (71) proposed the following correlation:

$$a = 48.7\ (u/\mu_a)^{.51} \tag{28}$$

where μ_a is the apparent viscosity. In general, in non-Newtonian liquids, the interfacial areas are much smaller than those of the Newtonian cases.

In stirred tanks, for a given gas flow rate, there is a critical impeller speed (or power input per unit volume) below which the interfacial area remains essentially constant (9). This was explained in the gas holdup section. The correlations are for the regions beyond the critical impeller speed. Calderbank and Moo-Young (4) proposed the following correlations:

$$a = .55(P/V)^{.4}u^{.5} \tag{29}$$

for coalescing liquid (clean, air-water system), and

$$a = .15(P/V)^{.7}u^{.3} \tag{30}$$

for non-coalescing liquid (air-electrolyte solution). Note that the recent investigation by Lopes de Figueriredo and Calderbank (72) showed that 'a' is proportional to $(P/V)^{.25}$ and $u^{.75}$.

Fig. 2 (from Ref. 64) shows the comparison of interfacial areas between different contacting devices as a function of power input. It is noted that bubble columns are comparable with stirred tanks. The advantage of the stirred tanks over the bubble columns is that a higher energy input can be applied to generate more interfacial area. Reactors equipped with two-phase jet nozzle spargers are shown to be far more efficient compared with the rest of the reactors.

Schuegerl, et al. (27), investigated the effect of yeast cell concentrations on interfacial area: the interfacial area increased with cell concentration up to 6 gm/L and then leveled off. Compared with the seed medium, the interfacial area increased over 100% when the cell concentration exceeded 6 gm/L.

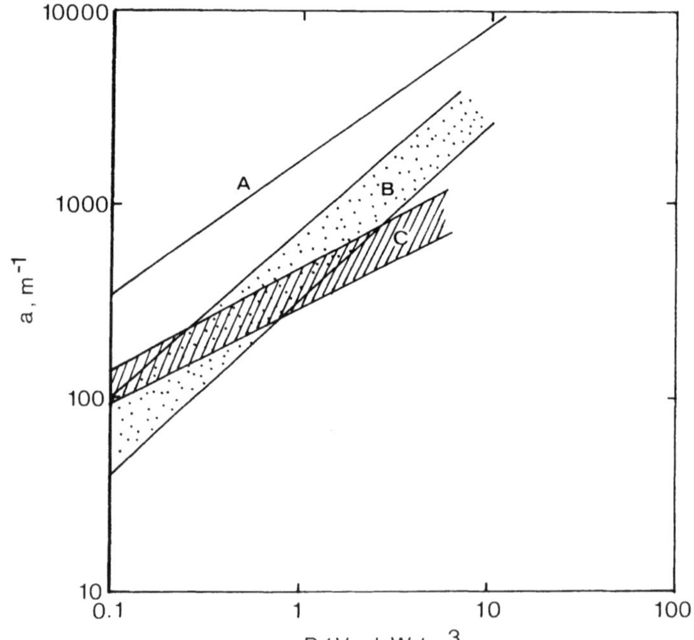

Figure 2. Effect of power input on interfacial area (from Ref. 64). A: column with jet sparger; B: stirred tank; C: bubble column.

C. *Volumetric Mass Transfer Coefficient*

1. *Stirred Tanks.* The volumetric mass transfer coefficient, k_La, is the ultimate parameter that describes the performance of aeration devices. It is, therefore, desirable to express k_La in terms of operational parameters and fluid properties. Reviews on k_La correlations were given by Van't Riet (73), Moo-Young and Blanch (4), Schuegerl (29), and Joshi, et al. (9).

There are basically two different types of correlations for k_La: one is based on power input per unit volume and the other is based on dimensional analysis. The correlation based on (P/V) has its root from Kolmogoroff's isotropic turbulence theory (53). The advantage of such approach is in its simplicity and wide applicability: for example, the same correlation can be used for both stirred tanks and bubble columns as long as the condition of isotropic turbulence prevails in the reactor. However, since the liquid properties are not included, it is rather difficult to find out their effects on k_La. The correlations based on dimensional analysis are more device dependent, but the liquid properties appear explicitly in the correlation, which is especially useful in the case of viscous media.

Van't Riet (73) proposed the following correlations from extensive literature data. For water (coalescing system),

$$k_La = 2.6 \times 10^{-2}\ (P/V)^{.4}\ u^{.5}\ \ s^{-1} \tag{31}$$

where (P/V) is in W/m^3 and u is in m/s. Eq. (31) is applicable for reactors of up to 2600 L, and (P/V) of between 500 and 10,000 W/m^3. For strong ionic (non-coalescing) solutions.

$$k_La = 2.0 \times 10^{-3}\ (P/V)^{.7}\ u^{.2}\ \ s^{-1} \tag{32}$$

As long as the vessel is properly stirred (indicated by a good circulation), vessel geometry, stirrer type and number, stirrer position, liquid height in tank, and sparger type do not affect the correlations geiven by (31) and (32). If (P/V) is less than 10^3 W/m^3, then the energy supplied by the sparged gas has to be taken into account. A liquid is considered strongly ionic, when, for example, NaCl concentration exceeds 10 gm/L. The prediction is accurate within 20% to 40%.

The constants and exponents in Eq. (31) and (32) are quite close to the original correlation by Calderbank and Moo-Young (74). Moo-Young and Blanch (4) gave a list of similar correlations and showed that the exponent for (P/V) ranges between 0.4-0.42 for water and 0.52-0.74 for electrolyte, while the exponent for u ranges 0.35 to 0.5 for water and 0.26 to 0.62 for electrolyte. Comparisons of Eqs. (31) and (32) with Eqs. (29) and (30), respectively, show that k_La is affected mainly by the interfacial area.

For low concentration single cell cultures such as bacteria and yeast, where the liquid properties are not appreciably different from those of water, Eq. (31) and (32) can be used for k_La predictions. Since the fermentation broth usually shows non-coalescing behavior due to the presence of surface active materials, Eq. (32) will give a better estimate. When the broth becomes coalescent due to a high concentration of antifoaming agent, Eq. (30) will be more appropraite.

Correlations originating from dimensional analysis can be divided into two types (29): those relating impeller Sherwood number with impeller Reynolds number and other dimensionless groups:

$$Sh_I = f(Re_I, Sc, \ldots) \qquad (33)$$

and those relating dimensionless volumetric mass transfer coefficient with modified power input and Schmidt number:

$$k_La(V/Q) = A\ (P/Q)^{*p}\ Sc^q \qquad (34)$$

where

$$(P/Q)^* = (P/Q)\ (\rho(\nu g)^{2/3})^{-1} \qquad (35)$$

The correlation of the type given by Eq. (33) was originated from Sideman, et al. (75), and followed by Perez and Sandall (76), and Yagi and Yoshida (43). Eq. (34) type correlation was proposed recently by Zlokarnik (77), and adopted by others (48,78).

A summary of k_La correlations are given in Table 1. Note that Zlokarnik's correlations on coalescing and non-coalescing systems are comparable with Eqs. (31) and (32). The same type of correlations are used for both the Newtonian and non-Newtonian liquids. However, in the non-Newtonian case, apparent viscosity, μ_a, has to be used in defining the Schmidt number, Sc^*. In power law fluids:

$$\mu_a = k(11N)^{n-1}((3n+1)/4n)^n \qquad (36)$$

where k and n are fluid consistency and behavior index, respectively, and N is impeller speed in rps.

There are two types of non-Newtonian broths in fermentation (4): in antibiotic fermentation, the non-Newtonian behavior originates from mycelia growth, whereas, in

Table 1. k_La Correlations for Stirred Tank.

	Correlation	System	Ref.
1	$k_La \frac{V}{Q} = 1.1 \times 10^{-4} [\frac{P}{V} \cdot \frac{1}{\rho(\nu g^4)^{1/3}}]^{0.8}$	Electrolyte solution	Zlokanik (77)
2	$k_La \frac{V}{Q} = 0.015 (\frac{P}{Q})^*$	Water	Zlokanik (77)
3	$k_La \frac{V}{Q} = 0.082 (\frac{P}{Q})^{*0.6} Sc^{*-0.3}$	CMC	Henzler (85)
4	$k_La \frac{V}{Q} = 0.045 (\frac{P}{Q})^{*0.5} Sc^{*-0.3}$	Glycerol, Glucose	Henzler (85)
5	$(\frac{k_LaD_I^2}{D}) = 21.2 (\frac{N_\rho D_I^2}{\mu_a})^{1.11} (\frac{\mu_a}{\rho D})^{0.5} (\frac{D_I u}{\sigma})^{0.45} (\frac{\mu_g}{\mu_a})^{0.69}$	Carbopol solution	Perez and Sandall (76)
6	$(\frac{k_LaD_I^2}{D}) = 0.06 (\frac{N\rho D_I^2}{\mu})^{1.5} (\frac{D_I N^2}{g})^{0.19} (\frac{\mu_a}{\rho D})^{0.5} (\frac{\mu_a u}{\sigma})^{0.6} (\frac{\mu_a D_I}{u})^{0.32} (1 + 2 D_e^{0.5})^{-0.67}$	CMC	Yagi and Yoshida (43)

polysaccharide fermentation, the non-Newtonian behavior is from the polymer produced by the organism. Ruess, et al. (48), stressed that, to use the correlation developed for homogeneous non-Newtonian model fluid such as CMC in heterogeneous fluid (such as mycelial culture broth), the Schmidt number of the supernatant has to be used.

In Yagi and Yoshida's correlations (Table 1), the same form is used for both the Newtonian and the Non-Newtonian liquids except that an additional term is added for the non-Newtonian case. The additional term is the Deborah number, De, defined by:

$$De = t_r/t_p \qquad (37)$$

where t_r represents the characteristic material relaxation time proposed by Prest, et al. (79), and t_p the characteristic process time. For stirred tanks with a given impeller speed N (in rps), t_p is expressed as:

$$t_p = 1/N \qquad (38)$$

Note that t_r is 0 for Newtonian fluid. The method for calculating t_r is given by Yagi and Yoshida (43).

The method of calculating the mechanical agitation power under aeration condition was summarized by Moo-Young, et al. (4), for both the non-Newtonian and the Newtonian liquids. In general, the ratio of gassed power vs. ungassed power, (P_g/P_o), is correlated with the aeration number, Na, defined by:

$$Na = Q/(ND_I) \qquad (39)$$

where Q, N, D_I are volumetric gas flow rate, impeller speed and impeller diameter, respectively. Dickey (145) recently reviewed different correlations and concluded that none is satisfactory. However, the compiled data showed that, for Na $\geq$ 0.1 (most of the industrial application), (P_g/P_o) ranged between 0.3 and 0.6 with a high density of data around 0.4.

Andrew (5) warned that the mass transfer data extracted from intensely agitated, small, laboratory stirred tanks are useless for the design of commercial or even the pilot scale reactors because they operate under a totally different hydrodynamics. Fig. 3 from Andrew shows the region where the k_La for large scale reactors lies. Note that below the power input of approximately 1 kW/m^3, the stirred tank behaves essentially as a bubble column.

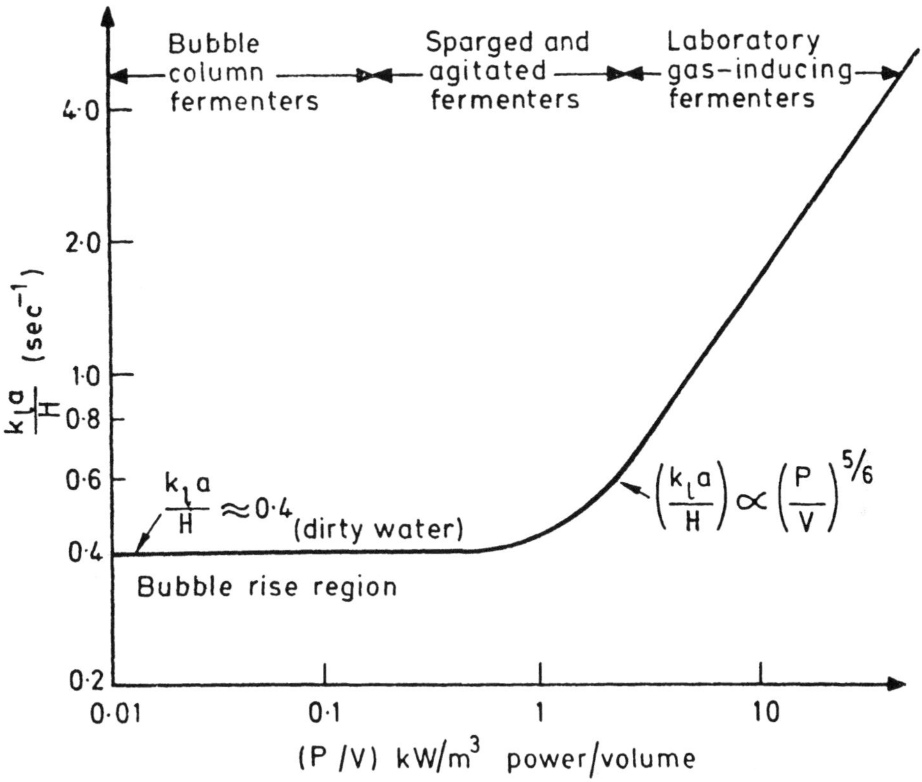

Figure 3. Effect of power input on volumetric mass transfer coefficient (from ref. 5).

2. *Bubble Columns.* Although various configurations exist for bubble columns (for example, those with internal draft tubes, baffles, or external loop), the correlation for k_La has been obtained for plain bubble columns without any internals. Similar to the case of stirred tanks, the correlations can be classified into two categories: those based on superficial gas velocity, and those based on dimensional analysis. Since the concentration driving force may not be constant throughout the column, a proper flow model has to be adopted to calculate k_La, especially for large and long columns. Comprehensive reviews on k_La in bubble columns are available from Shah, et al. (7), and Moo-Young and Blanch (4).

For non-viscous liquids (coalescing or non-coalescing), the following correlation has been widely used:

$$k_La = b\,u^m \tag{40}$$

where b and m are empirical constants. Note that k_La is in s^{-1} and u is in m/s. Wide range of available data (4,7) show that m is more or less constant (0.78 -1), whereas, b varies widely (0.24 - 1.45). Based on Higbie's penetration theory and Kolmogoroff's isotropic turbulence theory, Kastanek (80) showed that m varies between 0.78 and 0.82 and does not depend on sparger design. Shah, et al. (7), recommended to use m = 0.8. Comparisons of Eq. (40) with Eqs. (29) and (30) indicate some uncertainty regarding the value for m. The constant b is a function of liquid property and sparger design. Although b is slightly affected by the liquid type (b is slightly higher in electrolyte solutions compared with water), it is most significantly affected by the type of sparger employed: Deckwer (81) observed a three-fold increase in b when single hole sparger was replaced by a sintered plate sparger.

As long as the column exceeds minimum diameter (D_C = 15 cm), k_La does not depend on D_C. This was shown by Shah, et al. (7): Akita and Yoshida's correlation (26) which was obtained for D_C ranging between 15 to 60 cm, fits data well which were obtained from columns with D_C of 1 to 5.5 m. Shah, et al., also showed that the liquid velocity does not have a significant effect on k_La: only a slight increase in k_La was observed when the liquid velocity was varied from 5 cm/s to 10 cm/s. Based on air-water data in different columns, Deckwer (81) concluded that column geometry, the direction of liquid flow (counter current or co-current) or the direction of mass transfer (absorption or desorption) do not have significant effects on k_La. The discrepancy between different data sets could be attributed to differences in liquid properties (coalescing or non-coalescing) and the sparger design.

There have been steady developments in k_La correlations via dimensional analysis. Hughmark (82) seems to be the first to propose a correlation which takes into account of the liquid properties. Akita and Yoshida (26) later presented a dimensionless correlation based on extensive oxygen absorption data in various liquids including the viscous glycol solution. Hikita, et al. (65), developed a correlation that takes into account of ionic strength effect. Nakanoh and Yoshida (83) presented a correlation that is applicable to both the Newtonian and the non-Newtonian liquids.

A list of most recent correlations including descriptions of range of parameters and systems involved can be found from Shah, et al. (7). It appears that the most widely accepted model (4,7,10,81) is still that of Akita and Yoshida (26):

$$k_L a (D_c^{\,2}/D) = .6\ H^{1.1} (\nu/D)^{.5}\ \times$$

$$(gD_c^{\,2}\rho/\sigma)^{.62} (g\ D_c^{\,3}/\nu^2)^{.3} \tag{41}$$

where the gas holdup, H, can be calculated from Eq. (23). Note that the effect of electrolyte is considered when calculating H.

More explicit description of electrolyte effect is given in the correlation of Hikita, et al. (84):

$$k_L a = 14.9F(g/u)\ (u\mu/\sigma)^{1.76} (\mu^4 g/\rho\sigma^3)^{-.248}$$

$$\times\ (\mu_g/\mu)^{.243} (\mu/\rho D)^{-.604} \tag{42}$$

where the correction factor F depends on ionic strength, I, of the solution as follows:

$$\text{For } 0 < I < 1.0 \text{ gm-ion/L},\ F = 10^{.0681}$$

$$\text{For } I > 1.0 \text{ gm-ion/L},\qquad F = 1.114 \times 10^{.021I} \tag{43}$$

In general, the correlation of Hikita, et al. (84), gives a slightly higher k_La compared with that of Akita and Yoshida. Also, note that the column diameter is no longer included.

Eq. (41) requires separate calculation of the gas holdup, which is inconvenient. Nakanoh and Yoshida (83) combined Eqs. (23) and (41), and added an additional term so that a single correlation can be used for both the Newtonian and the non-Newtonian liquids:

$$k_L a (D_c^2/D) = 0.09(\nu_a/D)^{.5}(gD_c^2\rho/\sigma)^{.75}$$

$$\times\ (gD_c^3/\nu_a^2)^{.39}(u/gD_c)$$

$$\times\ ((1 + C\ (De)^m)^{-1} \qquad (44)$$

The Deborah number in Eq. (44) is defined by Eq. (37), where the characteristic process time, t_p, in bubble column is defined as:

$$t_p = d_s/v$$

$$= Hd_s/u \qquad (45)$$

where v is the average bubble rise velocity defined by Eq. (8). The values of C and m were 0.13 and 0.55, respectively, for CMC and sodium polyacrylate solutions, and the apparent viscosity, μ_a, for the non-Newtonian liquid is calculated by:

$$\mu_a = K(50u)^{n-1} \qquad (46)$$

where u is in cm/s, K and n are consistency and flow behavior index, respectively. Note that the Newtonian part (De = 0) is essentially the same as Eq. (41) except that the prediction is 25% higher because of a higher value used for the constant (it should have been 0.072 instead of 0.09 to be identical). As mentioned by the authors, the non-Newtonian part needs further experimental verification.

Another type of correlations for non-Newtonian fluids are those of Deckwer, et al. (7), and Henzler (85). For CMC solutions of 1 to 2 wt %, Deckwer, et al., gave the following correlation:

$$k_L a = 0.00315\ u^{.59}\mu_a^{-.84} \qquad (47)$$

where all units are in metric (u = m/s, μ_a = Pa-s). Eq. (47) was recommended when slug flow prevails. Note that Eq. (47) is a modification of Eq. (40). For the same CMC solution, Henzler gave the following correlation:

$$(k_L a/u)(\nu_a^2/g)^{1/3} =$$

$$.06\ (u/(g\ \nu_a)^{1/3})^{-.9}\ (\nu_a/D)^{-.5} \qquad (48)$$

This correlation is of the Zlokarnik (77) type as illustrated in the previous section. Henzler's correlation can be re-written as:

$$k_L a = 0.06\, g^{.633}\, D^{.5}\, u^{.1}\, \nu_a^{-.886} \tag{49}$$

Eqs. (47) and (49) show some discrepancy: while the exponents for the apparent viscosity are almost the same, those for the superficial gas velocity are quite different (.59 vs. .1).

Although several correlations for non-Newtonian fluids emerged in recent years, more experimental verifications, including tests on actual fermentation media, are necessary before we can select the one that can be used with confidence. Simpler correlations are attractive but the effects of different liquid properties on $k_L a$ cannot be readily assessed. The answer to the question: "How many variables need to be incorporated in the correlation?", can only be found from both the theory and experimental verifications.

As noted by Shah, et al. (7), and Deckwer (81), the sparger design has a strong effect on $k_L a$, and the correlations based on solid spargers (single hole or multi-holes) give too low an estimate when more efficient spargers (such as jet nozzles, or rubber nozzles) are used. More recently, Tojo and Miyanami (86) tested down jet spargers and correlated their data as follows:

$$k_L a = 0.029\, (P/V)^{.65} \tag{50}$$

Comparisons with Eqs. (31) and (32) show that the down jet sparger appears to be more effective. More work is necessary to characterize these spargers.

3. Factors Affecting $k_L a$. Most of the correlations presented in the previous sections are based on data on clean liquids of well defined compositions. The actual fermentation broth, however, contains various surface active materials of unknown concentrations (mostly proteins), organic acids and alcohols, and, invariably, antifoaming agents. Andrew (5) pointed out that due to the effects of surfactants and antifoaming agents, the mass transfer experiments made in tap water or salt solutions are of virtually no quantitative value for design of industrial scale biochemical reactors.

The most extensive study on the effect of antifoaming agents on $k_L a$ is that of Schuegerl, et al. (27), in a tower fermenter. When the antifoaming agent Ucolub (a water insoluble polyoxyethylene-propylene copolymer with a mean molecular weight of 2,500 and a kinematic viscosity of 240 cp at 50°C) was present in the amount of 0.1% in water, $k_L a$ values were lower up to u = 3 cm/s, but, beyond that, they were higher compared with clean water. When salt was added (1%), the superficial gas velocity where the transition occurs increased to approximately 5 cm/s. Since the fermenter was operated at superficial velocities higher than the transition value, the k_La in the actual run was lower than that of pure water. The presence of antifoaming agents suppressed the dependency of k_La on superficial gas velocity considerably. Separate determinations of the gas holdup showed the similar trend, which means that the observed effect is mainly due to change in the gas holdup.

Recent study by Paca and Ettler (87) on *Erythromycin* fermentation in a 300 L stirred tank reacter showed a trend similar to that of Schuegerl, et al. The sperm whale oil (antifoaming agent) added to water decreased the k_La, and the decrease became more pronounced with increase in (P/V). Although k_La data at low (P/V) values were not shown, an extrapolation shows that the transition does occur at rather low (P/V). In a simulated viscous medium, the effect of antifoaming agent was pronounced in decreasing k_La. However, in the actual fermentation media, k_La was higher presumably due to the presence of surface active materials and ionic nature of the solution.

As reviewed recently by Viesturs, et al. (88), there are a variety of antifoaming agents, both nautral (such as sunflower oil, cod-liver oil, etc.) and synthetic (silicones, detergents, organic polymers). It appears that the foaming tendency of a fermentation broth depends on the composition, the conditions of treatment (temperature used during sterilization) and also on the type of microorganism employed. Without the antifoaming agent, the fermentation broth is usually non-coalescing due to the various surface active materials, and these surfactants tend to stabilize the foams when they are formed. The antifoaming agents strongly adsorb on the interface and lower this stabilizing action, leading to break down or coalescence of foams (27). On a qualitative basis, the mechanism is known. What is needed is a quantitative description of this phenomena just as it has been done with surfactants (89). Fundamental studies on mixed surfactants-antifoaming agent systems are necessary.

Polymers such as CMC or PAA (Polyacrylamide) are often added in small amounts (100 ppm range) to fermentation media, which has the beneficial effect of decreasing power input per unit volume, and protecting microorganisms from shear deactivation (3). However, Ranade and Ulbrecht (90) showed that mass transfer can be severely reduced. In studying the oxygen absorption rate in a stirred tank under increasing pressure, Sato, et al. (91), showed that k_La increased with increase in pressure. It is likely that the pressure increase has some effect on the gas holdup.

III. MEASURING TECHNIQUES

Reliable correlations or mass transfer models cannot be obtained if they are based on inaccurate experimental data or data reduction processes. Therefore, improving the accuracy of the measurement itself, and refining the procedures involved in estimating the parameters have been of considerable interest. In this section, recent advances in experimental methods and data reduction procedures for obtaining mass transfer related parameters are presented.

A. Volumetric Mass Transfer Coefficient

Active research has been devoted to finding new methods or improving the existing methods. Reviews on earlier state of the art can be found from Tsao and Lee (2), Blanch (3), Lee and Tsao (92), and Van't Riet (73).

There are essentially two different methods for determining k_La: steady state method and dynamic method. In the steady state method, the liquid phase dissolved oxygen concentration is maintained constant either by reaction or by continuous liquid flow, and the k_La is calculated from an oxygen balance:

$$k_La = OUR/(c^* - c_L)_{LM} \tag{51}$$

where $(c^* - c_L)_{LM}$ is the log mean concentration driving force between the inlet and the outlet, and the oxygen uptake rate, OUR, can be estimated from gas phase oxygen balance (for example, in actual fermentation) or by back calculation from the oxidation product formed in the liquid phase (for example, sulfite oxidation method). The log mean concentration difference is more appropriate than the simpler form, $(c^* - c_L)$, because, often, the gas phase, i.e., bubbles in the reactor, cannot be assumed well-mixed, especially in non-coalescent systems. The saturation concentrations at the inlet and the outlet can be obtained from oxygen partial pressure measurements and the Henry's Law constants.

The dynamic method (sometimes, referred to as the non-steady state method, or the "gassing-out" method) is based on transient response of the liquid phase oxygen concentration after a sudden change in inlet gas concentration:

$$dc_L/dt = k_La(c^* - c_L) \tag{52}$$

where c_L is the dissolved oxygen concentration in the bulk liquid, which is usually measured with an oxygen probe. Although the application of Eq. (52) appears simple, there are a number of factors that have to be taken into account to evaluate k_La successfully. Refinements added are the considerations of probe dynamics and the gas phase dynamics (93), and the start-up effect (94).

Van't Riet (73) noted that although the steady state method is simpler to use and may yield more reliable data, very little attention has been paid by investigators. On applying the dynamic method in determining k_La in a stirred tank, he suggested the following. First order probe response can be used rather than the elaborate probe model (95), when the probe time constant (approximated as first order), t_p, is much less than $1/k_La$. The error in the evaluated k_La is less or equal to 6%, when

$$t_p \leq 1/k_La \tag{53}$$

Eq. (53) gives a guideline to estimate the maximum k_La which can be measured with sufficient accuracy, once the probe response time (the time required to reach 63% of the final value in a step response experiment) is measured. Another consideration is the gas phase residence time, t_g, represented by:

$$t_g = HV/(u\pi D_T^2/4) = Hh_T/u \tag{54}$$

where V, D_T, and h_T are aerated volume of fluid, tank diameter, and height of the fluid when aerated. If the t_g is relatively high compared with $1/k_La$, then considerable corrections have to be made, and, as a result, measurements lose accuracy.

Leeuwen (96) suggested that when measuring k_La in oxygen-consuming systems, the initial state has to be quasi-steady state: in other words, before making a step change in oxygen concentration in the gas inlet, the probe reading has to be constant for some time.

Heijnen, et al. (97), studied the influence of very small bubbles on k_La measurements in a viscous system. Since the small bubbles of less than 200 μm diameter move very slowly and, thus, quickly equilibriate with liquid, they behave as an additional oxygen source. As a result, the effective concentration driving force becomes higher, and, if this effect is not considered, the measured k_La by the dynamic method will be erroneous. To correct for this effect, the holdup of small bubbles has to be estimated accurately.

Andre, et al. (98), developed a dynamic method that is applicable to semi-batch fermentation involving solid substrates. The method involves unsteady state absorption or desorption of carbon dioxide: k_La is measured from the gaseous response of CO_2 concentration in the exit gas to a step change in the inlet gas composition. It was shown that at low k_La values, the choice of gas phase dynamics (either well-mixed or plug-flow) does not affect the evaluation. However, when k_La exceeds 0.02 s^{-1}, a proper model has to be used to obtain accurate results. The obtained k_La has to be corrected for the diffusivity difference between oxygen and carbon dioxide.

Linek, et al. (99), compared the dynamic method with the steady state method employing a glucose oxidase reaction. They reported that when a step change in inlet gas composition is made without interrupting aeration and agitation, erroneously low values of k_La were obtained for k_La values higher than 0.03 s^{-1}. When the step change was made after completely interrupting aeration and agitation and allowing all the bubbles to escape, the k_La values agreed well with those of the steady state method for k_La less than 0.08. However, for k_La values exceeding 0.08, the dynamic method yielded lower values. The transport of nitrogen was also taken into account in their evaluation.

In all of the previous dynamic methods, the gas phase was modeled as either well-mixed or plug flow. Chandrasekharan and Calderbank (100) considered a case where the gas phase dynamics is modeled as N tanks-in-series. They concluded that the physical methods with correct characterization of gas phase dynamics and residence time distribution, provide a reliable method for estimating k_La as long as equilibrium conditions between the gas and the liquid are not approached at all times.

Recently, computers have been increasingly used for monitoring and control of biochemical reactors, and, as a result, on-line or real-time estimation of k_La received considerable attention. Yoshida, et al. (101), introduced a method of continuously estimating k_La by the statistical approach. Analog computer was used for the estimation, but a digital computer may also be used. Another continuous measurement of k_La was introduced by Sobotka and Votruba (102), who used adaptive identification method. They reported that the method is especially advantageous when k_La is greater than 0.03 s^{-1}. The results of continuous measurement of k_La showed that, in an intensely agitated small scale reactors, the k_La is not independent of time as assumed by other investigators.

Vardar and Lilly (103) introduced a method for measuring k_La by frequency response technique commonly used in process control for identifying process variables. The electrode was modeled as a first order system with time delay and the probe response to a sinusoidal pressure input was used to construct

Bode diagram, from which k_La was obtained. The method has the advantage of eliminating errors originating from nitrogen flushing due to the dilution of gaseous oxygen and the start up effect (94).

Spriet, et al. (104), presented a method for continuously monitoring k_La with a computer. It was essentially a steady state method represented by Eq. (51) except that considerable corrections were made for each measured variable. Among them were the utilization of carbon dioxide concentration measurements to correct for the outlet flow rate, the barometric pressure corrections for gas phase oxygen concentrations measured by the paramagnetic analyzer, the temperature compensations for carbon dioxide measurements, and the oxygen solubility corrections considering salt content and the actual partial pressure. Comparisons with those obtained by dynamic method showed reasonable agreements. They observed that the accuracy of the static method depends mainly on the precision of the oxygen analyzer.

Recent work by Brooks, et al. (105), show that even with good quality instruments, the errors in the oxygen uptake rate (OUR) estimation based on gas phase concentration measurements can be as high as ± 100%, unless careful calibrations and proper corrections are made. As pointed out by Lee and Tsao (92), there is inherent instability with the dissolved oxygen probes due to the aging effect. The probe stability has been improved in a more recent design incorporating two cathodes in a single probe (106).

The dynamic method introduced by Chapman, et al. (107), appears to be convenient because it does not require the assumption regarding the gas phase dynamics, and k_La can be estimated continuously. The method involves measurements of both the liquid and the gas concentration responses following a step change in inlet gas concentration. The expression for k_La is given by:

$$k_La = \frac{(dC_L/dt)}{(Q/V_G)\int_0^t (1-C_o)dt - C_L\ (HeV_L/V_G + 1)} \tag{55}$$

where C_L and C_o are normalized responses of liquid concentration and outlet gas concentration, respectively, and Q, V_G, V_L and He are volumetric gas flow rate, gas phase volume, liquid volume, and Henry's Law constant, respectively. Since only the normalized responses are used, there is no need to use solubility data. However, accurate measurements on the gas holdup and the volumetric flow rate are required.

All of the methods described so far assume that the liquid phase is well mixed. However, this may not be true in large bubble columns. Experimental measurements (27,108) show that there are considerable distributions of concentrations in the liquid phase. Shioya and Dunn (109) presented a method of estimating k_La in bubble columns exhibiting concentration distributions in both the gas and the liquid phase. Also, a method for finding the optimum probe location was presented: in general, it is advantageous to place the probe in a position which gives a response which is the most sensitive to changes in k_La. Although the optimum position changes with the magnitude of k_La, in most realistic conditions, the position is between Z = 0.4 to 0.8, where Z is the normalized column height from bottom. The method requires known values of dispersion coefficients in both phases. The methods of obtaining dispersion coefficients are recently reviewed by Shah, et al. (7,110).

B. *Interfacial Area*

Although, for all practical purposes, k_La determination is sufficient for design and scale-up of biochemical reactors, separate measurements of interfacial area are necessary when it is desired to determine the underlying transport mechanism. The experimental techniques are recently reviewed by Shah, et al. (7), and Charpentier (1). A summary of the recent advances are given below.

1. *Physical Method.* The gas holdup and the Sauter mean bubble diameter are determined first, and then the interfacial area is calculated by Eq. (17). The gas holdup can be easily estimated by:

$$H = (h_T - h_L)/h_T \tag{56}$$

where h_T is the liquid height when aerated, and h_L is the liquid height without aeration. The direct height measurements are not very accurate (15% to 20% accuracy), and Charpentier (1) recommends manometer tapping method or conductivity level detector method for improving the accuracy.

The Sauter mean bubble diameter can be estimated from experimentally measured bubble size distributions by using Eq. (18). Basically, there are three different methods of determining bubble size: direct photography, light transmission method using optical probes, and two-electrode conductivity probe method. Recently, Buchholz, et al. (63), compared these three methods: direct photography has the advantage of measuring wide bubble sizes (0.1 to 20 mm), but it can only measure bubbles close to the wall; conductivity probes and the optical probes can be used in any positions inside the reactor, but the measurable bubble size spectrum is somewhat limited (0.6-20 mm for the conductivity probes and

1-20 mm for the optical probes). Although different methods give somewhat different bubble size distributions, the calculated Sauter mean diameters differ only slightly (7).

The conductivity probe is applicable only for conducting liquids. Recently, Burgress and Calderbank (111) developed a resistance probe method that can be used in non-conducting liquids. With a proper probe design, gas holdup, interfacial area and even the liquid side mass transfer coefficient can be estimated from the locally measured bubble size distribution and velocity.

Recently, Sriram and Mann (112) developed a new technique of measuring bubble size distribution by measuring the transient gas holdup after disconnecting the gas supply. This so-called "dynamic gas disengagement technique" may be quite useful because the method is rather simple and easy to apply.

2. *Chemical Method.* While the physical method enables measurements of local values, the chemical method yields the overally interfacial area. By selecting a suitable reactant, the chemical method gives volumetric mass transfer coefficient or interfacial area, or both. Intensive research has been devoted to this area as summarized by Charpentier (1). Most widely used reactions are sulfite oxidation and carbon dioxide absorption in alkali. Schumpe and Deckwer (113) compared the two methods and concluded that the chemical method be used at low conversions to obtain more consistent data. Recently Midoux, et al. (114), presented criteria to apply when selecting a reaction scheme to measure k_La, a, or both.

In general, the chemicals used for reaction make the system non-coalescent except when the reactant concentration is relatively small (for example, the Hydrazine method by Zlokarnik (77)). When the system is coalescent, the gas phase can be considered as completely mixed, whereas, in non-coalescent system, the gas phase may be between well-mixed and plug flow. Hassan and Robinson (115) proposed to make corrections for the concentration driving force for the non-coalescent systems.

The interfacial area measured by the chemical method appears to be consistently higher than that obtained by the physical method. Schumpe and Deckwer (113) showed that, in bubbly flow, the photographic method yielded an interfacial area which is 35% greater than that of the sulfite method. This could have been caused by the spatially dependent nature of the photographic method: measurements are possible only on locations close to the vessel wall (63). Sridhar and Potter (116) compared the light transmission method with the chemical method, and found a similar trend. They attributed the discrepancy to the increase in mass transfer rate in the impeller region. Charpentier (1) emphasized using both the chemical and the physical methods simultaneously: while the chemical method provides overall interfacial area immediately usable for design, the physical method gives local distributions which are necessary to deal competently with scale-up.

C. *Liquid Properties*

Important liquid properties in mass transfer studies are diffusivity, solubility and viscosity. The solubility and the viscosity in Newtonian liquids are relatively easy to measure, while the viscosity in non-Newtonian liquids and the diffusivity are more difficult to measure. Since viscosity measurements in non-Newtonian fermentation media were reviewed elsewhere (117,118) only the diffusivity and the solubility will be discussed in this section.

1. *Diffusivity*. Accurate diffusivity data are needed for predicting mass transfer, and to use correlations presented in the previous sections. The literature values on oxygen diffusivity vary widely under the same condition: at 25°C, it ranges from 1.87 to 2.6×10^{-5} cm^2/s (41). Although different experimental methods used might have caused some errors, this indicates clearly the difficulty involved in diffusivity measurements.

Different experimental methods and their relative accuracy have been reported (41). The experimental methods can be classified into two categories: those applicable to gases in general, such as the laminar jet method (119) or the stationary bubble method (120), and those specific to oxygen (121,122).

Recently, there have been some improvements in diffusivity measurement techniques. Sovoca and Prochazka (123) presented a method consisting of measuring the rate of absorption of a gas into a laminar film of liquid flowing radially over the surface of a horizontal disk. The proposed method preserves the advantages of the laminar jet method (short contact time, lesser sensitivity to vibrations and convection) and, at the same time, eliminates the end effect. De Block and Fortuin (120) have considerably improved the stationary bubble method. In the old method, the bubbles were allowed to decrease in size as the absorption progressed. By maintaining the bubble size constant, De Block and Fortuin could obtain higher accuracy.

Literature survey indicates that very little diffusivity data are available on viscous, non-Newtonian aqueous media, and furthermore, the reported values are not consistent. Diffusivity data in cell suspensions and in heterogeneous media such as cell flocs, and mycelial pellets are also needed to aid mass transfer studies.

2. *Solubility*. Significant studies came from Schumpe, et al. (124), and Quicker, et al. (125). It was shown (124) that the solubility of oxygen in electrolyte solutions can be expressed by:

$$\log(He/He_o) = \Sigma H_i I_i \tag{57}$$

where He and He_o are Henry's Law constants for electrolyte solution and water, respectively, and H_i and I_i are ion specific constants and ionic strength, respectively. The values of H_i are given in the reference. Note that the method is a slight modification of the method of Van Krevelen and Hoftijer (126).

Quicker, et al. (125), investigated the effect of sugars on the solubility of oxygen and carbon dioxide in fermentation media, and found that the solubility reduction is log-additive with respect to various compounds. The solubility in mixed electrolytes and sugars is expressed as follows:

$$\log(He/He_o) = \Sigma H_i I_i + \log(He_{sj}/He_o) \tag{58}$$

where He_{sj} is the Henry's Law constant for jth sugar in the solution. The method of calculating He_{sj} are given in the reference for glucose, lactose, and sucrose.

Improved experimental methods for determining solubilities of gases in general are given by Rivas and Prausssnitz (127) and by Armitage, et al. (128). A rapid oxygen solubility measurement by using oxygen probe has recently been introduced by Kaeppeli and Fiechter (129).

IV. AERATION DEVICES

A. *Aeration Devices Without Mechanical Agitation*

In the past few years, considerable research has been devoted to aeration devices that do not have mechanical agitation. It is recognized that the bubble columns (of various designs) have the advantage of simplicity and good aeration capability, especially when equipped with efficient spargers such as two phase nozzles.

Tower loop fermenters have been applied in cultivating various microorganisms by Schuegerl's group. In growing yeast on glucose or ethanol substrate (130), the interfacial area was found to increase 3 to 5 times, and k_L to decrease from .05 to .02 cm/s, as the cell concentration increased from 0 to approximately 20 gm/L. It was also found that the ethanol used as the substrate had a considerable influence on k_La and 'a': higher concentrations of ethanol gave higher mass transfer rates due to suppression of coalescence (131). In producing penicillin, the volumetric oxygen transfer coefficient was 4 to 5 times higher in pellet suspensions than in filamentous broth. In growing mycelial culture in 20 m long, 20 L tubular loop fermenter, Ziegler, et al. (132), reported that k_La was proportional to $(P/V)^{.4}u^{.4}$.

Large bubble columns have been studied by several investigators. In a 250 cm high, 66 cm diameter cocurrent bubble column, Alvaez-Cuenca and Nerenberg (108) tested both large gas (u = 4 - 28 cm/s) and liquid flow rates (liquid superficial velocity of 2.5 to 10 cm/s). It was found that more than 95% of the mass transfer occurred at lower 20% of the column, and the two zone model rather than the axial dispersion model described the system better. Kataoka, et al. (133), tested 550 cm diameter, 700 cm deep bubble column and found that the k_La determined agreed well with Akita and Yoshida's correlation, provided that perfect mixing is assumed for both the gas and the liquid phases. Botton, et al. (134) tested bubble columns operating at high gas throughput (up to u of 8 m/s): it was shown that, to achieve the same mass flux, the bubble columns require the same amount of energy but the size can be several times smaller, when compared with other equipment.

A U-tube aeration system was analyzed by Takamatsu, et al. (135): they found that optimal liquid velocity exists which gives minimum power consumption at a given gas velocity. Wick (136) applied cocurrent gas-liquid flow for continuous alcohol fermentation. Design aspects of fermenters with power introduced by aerating gas were investigated by Viesturs, et al. (137). Merchuk and Stein (138) and Lutmann, et al. (139), used distributed parameter model for analyzing airlift fermenters.

B. Stirred Tanks

In investigating the scale up of aerated stirred vessels, Lopes De Figueiredo and Calderbank (72) noted that, since k_La is proportional to the power dissipation to a power less than one and it is more dependent on superficial gas velocity, using multiple agitators and/or increasing the superficial gas velocity could be used for increasing k_La in large vessels. Also, they stressed that the gas phase dynamics has to be considered, because it is likely that well-mixing may not be achieved in large scale vessels.

In a subsequent study, Chandrasekharan and Calderbank (140) indicated that the scale-up with geometrical similarity results in a marked sacrifice in aeration performance, and that, for very large diameter vessels, multiple spargers rather than mechanical agitation might prove a better solution. Nienow and Lilly (141) have recently studied the power drawn by multiple impellers. Nishikawa, et al. (142), reported that, in a relatively large vessel, gas phase has to be assumed plug flow in calculating k_La.

Tanaka (143) studied the effect of apparent viscosity on k_La in high density plant cell culture (10 - 18 gm/L) which showed non-Newtonian behavior. When k_La was plotted as a function of apparent viscosity, two regions were identified: a region of mild decrease, and a region of steep decrease in k_La with increase in the apparent viscosity. The apparent viscosity where the transition occurred tended to increase with increase in power input. A similar trend was observed by Nishikawa, et al. (144), who used CMC as the fluid.

V. CONCLUSION

The capability of predicting volumetric mass transfer coefficients in conventional bubble columns and stirred tanks involving non-viscous media has been enhanced considerably over the past few years. Refinements have to come in the area of quantifying the effects of antifoaming agents, surfactants and cells on the interfacial area and the liquid phase mass transfer coefficient. More investigations are needed on bubble dynamics and mass transfer characteristics in bubble column reactors equipped with two phase jet nozzle spargers.

As for the aeration in non-Newtonian fermentation, we are still at the beginning stage. Fundamental data on diffusivity, solubility, and rheology are needed together with experimental confirmation of existing models or correlations, and improvements upon them. In stirred tanks, more experiments need to be done under flow conditions that resemble large scale reactors.

NOMENCLATURE

a interfacial area per unit volume (L^{-1})

b constant

c^* saturation concentration (ML^{-3})

c_L liquid bulk concentration (ML^{-3})

C constant

C_o normalized outlet gas concentration

C_L	normalized liquid concentration
d	bubble diameter (L)
d_s	Sauter mean bubble diameter (L)
d	mean bubble diameter (L)
D	diffusivity (L^2T^{-1})
D_C	column diameter (L)
De	Deborah number defined in Eq. (37)
D_i	impeller blade width (L)
D_I	impeller diameter (L)
D_T	tank diameter (L)
D_W	diffusivity in water (L^2T^{-1})
e	energy dissipation per unit volume
F	constant in Eq. (42)
g	gravitational acceleration (LT^{-2})
h_T	liquid height when aerated (L)
h_L	liquid height without aeration (L)
H	gas holdup
H_i	ion specific constant
He	Henry's Law constant
He_o	Henry's Law constant of base solution
He_{sj}	Henry's Law constant for jth sugar in solution
I	ionic strength (ML^{-3})
k	consistency index
k_L	liquid film mass transfer coefficient (LT^{-1})

K constant

K' constant

l micro length scale (L)

L length scale (L)

m constant

n constant or behavior index

N impeller speed (T^{-1})

Na aeration number defined by Eq. (39)

OTR oxygen transfer rate

OUR oxygen uptake rate

p constant

q constant

(P/V) power per unit volume

Q volumetric gas flow rate (L^3T^{-1})

Re Reynolds number defined by Eq. (7)

Re_I Impeller Reynolds number ($ND_I{}^2/\nu$)

s surface renewal rate (T^{-1})

Sc Schmidt number (ν/D)

Sh_I impeller Sherwood number (k_LaD_I/D)

t_e exposure time (T)

t_g gas phase residence time (T)

t_p characteristic process time (T)

t_r characteristic material relaxation time (T)

T absolute temperature

u	superficial gas velocity (LT^{-1})
U	bubble free rise velocity (LT^{-1})
v	characteristic velocity (LT^{-1})
V	reactor volume (L^3)
V_G	gas volume (L^3)
V_L	liquid volume (L^3)
We	Weber number
δ	film thickness
ρ	liquid density (ML^{-3})
μ	liquid viscosity ($ML^{-1}T^{-1}$)
μ_a	apparent viscosity ($ML^{-1}T^{-1}$)
μ_g	gas viscosity ($ML^{-1}T^{-1}$)
σ	surface tension (MT^{-2})
ν	kinematic viscosity (L^2T^{-1})
ν_a	apparent kinematic viscosity (L^2T^{-1})
τ	shear stresss ($MT^{-2}L^{-1}$)

REFERENCES

1. Charpentier, J.C. *Trans. Inst. Chem. Eng. 60,* 131 (1982).
2. Tsao, G.T. and Lee, Y.H. *Ann. Reports Fermentation Processes 1,* 115 (1977).
3. Blanch, H.W. *Ann. Reports Fermentation Processes 3,* 47 (1979).
4. Moo-Young, M. and Blanch, H.W. *Adv. Biochem. Eng. 19,* 1 (1981).
5. Andrew, S.P.S. *Trans. Inst. Chem. Eng. 60,* 3 (1982).
6. Finn, R.K. *NATO Advanced Study Institute Series E. 51,* 445 (1981).
7. Shah, Y.T., Kelkar, B.C., Godbole, S.P., and Deckwer, W.-D. *AIChE J. 28,* 353 (1982).

8. Charpentier, J.C. *Adv. Chem. Eng. 11*, 1 (1981).
9. Joshi, J.B., Pandt, A.B., and Sharma, M.M. *Chem. Eng. Sci. 37*, 813 (1982).
10. Van Landeghem, H. *Chem. Eng. Sci. 35*, 1912 (1982).
11. Shah, Y.T. "Gas-Liquid-Solid Reactor Design," McGraw Hill (1979).
12. Whitman, W.G. *Chem. Met. Eng. 29*, 146 (1923).
13. Higbie, R. *Trans. Am. Inst. Chem. Eng. 31*, 365 (1935).
14. Danckwerts, P.V. *Ind. Eng. Chem. 43*, 1460 (1951).
15. Lee, Y.H. and Luk, S. *Ind. Eng. Chem. Fundam.*, in press (1982).
16. Fortescue, G.E. and Pearson, R.A. *Chem. Eng. Sci. 22*, 1163 (1967).
17. Lamont, L.C. and Scott, D.S. *AIChE J. 16*, 513 (1970).
18. Theofanous, T.G., Houze, R.N., and Brumfield, L.K. *Int. J. Heat Mass Transfer 19*, 613 (1976).
19. Hinze, J.O. "Turbulence," McGraw Hill, New York (1950).
20. Henstock, W.H. and Hanratty, T.J. *AIChE J. 25*, 122 (1979).
21. Petty, C.A. and Wood, P.E. *Chem. Eng. Sci. 7*, 261 (1980).
22. Ruckenstein, E. *Chem. Eng. Commun. 10*, 77 (1981).
23. Chan, W.C. and Scriven, L.E. *Ind. Eng. Chem. Fundam. 9*, 1114 (1970).
24. Levec, J. and Pavko, S. *Chem. Eng. Sci. 34*, 1159 (1979).
25. Kawase, Y. and Ulbrecht, J. *J. Canad. J. Chem. Eng. 60*, 87 (1982).
26. Akita, K. and Yoshida, F. *Ind. Eng. Chem. Proc. Des. Dev. 12*, 76 (1973).
27. Schuegerl, K., Leucke, J., Lehman, J., and Wagner, F. *Adv. Biochem. Eng. 8*, 60 (1978).
28. Franz, K., Buchholz, R., and Schuegerl, K. *Chem. Eng. Commun. 5*, 165 and 187 (1980).
29. Schuegerl, K. *Adv. Biochem. Eng. 19*, 71 (1981).
30. Motarjemi, M. and Jameson, G.J. *Chem. Eng. Sci. 33*, 1415 (1978).
31. Raymond, D. and Zieminski, S.A. *AIChE J. 17*, 57 (1971).
32. Akita, K. and Yoshida, F. *Ind. Eng. Chem. Proc. Des. Dev. 13*, 84 (1974).
33. Astarita, G. "Mass Transfer with Chemical Reaction," Elsevier Press (1967).
34. Phillips, K.L. *Canad. J. Chem. Eng. 51*, 371 (1973).
35. Phillips, K.L. *Chem. Eng. Sci. 34*, 1083 (1979).
36. Niiyama, H. and Smith, J.M. *AIChE J. 23*, 592 (1977).
37. Sylvester, D. and Dianat, S. *Ind. Eng. Chem. Proc. Des. Dev. 19*, 201 (1980).
38. Sobotka, M., Votruba, J., and Prokop, A. *Biotech. Bioeng. 23*, 1193 (1981).

39. Donaldson, T.L. and Palmer, H.J. *AIChE Symp. Ser. 75,* 190 (1978).
40. Lozowski, D., Langa, J., Andrews, G.F., and Stroeve, P. *Chem. Eng. Commun. 6,* 349 (1980).
41. St-Dennis, C.E. and Fell, C.J.D. *Canad. J. Chem. Eng. 49,* 885 (1971).
42. Hikita, H., Asai, S., and Azuma, Y. *Canad. J. Chem. Eng. 56,* 371 (1978).
43. Yagi, H. and Yoshida, F. *Ind. Eng. Chem. Proc. Des. Dev. 14,* 488 (1975).
44. Brignole, E.A. and Echarte, R. *Chem. Eng. Sci. 36,* 695 (1981).
45. Sovova, H. *Coll. Czech. Chem. Comm. 41,* 3715 (1976).
46. Greif, R., Cornet, I., and Kappesser, R. *Int. J. Heat Mass. 15,* 593 (1972).
47. Zandi, I. and Turner, C.D. *Chem. Eng. Sci. 25,* 517 (1970).
48. Reuss, M., Debus, D., and Niebelschuetz, H. *Ferment. React. Capteurs Autom. Colloq.,* 49 (1980).
49. Herschcovich, F., Merchuk, J.C., and Tamir, A. *Chem. Eng. Sci. 37,* 637 (1982).
50. Jalan, V.M., Tham, M.K., and Gubbins, K.E. *Canad. J. Chem. Eng. 50,* 85 (1972).
51. Navari, R.M., Gainer, J.L., and Hall, K.R. *AIChE J. 17,* 1028 (1971).
52. Gubbins, K.E., Bhatia, K.K., and Walker, R.D., Jr. *AIChE J. 12,* 548 (1966).
53. Calderbank, P.H. *The Chem. Eng. 45,* CE209 (1967).
54. Baykara, Z.S. and Ulbrecht, J.J. *Biotech. Bioeng. 20,* 287 (1978).
55. Bailey, J.E. and Ollis, D.F. "Biochemical Engineering Fundamentals," McGraw Hill (1977).
56. Moryakov, V.S., Nikolaev, N.A., and Nikolaev, A.M. *Int. Chem. Eng. 14,* 623 (1974).
57. Ueyama, K., Morooka, S., Koide, K., Kaju, H., and Mlyauchi, T. *Ind. Eng. Chem. Proc. Des. Dev. 19,* 592 (1980).
58. Schuegerl, K., Lueke, J., and Oels, U. *Adv. Biochem. Eng. 7,* 1 (1977).
59. Koide, K., Kato, K., Tanaka, Y., and Kuboda, H.J. *Chem. Eng. Japan 1,* 51 (1961).
60. Otake, T., Tone, S., Kuboi, R., Takkahashi, Y., and Nakao, K. *Int. Chem. Eng. 21,* 72 (1981).
61. Rice, R., Tupperainen, J.M.I, and Hedge, R.M. *Canad. J. Chem. Eng. 59,* 677 (1981).
62. Kawase, Y. and Ulbrecht, J.J. *Ind. Eng. Chem. Proc. Des. Deve. 20,* 636 (1981).

63. Buchholz, R., Zakrzewski, W., and Schuegerl, K. *Int. Chem. Eng. 21*, 180 (1981).
64. Nagel, O., Hegner, B., and Kuerten, H. *Int. Chem. Eng. 21*, 161 (1981).
65. Hikita, H.S., Asai, K., Tanigawa, K., Segawa, K., and Kitao, M. *Chem. Eng. J. 20*, 59 (1980).
66. Mersmann, A. *Ger. Chem. Eng. 1*, 1 (1978).
67. Sridhar, T. and Potter, O.E. *Ind. Eng. Chem. Fundam. 19*, 21 (1980).
68. Hassan, I.T.M. and Robinson, C.W. *AIChE J. 23*, 48 (1977).
69. Einsele, A. and Finn, R.K. *Ind. Eng. Chem. Proc. Des. Dev. 19*, 600 (1980).
70. Hughmark, G.A. *Ind. Eng. Chem. Proc. Des. Dev. 19*, 638 (1980).
71. Schumpe, A. and Deckwer, W.-D. Paper presented at ACS Meeting, Atlanta (1981).
72. Lopes de Figueiredo, M.M. and Calderbank, P.H. *Chem. Eng. Sci. 34*, 1333 (1979).
73. Van't Riet, K. *Ind. Eng. Chem. Proc. Des. Dev. 18*, 357 (1979).
74. Calderbank, P.H. and Moo-Young, M. *Chem. Eng. Sci. 16*, 39 (1961).
75. Sideman, S., Hortacsu, H., and Fulton, J.W. *Ind. Eng. Chem. 58*, 32 (1966).
76. Preres, J.F. and Sandall, O.C. *AIChE J. 20*, 770 (1974).
77. Zlokarnik, M. *Adv. Biochem. Eng. 8*, 133 (1978).
78. Judat, H. *Int. Chem. Eng. 21*, 188 (1981).
79. Prest, W.M., Porter, R.S., and O'Reilly, J.M. *J. Appl. Polym. Sci. 14*, 2697 (1970).
80. Kastanek, F. *Coll. Czechoslov. Chem. Commun. 42*, 2491 (1977).
81. Deckwer, W.-D. ACS Symp. Ser. Chemical Reactors, Chapt. 22, 213 (1981).
82. Hughmark, G.A. *Ind. Eng. Chem. Proc. Des. Dev. 6*, 218 (1967).
83. Nakanoh, M. and Yoshida, F. *Ind. Eng. Chem. Proc. Des. Deve. 19*, 190 (1980).
84. Hikita, H.S., Asai, K., Tanigawa, K., Segawa, K., and Kitao, M. *Chem. Eng. J. 22*, 61 (1981).
85. Henzler, H.J. *Chem. Ing. Techn. 52*, 643 (1980).
86. Tojo, K. and Miyanami, K. *Chem. Eng. J. 24*, 89 (1982).
87. Paca, J. and Ettler, P. *Prague Inst. Chem. Tech. E50*, 205 (1979).
88. Viesturs, U.E., Kristapsons, M.Z., and Levitans, E.S. *Adv. Biochem. Eng. 21*, 169 (1982).
89. Kwase, Y. and Ulbrecht, J.J. *Canad. J. Chem. Eng. 60*, 187 (1982).

90. Ranade, V.R. and Ulbrecht, J.J. *AIChE J. 24*, 796 (1978).
91. Sato, S., Mukataka, S., Kataoka, H., and Takahashi, J. *J. Ferment. Tech. 59*, 221 (1981).
92. Lee, Y.H. and Tsao, G.T. *Adv. Biochem. Eng. 13*, 35 (1979).
93. Dang, N.D.P., Karrer, D.A., and Dunn, I.J. *Biotech. Bioeng. 19*, 853 (1977).
94. Linek, V. and Vacek, V. *Biotech. Bioeng. 19*, 983 (1977).
95. Linek, V. *Biotech. Bioeng. 14*, 285 (1972).
96. Leeuwen, C. *Biotech. Bioeng. 21*, 2125 (1979).
97. Heijnen, J.J., Van't Riet, K., and Wolthuis, A.J. *Biotech. Bioeng. 22*, 1945 (1980).
98. Andre, G., Moo-Young, M., and Robinson, C.W. *Biotech. Bioeng. 23*, 1611 (1981).
99. Linek, V., Hovorka, F., and Holecek, O. *Biotech. Bioeng. 23*, 1467 (1981).
100. Chandrasekharan, K. and Calderbank, P.H. *Chem. Eng. Sci. 35*, 1473 (1980).
101. Yoshida, T., Ohasa, S., and Taguchi, H. *Biotech. Bioeng. 22*, 201 (1980).
102. Sobotka, M. and Votruba, J. *Collect. Czechoslov. Chem. Commun. 45*, 2266 (1980).
103. Vardar, F. and Lilly, M.D. *Biotech. Bioeng. 24*, 1711 (1982).
104. Spriet, J.A., Botterman, J., De Buyser, D.R., De Visscher, P.L., and Vandamme, E.J. *Biotech. Bioeng. 24*, 1605 (1982).
105. Brooks, J.D., MacLennan, D.G., and Barford, J.P. *Biotech. Bioeng. 24*, 847 (1982).
106. Taylor, R. Paper presented at Water Pollution Control Meeting, October, Houston (1980).
107. Chapman, C.M., Gibiaro, L.G., and Nienow, A.W. *Chem. Eng. Sci.*, 891 (1982).
108. Alvarez-Cuenca, M. and Nernberg, M.A. *AIChE J. 27*, 66 (1981).
109. Shioya, S. and Dunn, I.J. *Chem. Eng. Sci. 33*, 1529 (1978).
110. Shah, Y.T., Stiegel, G.J., and Sharma, M.M. *AIChE J. 24*, 369 (1978).
111. Burgress, J.M. and Calderbank, P.H. *Chem. Eng. Sci. 30*, 743 (1975).
112. Sriram, K. and Mann, R. *Chem. Eng. Sci. 32*, 571 (1977).
113. Schumpe, A. and Deckwer, W.-D. *Chem. Eng. Sci. 35*, 221 (1980).
114. Midoux, N., Laurent, A., and Charpentier, J.C. *AIChE J. 26*, 157 (1980).

115. Hassan, I.M. and Robinson, C.W. *Chem. Eng. Sci. 35,* 1277 (1980).
116. Sridhar, T. and Potter, O.E. *Chem. Eng. Sci. 33,* 1347 (1978).
117. Charles, M. *Adv. Biochem. Eng. 8,* 1 (1978).
118. Metz, B., Kosen, N.W.F., and Can Suijdam, J.C. *Adv. Biochem. Eng. 11,* 103 (1979).
119. Brignole, E.A. and Echarte, R. *Chem. Eng. Sci. 36,* 695 (1981).
120. De Block, W.J. and Fortuin, M.J. *Chem. Eng. Sci. 36,* 1687 (1981).
121. Goldstick, T.K. and Fatt, I. *Chem. Eng. Prog. Sym. Ser. 66,* 101 (1970).
122. Greif, R., Cornet, I., and Kappesser, R. *Int. J. Heat Mass Transfer 15,* 593 (1972).
123. Sovova, H. and Prochazka, J. *Chem. Eng. Sci. 31,* 1091 (1976).
124. Schumpe, A., Adler, I., and Deckwer, W.-D. *Biotech. Bioeng. 20,* 145 (1978).
125. Quicker, G., Schumpe, A., Koenig, B., and Deckwer, W.-D. *Biotech. Bioeng. 23,* 635 (1981).
126. Dackwerts, P.V. "Gas-Liquid Reactions," McGraw Hill (1970).
127. Rivas, O.R. and Praussnitz, J.M. *Ind. Eng. Chem. Fundam. 18,* 289 (1979).
128. Armitage, A.D., Linford, R.G., and Thornhill, D.G.T. *Ind. Eng. Chem. Fundam. 17,* 362 (1978).
129. Kaeppeli, O. and Fiechter, A. *Biotech. Bioeng. 23,* 1897 (1981).
130. Buchholz, H., Luttmann, R., Zakrzeweki, W., and Schuegerl, K. *Eur. J. Appl. Micro. Biotech. 11,* 89 (1981).
131. Buchholz, H., Luttmann, R., Zakrzewski, W., and Schuegerl, K. *J. Chem. Tech. Biotech. 31,* 435 (1981).
132. Ziegler, H., Dunn, I.J., and Bourne, J.R. *Biotech. Bioeng. 22,* 1613 (1980).
133. Kataoka, H., Takeuchi, H., Nakao, K., Yagi, H., Tadaki, T., Otake, T., Miyauchi, T., Washimi, K., Watanabe, K., and Yoshida, F. *J. Chem. Eng. Japan 12,* 105 (1979).
134. Botton, R., Cosserat, D., and Charpentier, J.C. *Chem. Eng. J. 20,* 87 (1980).
135. Takamatsu, T., Shioya, S., and Nakatani, H. *J. Ferment. Technol. 59,* 287 (1981).
136. Wick, E. *Canad. J. Chem. Eng. 59,* 297 (1981).
137. Viesturs, U.E., Sturmanis, I.A., Kriskis, V.V., Prokopenko, V.D., and Erickson, L.E. *Biotech. Bioeng. 23,* 1171 (1981).

138. Merchuk, J.C. and Stein, Y. *Biotech. Bioeng. 23,* 1309 (1981).
139. Lutmann, R., Buchholz, H., Zakrzewski, W., and Schuegerl, K. *Biotech. Bioeng. 24,* 817 (1982).
140. Chandrasekharan, K. and Calderbank, P.H. *Chem. Eng. Sci. 36,* 819 (1981).
141. Nienow, A.W. and Lilly, M.D. *Biotech. Bioeng. 21,* 2341 (1979).
142. Nishikawa, M., Nakamura, M., Yagi, H., and Hashimoto, K. *J. Chem. Eng. Japan 14,* 219 (1981).
143. Tanaka, H. *Biotech. Bioeng. 24,* 425 (1982).
144. Nishikaw, M., Nakamura, M., and Hashimoto, K. *J. Chem. Eng. Japan 14,* 227 (1981).
145. Dickey, D.S. AIChE 72nd Annual Meeting, San Francisco, Paper 116d, November (1979).

CHAPTER 6

ENGINEERING AT THE MICROORGANISM SCALE

H. R. Bungay
M. L. Bungay

Rensselaer Polytechnic Institute
Troy, New York

C. N. Haas

Illinois Institute of Technology
Chicago, Illinois

I. INTRODUCTION

In previous contributions to this series, Tsao and Lee (1977), Blanch (1979), and Yoshida (1982) reviewed aeration. These papers mentioned micro-scale effects and noted the importance of advancing from empirical approaches to a more fundamental understanding of mass transfer. Some years ago, Donovick (1960), estimated the amount of cell surface in a fermentation and calculated rates of transfer per square micrometer. This was one of the first attempts to use engineering analysis on a micro scale, and began a shift in thinking away from the lumped, black box approach.

Today, practical design of fermentation uses the relationship:

transfer rate = coefficient X area X driving force.

Because the areas are difficult to measure, a Kla term is used to handle both area and coefficient. The Kl indicates an appreciation that liquid film resistances predominate over gas film resistances at an interface with such sparingly soluble gases as oxygen. The driving force for oxygen transfer is taken as the difference between actual concentration in the liquid and the concentration of oxygen if in equilibrium with

ISBN 0-12-040306-4

the inlet gas. Engineers make good use of such simple flux equations, but little additional understanding of the system is gained from such oversimplification. Ultimately, we need to know actual mechanisms and to employ rational coefficients instead of variable "constants" that depend on each individual situation.

The size range of microbial cells, microbial aggregates, and gas bubbles is reasonable for considering micro effects. Eddies in fermentation fluids are roughly of these dimensions, and there would be little to gain from a smaller scale of reference. Although certain aspects of surface activity are understood on a molecular level, the microorganism scale seems more appropriate for advancing the status of practical processes.

Gradients of oxygen, nutrients, and product concentrations are established by microbial metabolism. These gradients extend from the cell surface toward the bulk of the medium, but there may be major inhomogenieties in the medium when distances from the sources and sinks for gases, nutrients, and products are considered. Thus the location of an individual cell with respect to other cells and with respect to the overall system is important. Furthermore, suspended cells are not stationary. They may be in a region rich in oxygen near a bubble or an air/water interface and later be swept by circulation of the fluid to a new location remote from any oxygen source.

Non-uniform gradients are expected between metabolizing cells or aggregates, but the gradients may be distorted further because the cells or aggregates themselves are not homogeneous. For example, a filamentous cell grows by extension of the tip and has varying rates of metabolic activity along its length. Also, an aggregate such as a slime will often vary in thickness, and the microbial population will seldom be uniformly distributed throughout the slime. In other words, there may be regions of high or of low metabolic activity within an aggregate or film.

Oxygen transfer studies are no longer restricted to gross measurements of average rates to the medium. Instrumentation is now available, in the form of various microelectrodes, for detecting gradients of specific ions, conductivity, organic compounds, and even enzymes and biochemicals. Oxygen microelectrodes with tip sizes from one to several microns have sensing zones approximately the size of bacteria. Oxygen gradients near and within interfacial films, microbial slimes, colonies, microbial pellets, and flocs have been measured.

II. AGGREGATION

Individual cells, groups of just a few cells, and aggregates all have practical importance. Attachment of any of these to a surface markedly affects survival and mass transfer. In a flowing system, attached organisms can persist, while suspended cells leave with the effluent. Non-living suspended particles may serve as attachment sites for cells aiding their persistence by more rapid sedimentation, thereby escaping effluent withdrawal. Sizes and states of aggregation are shown in Fig. 1. Obviously these can be major factors in determining rates of mass transfer.

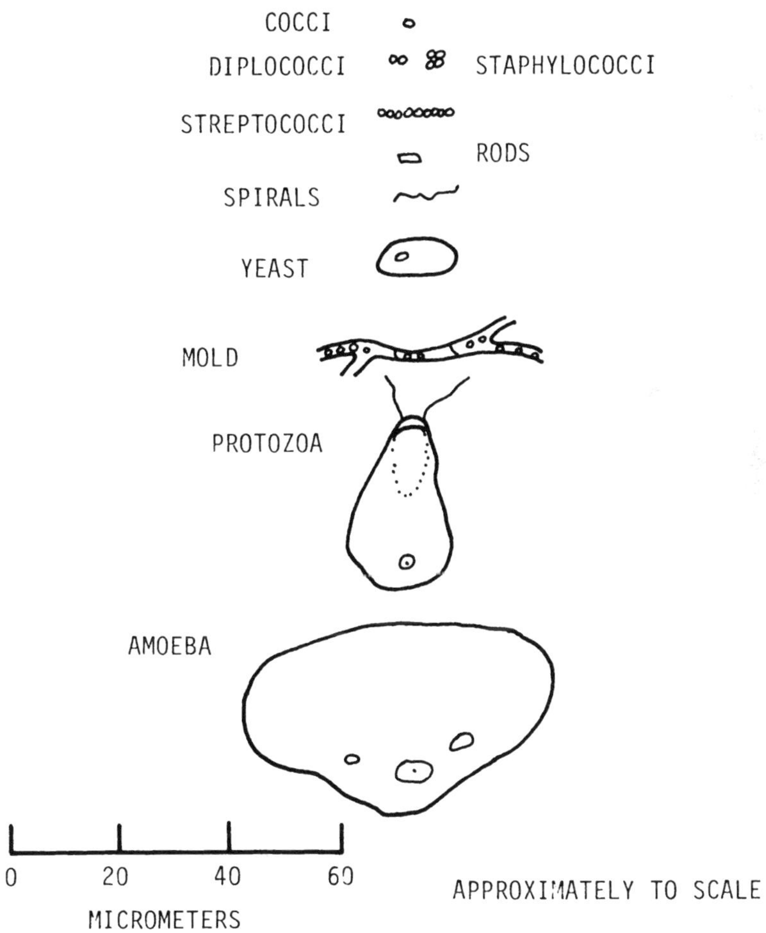

Figure 1. Microbial entities. (Figure 1 continued next page).

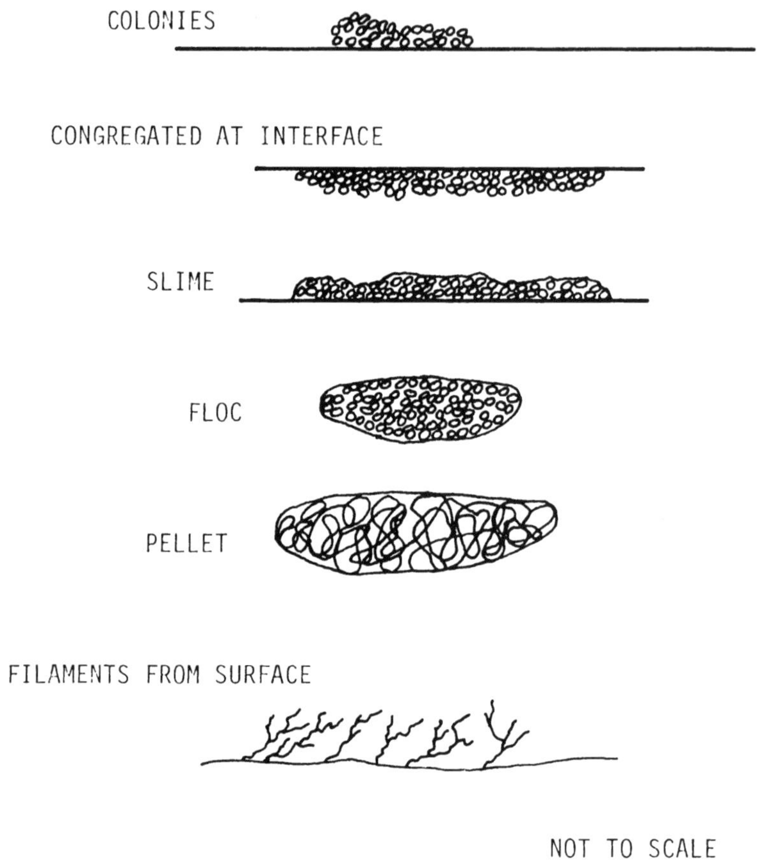

Figure 1. (Continued) Microbial entities.

Attachment of microbial cells to surfaces is so important that much greater understanding of mechanisms and forces is warranted. Goulder (1977), found that in estuaries with high concentrations of suspended solids, most of the bacteria were attached to particles. This suggests that most of the breakdown of organic matter resulted from attached organisms. Some of the literature about attachment is listed in Table 1. Microbial slime is similar in composition to capsular material. Sometimes particulate matter is embedded in the slime, and there may be a sheath with encrustation or embedded

Table 1. Attachment of Microorganisms to Surfaces.

Reference	Comments
Danielli and Davies, 1951	Review of biochemistry at interfaces.
Heukelelian and Heller, 1940	Adding glass beads to E. coli allowed growth on more dilute media because substrates adsorbed.
Zobell, 1943	Discussion of basic mechanisms.
Munson and Bridges, 1964	Considerable force is required to hold a bacterial cell on the wall of a vigorously stirred vessel.
Larsen and Dimmick, 1964	Density of attached cells is characteristic of the types of organisms. Silicone coating delays buildup of organism layer.
Meadows, 1965	Ionic strength critical for marine or fresh water organisms to remain attached.
Meadows and Anderson, 1966	Counted organisms growing on sand grains. Found in patches with much bare surface.
Daniels, 1972	Review of the literature.
Ou and Alexander, 1974	Magnesium ion lessens number of filamentous microbes attached to glass beads. Smaller beads promoted more rapid colonization.
Hendricks, 1974	Observed sorption and growth of heterotrophic bacteria on glass immersed in river water. Doubling time was about 2 hr until film stabilized, then dropped to about 200 hr.
Fletcher and Floodgate, 1977	Electron microscopy of attaching cells. Noted polysaccharide coating surface.
Ash, 1979	Review emphasizes ores, cell-to-cell forces, and whole-cell immobilization.
Marshall, 1978	Surface tension and adsorption when solids are immersed in dilute microbial suspensions in artificial sea water.
Gerson and Zajic, 1979	Analyzed adhesion in terms of cell and solid surface considering free energy, zeta potential, electrostatic forces.
Meadows and Anderson, 1979	Review of literature about microbial growth at interfaces.

mineral matter formed by the microorganisms. There seems to be a common misconception that slimes are about as rigid as a gelatin dessert. In fact, many slimes are quite fluid and protozoa can swim through them in search of prey as the slime closes in their wake. Slimes are inhabited by the organisms that formed them as well as other organisms that seek protection or food.

Cells may have a polymeric coating of hydrophillic polysaccharides that enables them to adhere to each other or to non-biological surfaces. This coating, referred to as a glycocalyx, may not be present on some microbial cells, and an environment in which attachment is of little advantage may select strains or mutants having no glycocalyx. In fact, information about mechanisms for adhesion was slow to develop because common laboratory culturing techniques selected for organisms that did not expend energy in synthesizing an unneeded attachment polysaccharide. Some microscopic forms form a "holdfast" structure to facilitate attachment to surfaces.

The presence of water greatly complicates adhesion of cells to a surface. The bond must resist hydrolysis while forming in an aqueous environment. Much of the study of adhesion has been with surface active substances, and non-living processes such as ore flotation were the focus of research. Solids with hydrophillic surfaces participate readily in adsorbing microbial cells, but hydrophobic materials such as polyethylene and Teflon resist cellular attachment and easily slough weakly adsorbed cells. Papers edited by Drost-Hansen and Clegg (1979), deal with cell-associated water in systems of complexity ranging from pure colloids to zoogeography.

Ionic binding of cells involves the counter ions in the Stern layer adjacent to a surface, and ions and charges must shift to maintain electrical neutrality as the new bonds form. A discussion of surface potential is beyond the scope of this brief treatment, but there is a wealth of related material in books about corrosion or electrochemistry. Of course, hydrogen ions and hydroxyl ions usually play dominant roles at surfaces, so pH is a major factor in adhesion. Ionic bonding may be important in the initial stages of attachment, but stronger bonds are needed for fastening microbial communities. Although a hydrogen bond is relatively weak, there can be such great numbers that hydrogen bonding provides strong attachment. Additional forces from adsorption of functional groups to the surface can securely anchor microbial slimes or colonies. Nevertheless, when an attached aggregate grows to large size and is torn from the surface by shear forces, a bare surface is usually exposed. For example , trickling filter slimes seem to be stronger internally than are their bonds to a surface because intact biological material is torn away by hydraulic forces to leave a surface that starts over in the attachment of new organisms.

As slime grows thicker, the region attached to the solid support becomes remote from nutrients and sources of oxygen. It may be postulated that this layer becomes anaerobic and that organisms lyse to weaken their ability to attach or the polymers may be attacked by lytic enzymes. However, the concentration of dissolved oxygen tends not to drop much lower than the critical oxygen concentration because metabolic rates tend to be very slow below this point, and there may be no truly anaerobic region. Let us advance the speculation that there are organisms producing slime polymers while other organisms live in and eat slime. In the lower layers, slime production rate may be curtailed by being far from the nutrient supply so that slime consumption leads to a region too weak to maintain attachment.

There have been many reports about the performance of microbial films as in waste treatment with trickling filters. A selection of papers is in Table 2. Microorganisms affect the binding of soil particles, a phenomenon of special concern in the erosion of farm lands. McCalla (1946) measured the

Table 2. Microbial Film Systems.

Reference	*Comments*
Sanders, 1966	*Related oxygen and substrate uptake to slime growth.*
Kornegay and Andrews, 1969	*Correlated film formation with substrate concentration.*
Sanders, et al., (1971)	*Microelectrode measurements of oxygen concentrations near and within slimes.*
Hamer, 1973	*Discussion of fouling of heat transfer surfaces and mass transfer limitations from film formation.*
Atkinson and Daoud, 1970	*Diffusion and biochemical processes for modeling film systems.*
Atkinson and Davies, 1974	*Derivation of relations of growth rate and substrate utilization to film properties and substrate concentration.*
Atkinson and Knight, 1975	*Considered total biomass in a film reactor.*

(Table 2 continued on next page)

Table 2. Microbial Film Systems (continued).

Reference	*Comments*
Howell and Atkinson, 1976	*Discussed "ideal" film thickness.*
Atkinson and Rahman, 1976	*Used plastic packing and correlated process parameters. Measured wetted areas.*
Atkinson and Rahman, 1978	*Additional correlation including hold up and packing surface.*
Atkinson and Rahman, 1979	*Used shape parameters to extend film analysis to microbial flocs.*
Atkinson, et al., 1979	*Devised means for fabricating supports of any desired size and shape. Listed related effects.*
Pawlowsky, et al., 1973	*Upset of a mixed culture with wall growth could cause return to a different steady state. Wall growth improves resistance to shocks.*
Characklis, 1973	*Review of effects of slimes in water systems and of control methods.*
Hoehn and Ray, 1973	*Effect of film thickness on substrate removal.*
Gerchakov and Sallman, 1977	*Review of corrosion in sea water. Formation of microbial films is one of factors.*
Jennings, et al., 1976	*Math model of biological film reactor.*
Characklis, 1981	*Development of biofilms, effects on heat and mass transfer, metabolism, and polymer formation.*
La Motta, 1976	*Used effectiveness factor for substrate removal by aerobic slimes. Estimated apparent diffusivities.*
Livansky and Prokes, 1972	*Microbial respiration, pH, hydration of carbon dioxide, and gradients affect transport rates. Algae in phosphate buffer were studied.*
Rittman and McCarty, 1980	*Minimum concentration of acetate to maintain a steady state biofilm was 0.66 mg/L.*
Rittman and McCarty, 1981	*Math model of steady state fluxes for biofilm.*

stability of lumps of soil when struck by water drops and found that success in soil stabilization decreases in the following order: fungi, actinomycetes, certain bacteria, yeasts, and most common bacteria. Nutrition is important because growth on glucose was more stabilizing than growth on decaying organic matter. When soil flora were added to the pure cultures, fungi were no longer excellent stabilizers. Further tests with various microbial species by Watson and Stojanovic (1963) showed that species that were good stabilizers in pure cultures were less effective in mixed cultures. Although concentration of polysaccharide correlated with stability, hexoseamine level did not.

Flocculation of microorganisms has practical importance in waste treatment because flocculated organisms are easier to collect by sedimentation from various streams in a waste treatment plant. Commercial fermentation processes utilize flocculated cultures to maintain high cell density for more rapid bioconversion. For example, tower fermentation with flocculated yeast has a conical section above the fermentaion section so that rising fluid has less vertical velocity and flocs can separate and settle back. High cell density permits ethanol fermentation in 3 to 4 hours compared to days in conventional batch fermentation.

There are several ways that flocculation can occur, and colloidal phenomena may be important. Many microorganisms exude polymers, predominantly of carbohydrates, that form capsules or slimes. As organisms aggregate, settling times are shortened so that collection or retention by sedimentation becomes feasible. Selected references about flocculation are in Table 3.

Table 3. Flocculation.

Reference	*Comments*
Nakamura, 1961a	*Coagulated microbial suspensions with various cations.*
Nakamura, 1961b	*Calcium hydroxide, chloroform, and salts of cationic detergents were good coagulating agents.*
Curds, 1963	*Carbohydrate polymer from protozoa flocculated india-ink particles.*
Chester, 1965	*Effects of various substances on flocculated yeast were measured. Calcium ion promoted floc formation.*
Crabtree, et al., 1966	*Zoogleal flocs studied by affecting cell metabolism. Inhibiting esterases helped flocs.*
Busch and Stumm, 1968	*Flocculated bacteria with anion and non-ionic polymers. Mechanisms are specific adsorption and bridging.*

(Table 3 continued on next page)

Table 3. Flocculation (continued).

Reference	Comments
Tezuka, 1969	Diluting minerals in flocs caused dispersion. Flocs reformed if ions were added back.
Pavoni, et al., 1972	Explained floc formation in terms of polymers, binding, and bridging.
Pavoni, et al., 1973	Studied algal flocs by removing and adding back their polymers.
Williams and Wiseman, 1973	Kinetics of floc formation related to calcium ion.
Paerl, 1973	Aggregation of particles in Lake Tahoe resulted from bacterial secretions more than from attachment structures.
Tenney and Verhoff, 1973	Compared chemical and autoflocculation.
Daniels, 1974	Review of processes, designs, and various flocculants for water and waste treatment.
Unz and Farrah, 1974	Determined composition of zoogleal polymer. Strains with rough colonies formed flocs readily, but those with smooth colonies did not.
Hahn and Eppler, 1976	Rate of aggregation was measured for some bacteria and alga. Salt conc.and pH determined rates, but there was some aggregation in distilled water.
Forster, 1976	Uptake of metals by activated sludge was studied and role of poly-beta-hydroxybutyric acid discussed. Polymers were characterized.
Nakamura, et al., 1976	Screened microorganisms for ability to flocculate yeast. Less than 10% could.
Stewart, et al., 1976	Characterized flocculating species by electron microscopy and biochemical analysis. Noted "hairy" cell surface.
Treweek and Morgan, 1977	Flocculation of E. coli with polyethyleneimine has charge neutralization and bridging. Surface charge sites are covered.

Whereas slimes and zoogleal flocs are continuous, pellets may be networks of filaments resembling, on a tiny scale, a tangle of fishing line. Fluid can flow through the pellet but only at a slow rate, since the open spaces have high resistances because of their sizes. Hydraulic deformation of a filamentous floc may squeeze out interstitial water and thus aid mass transfer. By contrast, the viscosity of slime retards convection and slows molecular diffusion.

Blakebrough (1967) discussed mass transfer and scale up in several microbial situations: 1. a microbial cell in contact with liquid with consideration of film resistances and convective transport, 2. suspended cell aggregates or filaments, and 3. systems in which organisms grow as a film attached to a supporting medium. Oxygen and nutrients diffuse to the organisms, and products of metabolism diffuse away. It is claimed that gradients in the bulk of the medium are negligible for well-mixed, low viscosity systems, but thicker media can have a significant mass transfer barrier in the bulk. Models based on equal concentration throughout the bulk had steep gradients within clumps of organisms or in films formed by cells. Although the general concepts of Blakebrough's analysis have merit, actual measurements on real systems are quite different from his postulates. Fig. 2 shows the hypothetical gradients in a spherical floc and an actual

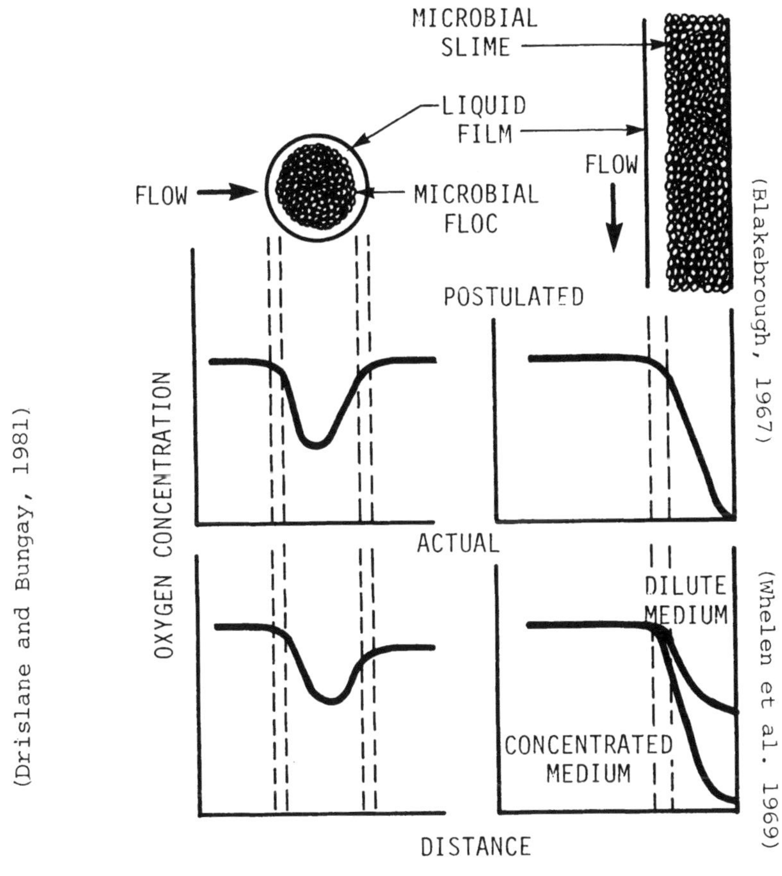

Figure 2. Oxygen profiles.

profile of dissolved oxygen. The downstream oxygen concentration is lower than that upstream for a real floc because some oxygen is depleted and not replaced by diffusion quickly enough. The postulated and real profiles for a microbial film are also shown in Fig. 2. Here the difference can be very great for a film in contact with dilute medium because there is little demand for oxygen, and this demand is easily satisfied by diffusion. With concentrated medium, the dissolved oxygen deep in the film does not fall to zero because metabolic rates and oxygen demand drop drastically as a critical level of dissolved oxygen is reached.

III. INTERFACES

Microbial cells have surface activity that attracts them to interfaces. Greater concentration of virus particles because of attraction to an air-water interface was measured by Baylor (1977a,b), and droplets could have 30 to 50 times the concentration of bacteria or viruses as the water from which the droplet was formed. Adsorption of viruses to biological and to non-biological surfaces has been reviewed by Bitton (1975). Various bacteria used as indicators of water quality have been more concentrated at the surfaces of natural waters than in the bulk (Dutka and Kwan, 1978). In studies of the North Atlantic Ocean, Sieburth, et al. (1976), found more organic compounds, bacteria, and amoeba at the surface than in the waters beneath.

Insight into mechanisms for impairment of mass trasnfer by surface active agents comes from liquid-liquid extraction studies with droplets by Garner and Skelland (1956). The internal fluid circulation of a droplet can be observed to decline with addition of surface active agents, and circulation may stop completely. With little or no convection, mass transfer by diffusion only in the droplet is much slower, about 1/12th of the original value.

Eddy circulation was noted by measuring oxygen concentration just inside an air-water interface with a microelectrode (Bungay, et al., 1973). Tsao and Lee (1974) and Lee, et al., (1980) extended this approach for investigations of the effects of surface active agents. Eddy circulation of liquid at the interface was greatly reduced by the formation of a "skin" produced by a high concentration of surface active molecules, but there was good eddy circulation just below the skin. As eddy concentration of oxygen approached the concentration of the main bulk of fluid, the oxygen microelectrode could no longer detect eddies.

Marshall and Cruickshank (1973) showed that *Flexibacter aurantiacus* and *Hyphomicrobium vulgare* were oriented perpendicular to air-water, oil-water, or solid-water interfaces. It was suggested that hydrophobic portions of the cell were rejected by water. With solids, an extracellular adhesive material, probably polysaccharide, aided attachment.

Measurement of a "skin" has been performed with bacteria oriented at an interface. The somewhat weak surface activity of cells is shown by measurements with *Pseudomonas ovalis* of a skin with a thickness of about 40 micrometers (Bungay and Masak, 1981) that is much less than the reported thickness of 240 micrometers for a detergent film at an air/water interface (Tsao and Lee, 1974). Negligible film thickness was noted until the concentration of bacteria in the main bulk reached about 10^7 per ml. Although there has been controversy about whether overall mass transfer rates of oxygen can be enhanced by the presence of actively respiring organisms right in the interfacial film, (Tsao, 1977, Yoshida and Yagi, 1977, Linek and Benes, 1977) measurements with a microelectrode indicated slower transfer (Bungay and Masak, 1981). Oxygen is a small molecule that diffuses much more rapidly than an organic molecule such as glucose. This means that a barrier film of microorganisms at an air/water interface retards nutrient transfer so that the advantage of proximity to the oxygen supply cannot be fully exploited. This also illustrates the fallacy of focusing exclusively on oxygen transfer rates because the diffusion of a substrate or product may be rate controlling. However, the biochemical engineer has often faced the problem of supplying oxygen to a highly aerobic process, thus the fixation on oxygen transfer.

Among the many papers about the effects of surface active agents on mass transfer are those by McKeown and Okun (1963), Mancy and Okun (1965), Eckenfelder and Barnhart, 1961, Bull and Kemp, (1971), Greenhalgh, et al. (1975), Dominguez, (1980), Mukerjee, (1973), and Yagi and Yoshida, (1974).

Real systems have many different surface active agents present. Little is known about their interactions at an interface, and even less is known about how microbial cells form at an interface when lipids, protein, and the like are packed in the same region. Consumption and excretion by the cells can alter the amounts of surface active substances. The effects of mixtures of cells and surface active agents on mass transfer should be a fertile research area.

An interface that has persisted for some time may approach steady state. The surface of a body of water or of a culture fluid in a container are examples. However, interfaces at bubbles are highly dynamic. For a very brief time, a recently-formed bubble may have a clean surface. Within a few hundredths of a second, surface active agents migrate to a bubble and markedly change its properties. A clean bubble is very flexible and has little resistance to gas transfer. A polluted bubble surface is relatively rigid and has much more resistance to penetration by gases. The bubble rises and expands as hydrostatic pressure lessens. If there is sufficient shear, a single bubble may subdivide. Fluid flow around the bubble affects its shape and may orient the surface active materials. Microbial cells in or near the surface active layer of a bubble have complicated and poorly understood mass transfer.

Vandenberg (1972) studied the effect of solid particles on the gas-liquid interfacial area in a stirred tank and found that up to 4 per cent by weight of particles in the diameter range of 75 to 600 micrometers had no effect. As particles can reduce rates of gas transfer, it is interesting that disturbance of the area is not the reason. Wang, et al. (1982), added cellite microbeads to penicillin fermentations and found that mass transfer improved because the system became less viscous when growth occurred on the beads.

There has been speculation about the effects of depth on organisms in tall tanks. There may be a rapid change from small pressure to several atmospheres in a few seconds as elements from the top are swept to the bottom of a vigorously agitated tank. Pressure increases gas solubility but decreases bubble size and with it the important area for mass transfer, and small bubbles move slowly with thicker fluid films that resist mass transfer. Pressure increases on the entire tank have not given significant improvements in mass transfer unless the gas flow rate was increased. Hydrostatic pressure is important in culturing organisms found in the great depth of the oceans, and the deep-shaft fermenter is an attempt to improve oxygen transfer. A fermenter of the type shown in Fig. 3 introduces air at a point below its eventual exit. The circulation pattern carries the gas downward to regions of great hydrostatic pressure that increase gas solubility. Spent air plus carbon dioxide buoy up the fluid for flow without mechanical agitation. While this concept is interesting and deserves research, there are practical limitations for capacity of a fermenter constructed of long lengths of vertical pipe.

Some microorganisms are almost always freely suspended while others aggregate or attach to a surface. Attachment imparts great directionality to mass transfer. It would seem wise to operate commercial processes with freely suspended cells in order to maximize mass transfer, but other factors may be more important. For example, a reactor can be packed with a material to which cells are attached to maintain a very high cell concentration. Cells immobilized in gel particles can be much more dense than cells in suspension. There are several reasons why they function when so closely packed; there is an advantage of fluid flowing past immobilized cells to flush out wastes that limit growth in batch cultures. High cell density can also be achieved by separating and recycling cells from the effluent. Mass transfer on a per cell basis could be terribly inefficient for densely packed cells, but their sheer numbers lead to high overall rates. There are also cases where activity of closely packed cells is higher than that of the same type of cells in a relatively thin suspension. For example, ethanol production by immobilized cells can be surprisingly high on a per cell basis (Chen and Tsao, 1982).

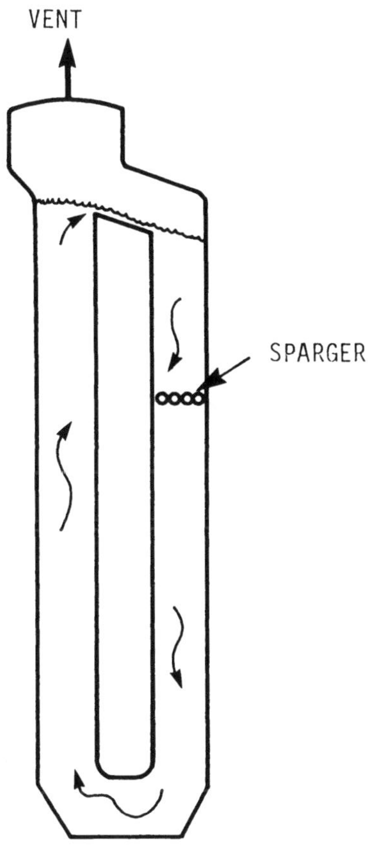

Figure 3. Deep shaft fermenter.

Mass transfer for immobilized cells can have serious problems. The essence of immobilization is to have high rates, but oxygen supply to a cell paste inside of beads or to dense cell attachments on a supporting surface may be limiting. Furthermore, carbon dioxide may be evolved rapidly to burst beads or to push cell aggregates away from their support.

Chemostats operated at high dilution rate strongly favor attached organisms. When the dilution rate excedes the maximum specific growth rate coefficient for suspended cells, loss is greater than growth so washout occurs. Attached cells are not washed out and persist.

Attachment may be initiated because the concentration of a nutrient adsorbed at a surface is greater than its concentration in the medium. After colonies coat the surface, there is no reason why adsorption of nutrients would differ much from that for freely suspended cells. On the other hand, colonies or slimes on the benthos of a stream or lake can benefit from nutrient particulate matter that settles.

IV. RELATIVE MOTION OF CELLS AND MEDIUM

When cells are attached or entrapped, the medium may flow past at a very great rate. However, suspended cells tend to have little motion relative to the medium. Flagellated cells force their way through the medium either at random or in response to tactile forces, concentration gradients, or to illumination. For example, a *Spirillum* with a flagellum at each end may be imprisoned in a tube. It swims to one end, and striking the surface causes inactivation of one flagellum and the other starts so that it reverses. It will traverse the capillary tube many times. Other motile organisms seek light or higher concentrations of nutrients. Even so the velocities are small.

The centrifugal forces of mixing have little effect on cells because their densities are so nearly the same as that of the medium. Even at the tips of rapidly rotating impeller blades, the centrifugal force could impart motion of only the same order as that of Brownian motion. Gravitational forces are small; yeast cells only sediment at a rate of roughtly 0.6 cm/hr, and this is much faster than the sedimentation rate of bacterial cells. However, there is a phenomenon that causes cells to move at rates of nearly a millimeter per second. Fermentation broths, silt particles, fine suspensions of ion exchange resins, and the like, can be stirred in a beaker, and within a few seconds of removing the stirrer, a clear annulus forms concentric with the walls. This is analogous to the "tubular pinch zone" observed for particles in Poiseulle flow in a tube (Segre and Silberberg, 1961). A particle released near the wall migrates toward the center while one starting at the center moves toward the wall. The region of zero migration is an annulus. There the forces due to streamlines are equal to the Magnus forces caused by particle rotation. Oliver (1962) showed that a particle drilled and packed with heavier material eccentricly to prevent rotation moved to the center and formed no annulus. Forces on a particle in a shear field are shown in Fig. 4. The different velocities at different points create unequal pressures by the Bernouli principle to cause migration. The effect increases as the cube of the particle diameter until the particle becomes so large that the velocity field is distorted. For particles less than one micron in diameter, the migration effect is small. Microbial cells in the range of two to several hundred microns are affected strongly, and larger particles have diminishing migration effect because their size distorts the streamlines. This simplified explanation based on the Bernouli principle is a useful way of interpreting particle migration data (Bungay and Wiggert, 1967).

Sedimentation of particles in inclined tubes has been studied by Acrivos and Herbolzheimer (1979). The suspension is more dense than the clear fluid, thus buoyant forces create additional motion during separation. The more concentrated particles slide down along the walls of the tube as the clearer fluid rises. Sedimentation is more rapid than in conventional gravity settling, and there are very complicated hydrodynamics and relative motion between particles and fluid.

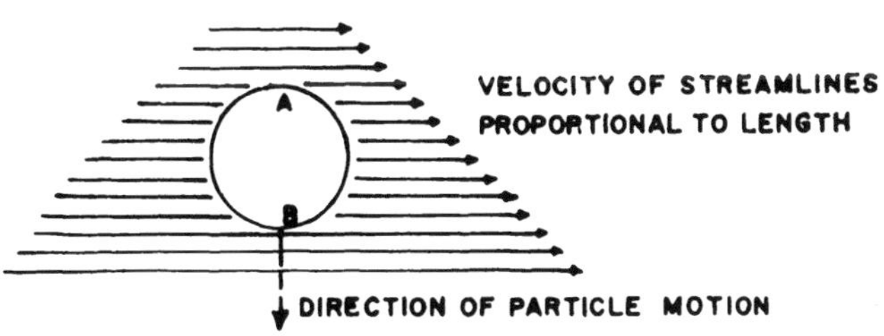

$$F_A = H - \frac{V_A^2}{2g} \qquad V_B > V_A$$

$$F_B = H - \frac{V_B^2}{2g} \qquad \therefore F_A > F_B$$

Figure 4. Development of forces on a suspended body.

In an intriguing report, Elmayergi and Moo-Young (1973) relate that mass transfer of oxygen, glucose consumption, and other growth parameters, all increased when a polymer was added to the medium of a fermentation with *Aspergillus niger* pellets. The enhancement may have resulted from better hydrodynamic slip between the pellets and the medium, lowering the thickness of films through which mass was transferred.

Hydrodynamic effects on cell membranes have been analyzed by Jain and Maldarelli (1982), and a book edited by Sorensen (1979) elaborates on interface phenomena. The biological processes described include deformation of red blood cells, rhythmic motion of red blood cells, breakdown of lipid bilayers, and phagocytosis. This type of analysis should be extended further to microbial systems.

V. SUBSTRATE AND MASS TRANSFER LIMITATIONS

Although this paper takes a different approach to mass transfer, some selected references in Table 4 provide some perspective on the more usual thinking in this area.

VI. FLUID TRANSPORT WITHIN CELLS

Movement of fluids through plants and animals is a fascinating topic that involves both fluid mechanics and biology (Vogel, 1981). Some general principles are: 1. large conduits are best for moving fluids to minimize frictional resistance; 2. small vessels are best for mass transfer through the walls to reduce diffusional paths and thickness of boundary films; and 3. viscosity effects vary because non-Newtonian fluids are moving with velocity and shear gradients (LaBarberra and Vogel, 1982).

Fluid transport within microorganisms is also complicated, but protoplasmal streaming and diffusion are predominant for tiny organisms. There may be some internal reaction to the spinning of flagella fastened to the cell, but there is no other internal pumping in bacteria, yeasts, and molds. Various microorganisms do have impulsive motion that must have some effect on the circulation of internal fluid. As examples, the thrusting motion of hydra must agitate their cytoplasm, and the cytoplasm of amoebae must mix as they glide across a surface.

So little is known about the mechanics of internal flow in microorganisms that no review is possible. Some nondestructive technique would be needed to visualize and quantify internal flow in tiny life forms.

Table 4. Mass Transfer.

Reference	Comments
Baillod and Boyle, 1970	Applied reaction modulus concepts and concluded that cells grow to a size that is at onset of diffusional limitations.
Calderbank and Jones, 1961	Used ion exchange particles as analog of transfer to dispersed organisms.
Massaro and Fatt, 1969	Theory of oxygen transfer to spherical cells based on region dependent on oxygen tension, region of constant oxygen, and a non-consuming membrane.
Conner and Gainer, 1970	Oxygen diffusivities in various systems.
Greenwood and Goodman, 1964	Lowering oxygen level around soil particles had little effect on aerobic organisms until critical level was reached.
Gulevich, et al., 1968	Effect of velocity on substrate removal determined with a rotating disc system.
Ghosh and Ghose, 1978	Effect of oxygen conc. on rate of gluconic acid production.
Reference	Comments
Weisz, 1973	Applied reaction modulus concepts to physiology and concluded that characteristic size of cells corresponds to onset of diffusional limitation.
Oels, et al., 1976	Detailed investigation of aeration and antifoam effects on single-cell protein production in a bubble column fermenter.
Rahman and Rahman, 1979	Model of substrate-inhibited systems considering diffusional resistances to spherical particles.
Haas, 1981	Thiele modulus analysis of floc size.

VII. CELLULAR INTERACTIONS

Microbial ecology, symbiosis, and synergistic metabolism have been reviewed several times, but little has been said about their roles in mass transfer, Haas, et al., (1981), Bull and Slater, ed. (1982). There are some extremely interesting associations where one organism lives within the cells of another. One example with great practical potential is the water fern Azolla that harbors an alga which harnesses sunlight for the fixation of nitrogen from the atmosphere. This fern is one of the most efficient nitrogen fixers with rates much better than those of alfalfa and other legumes and many times the rate of free blue-green algae. Certainly there are strong mutual effects on mass transfer for the plants cells and their intracellular alga. Another interaction with profound effects on mass transfer is predation. When one organism is ingested by another, its demand for and access to oxygen are reduced, and death may soon occur. The surviving prey face less competition for oxygen and nutrients, thus benefitting from the presence of predators. Parasitism affects mass transfer in several ways: organisms attached to a host may form a physical barrier to mass transfer and those inside a host may change metabolic rates.

Phage must have effects on mass transfer. Virulent phage redirect cellular metabolism to the synthesis of more phage until the cell dies. Such a cell has differing demands for oxygen and nutrients as it dies and disintegrates. Temperate phage attach to the chromosome and may persist in an inactive form through many cell generations. Even so there may be a small effect on the mass transfer of an infected cell because its requirements for synthesis have been modified slightly. Other microbial associations have stimulations or inhibitions of one on another, and mass transfer rates are likely to be affected.

Spatial order in microbial ecosystems has been reviewed by Wimpenny (1981). The structure of ecosystems affects how microorganisms interact, and spatial distribution, as mentioned previously, is a major factor in establishing gradients of oxygen, nutrients, and products. Stotzky (1967) concluded that clay minerals exert a marked influence on the activity, ecology, and population dynamics of microorganisms, especially in soil. Physical-chemical parameters of the clay are important, and surface characteristics must control attachment.

VIII. CONCLUSION

On a practical level, consideration of microscopic effects lends insight to mixing and aeration. Mixing is important because it creates velocity gradients so that cells can migrate away from fluid elements that they have depleted of oxygen and nutrients. Gentle mixing has little effect because the gradients are weak. Creating velocity gradients should be an aim of fermentation engineers, but excessive mixing is costly and may damage cells. Size, strength, and type of aggregation of cells must not be ignored, and may perhaps be improved by biologists to better suit the needs of engineers.

The traditional concept that mass transfer rate is dependent on area, a driving force, and diffusion constants is not of much value on a microscopic scale, where there are abrupt changes in cell dimensions and there are variations in concentrations of nutrients and dissolved gases. Aggregation, orientation in respect to fluid streamlines, and location of cells can play important roles in establishing rates of mass transfer. While there are formidable challenges in developing a coherent and integrated frame of reference, pointing out the complexities of mass transfer at the cellular level should be a stimulus to research.

REFERENCES

Acrivos, A., and E. Herbolzheimer. *J. Fluid Mech. 92:* 435 (1979).

Ash, S.G. *Spec. Publ. Soc. Gen. Microbiol. 2,* 57 (1979).

Atkinson, B., and M.E. Abdel Rahman Ali. *Trans. Inst. Chem. Eng. 54,* 239 (1976).

Atkinson, B., and M.E. Abdel Rahman Ali. *Water Research 12,* 147 (1978).

Atkinson, B., G.M. Black, P.J.S. Lewis, and A. Pinches. *Biotech. Bioeng. 21,* 193 (1979).

Atkinson, B., and I.S. Daoud. *Trans. Inst. Chem. Eng. 48,* 245 (1970).

Atkinson, B., and I.J. Davies. *Trans. Inst. Chem. Eng. 52,* 248 (1974).

Atkinson, B., and A.J. Knights. *Biotech. Bioeng. 17,* 1245 (1975).

Atkinson, B., and F. Ur-Rahman. *Biotech. Bioeng. 21,* 221 (1979).

Baillod, C.R., and W. C. Boyle. *ASCE, J. Sanit. Eng. Div. 96,* 525 (1970).

Baylor, E.R., V. Peters, and M.B. Baylor. *Sci. 197,* 763(1977).

Baylor, E.R., M.B. Baylor, D.C. Blanchard, I.D. Syzdek, and C. Appel. *Science 198*, 575 (1977).
Bitton, G. *Water Research 9*(5-6), 473 (1975).
Blakebrough, N. *Brit. Chem. Eng. 12*(1), 78 (1967).
Blanch, H.W. *Ann. Reports Ferm. Proc. 3*, 47 (1979).
Bull, A.T., and J.H. Slater, ed. "Microbial Interactions and Communities," Vol. 1, Academic Press (1982).
Bull, D.N., and L.L. Kempe. *Biotech. Bioeng. 13*, 529 (1971).
Bungay, H.R., and J.M. Wiggert. *Biotechnol. Bioengr. 7*, 322 (1965).
Bungay, H.R., M.Y. Huang, and W.M. Sanders. *AIChE Jour. 19*, 373 (1973).
Bungay, H.R., and R.D. Masak. *Biotechnol. Bioengr. 23*, 1155 (1981).
Busch, P.L., and W. Stumm. *Environ. Sci. Technol. 2*, 49 (1968).
Calderbank, P.H., and S.J.R. Jones. *Trans. Inst. Chem. Eng. (London) 39*, 363 (1961).
Characklis, W.G. *Water Res. 7*, 1249 (1973).
Characklis, W.G. *Biotech. Bioeng. 23*, 1923 (1981).
Chen, L.F., and G.T. Tsao, Paper presented at Am. Chem. Soc., Kansas City (1982).
Chester, V.E. *Soc. Chem. Ind. Monograph 19*, 59 (1965).
Conner, E.D., and J.L. Gainer. *CEP Symp. Series 99*, Vol. 63, 72 (1970).
Crabtree, K., W. Boyle, E. McCoy, and G.A. Rohlich. *J. Water Poll. Control Fed. 38*, 1968 (1966).
Curds, C.R. *J. Gen. Microbiol. 33*, 357 (1963).
Danielli, J.F., and J.T. Davies. *Adv. in Enzymol. 11*, 35 (1951).
Daniels, S.L. *Devel. Ind. Micro. 13*, 211 (1972).
Daniels, S. *AIChE Symp. Ser. 70*(136), 266 (1974).
Dominguez, J.G. *J. Dispersion Sci. Technol. 1*(2), 233 (1980).
Donovick, R. *Appl. Microbiol. 8*, 117 (1960).
Drost-Hansen, W., and J.S. Clegg, ed. "Cell-Associated Water," Academic Press (1979).
Dutka, B.J., and K.K. Kwan. *Canad. J. of Microbiol. 24*, 187 (1978).
Eckenfelder, W.W., and E.I. Barnhart. *AIChE J. 7*, 631 (1961).
Elmayergi, H., and M. Moo-Young *Biotechnol. Bioeng. Symp. 4*, 507 (1973).
Fletcher, M., and G.D. Floodgate. *Soc. Appl. Bacteriol. Tech. Ser. 10*, 101 (1976).
Forster, C.F. *Water SA 1976, 2*(3), 119 (1976).
Garner, F.H., and A.H.P. Skelland. *Ind. Eng. Chem. 48*, 51 (1956).
Gerchakov, S., and B. Sallman. NTIS AD-A045815 (1977).

Gerson, D.F., and J.E. Zajic. *ACS Symp. Ser. 106,* 29 (1979).
Ghosh, P., and T.K. Ghose. *J. Ferment. Technol. 56*(2), 139 (1978).
Goulder, R. *J. Appl. Bact. 43,* 399 (1977).
Greenhalgh, S., W.J. McManamey, and K.E. Porter. *J. Appl. Chem. Biotechnol. 25,* 143 (1975).
Greenwood, D.J., and D. Goodman. *J. Sci. Food Agr. 15,* 579 (1964).
Gulevich, W., C.E. Renn, and J.C. Liebman. *Environ. Sci. Technol. 2,* 113 (1968).
Haas, C.N. *Jour. Envir. Engr. Div. of Am. Soc. Civil Engr. 107,* 269 (1981).
Haas, C.N., H.R. Bungay, and M.L. Bungay. *Ann. Rep. Ferm. Proc. 4,* 1 (1981).
Hahn, H.H., and B. Eppler. *Colloid Interface Sci. Proc. Int. Conf. 50*(4), 125 (1976).
Hamer, G. *Biotech. Bioeng. 54,* 565 (1973).
Hendricks, C.W. *Appl. Microb. 28*(4), 572 (1974).
Heukelekian, H., and A. Heller. *J. Bacteriol. 40,* 547 (1940).
Hoehn, R.C., and A.D. Ray. *J. Water Poll. Control Fed. 45,* 2302 (1973).
Howell, J.A., and B. Atkinson. *Biotech. Bioeng. 18,* 15 (1976).
Jain, R.K., and C. Maldarelli. Paper presented at 4th Intl. Conf. of Physico-Chemical Hydrodynamics, New York (1982).
Jennings, P.A., V.L. Snoeyink, and E.S.K. Chian. *Biotech. Bioeng. 18,* 1249 (1976).
Kornegay, B.H., and J.F. Andrews. *Purdue Univ. Eng. Bull. Ext. Ser. 1967, 129*(2), 620 (1967).
LaBarbera, M., and S. Vogel. *Am. Scientist 70,* 54 (1982).
LaMotta, E.J. *Environ. Sci. Technol. 10,* 765 (1976).
Larsen, D.H., and R.L. Dimmick. *J. Bacteriol. 88,* 1380 (1964).
Lee, Y.H., G.T. Tsao, and P.C. Wankat. *AIChE J. 26,* 1008 (1980).
Linek, V., and P. Benes. *Biotechnol. Bioengr. 19,* 565 (1977).
Livansky, K., and B. Prokes. *Collect. Czech. Chem. Commun. 42*(1), 180 (1977).
Mancy, K.H., and D.A. Okun. *J. Water Poll. Control Fed. 37,* 212 (1965).
Marshall, K.C., and R.H. Gruickshank. *Arch. Mikrob. 91*(1), 29 (1973).
Marshall, K.C. *Life Sci. Res. Rep. 13,* 281 (1978).
Massaro, T.A., and I. Fatt. *Bull. Math Biophys. 31,* 327 (1969).
McCalla, T.M. *Soil Sci. Soc. Am. Proc. 11,* 260 (1946).
McKeown, J.J., and D.A. Okun. *Air Water Pollution 5,* 113 (1963).
Meadows, P.S. *Nature 207,* 1108 (1965).
Meadows, P.S., and J.G. Anderson. *Nature 212,* 1059 (1966).

Meadows, P.S., and J.G. Anderson. *Prog. Ind. Microbiol. 15,* 207 (1979).
Mukerjee, A. NTIS PB 229560 (1973).
Munson, R.J., and B.A. Bridges. *J. Gen. Microbiol. 37,* 411 (1964).
Nakamura, H. *J. Biochem. Microbiol. Technol. Eng. 3,* 395 (1961a).
Nakamura, H. *Mikrobiologiya 30,* 628 (1961b).
Nakamura, J., S. Miyashiro, and Y. Hirose. *Agric. Biol. Chem. 40,* 377 (1976).
Oels, U., T. Schuegerl, and J. Todt. *Chem. -Ing. Tech. 48*(1), 73 (1976).
Oliver, D.R. *Nature 194,* 1269 (1962).
Ou, L.T., and M. Alexander. *Arch. Microb. 101*(1), 35 (1974).
Paerl, H.W. *Science 180,* 496 (1973).
Pavoni, J.L., M.W. Tenney, and W.F. Echelberger, Jr. *J. Water Poll. Control Fed. 44,* 414 (1972).
Pavoni, J.L., M.W. Tenney, and W.F. Echelberger, Jr. *Eng. Bull. Purdue Univ. Eng. Ext. Ser. 1973, 140,* 957 (1973).
Pawlosky, U., J.A. Howell, and C.T. Chi. *Biotech. Bioeng. 15*(5), 905 (1973).
Rahman, F., and A. Rahman. *Pak. J. Biochem. 12*(2), 82 (1979).
Rittman, B.E., and P.L. McCarty. *Biotech. Bioeng. 22*(11), 2343 (1980).
Rittman, B.E., and P.L. McCarty. *Biotech. Bioeng. 22,* 2359 (1980).
Sanders, W.M. *Air Water Poll. 10,* 253 (1966).
Sanders, W.M., H.R. Bungay, and W.H. Whelan. *AIChE Symp. Ser. 107,* 69 (1971).
Santoro, T., and G. Stotzky. *Can. J. Micro. 14,* 299 (1968).
Segre, G., and A. Silberberg. *Nature 189,* 209 (1961).
Sieburth, J.M., P.J. Willis, K.M. Johnson, C.M. Burney, D.M. Lavoie, K.R. Hinga, D.A. Caron, F.W. French III, P.W. Johnson, and P.G. Davis. *Science 194,* 1415 (1976).
Sorensen, T.S., ed. "Dynamics and Instability of Fluid Interfaces," Springer-Verlag, New York (1979).
Stewart, G.G., I.F. Garrison, T.E. Goring, M. Meleg, P. Pipasts, and I. Russell. *Kemia-Kemi 3,* 465 (1976).
Stotzky, G. *Trans. N.Y. Acad. Sci. 30,* 11 (1967).
Tenney, M.W., and F.H. Verhoff. *Biotech. Bioeng. 15,* 1045 (1973).
Tezuka, Y. *Appl. Microbiol. 17,* 222 (1969).
Topiwala, H.H., and G. Hamer. *Biotech. Bioeng. 13,* 919 (1971).
Treweek, G., and J.J. Morgan. *J. Colloid Interface Sci. 60,* 258 (1977).
Tsao, G.T., and D.L. Lee. *AIChE J. 21,* 979 (1975).
Tsao, G.T. *Biotech. Bioeng. 19,* 557 (1977).

Tsao, G.T., and Y.H. Lee. *Ann. Reports Fermentation Processes 1,* 115 (1977).
Unz, R.F., and S.R. Farrah. NTIS PB-235023 (1974).
Vandenberg, H.J. *Adv. Chem. Ser. 109,* 240 (1972).
Wang, D.I.C., J. Meier, and K. Yokoyama. Paper presented at Am. Ch. Soc., Kansas City (1982).
Vogel, S. "Life in Moving Fluids: The Physical Biology of Flow," Willard Grant Press (1981).
Watson, J.H., and B.J. Stojanovic. *Soil Sci. 100,* 57 (1965).
Weisz, P.B. *Science 179,* 433 (1973).
Whelan, W.J., H.R. Bungay, and W.M. Sanders. *Environ. Sci. Technol. 3,* 1297 (1969).
Williams, N.J., and A. Wiseman. *Biochem. Soc. Trans. 1,* 291 (1973).
Wimpenney, J.W.T. *Biol. Rev. 56,* 295 (1981).
Yagi, H., and F. Yoshida. *J. Ferm. Tech. 52*(12), 905 (1974).
Yoshida, F., and H. Yagi. *Biotechnol. Bioengr. 19,* 561 (1977).
Yoshida, F. *Ann. Rep. Ferm. Proc. 5,* 1 (1982).
Zobell, C.E. *J. Bacteriol. 46,* 39 (1943).

CHAPTER 7

ENZYMES OF INDUSTRIAL INTEREST--TRADITIONAL PRODUCTS

K. Aunstrup

Novo Industries
Bagsvaerd, Denmark

In the turmoil of biotechnological dreams and promises, which we have seen over the last five years, the enzyme industry has enjoyed a steady but undramatic growth. The traditional enzymes have been improved, new applications have been found and a few new and valuable products have been marketed. It is to be expected, that in the coming five years, we will see a much more dramatic development of this industry. New products developed by the use of gene splicing will be marketed. For example, a microbial production of calf rennet, and the tremendous increase in biotechnological research, which has taken place all over the world, is bound to result in some untraditional and successful use of enzymes.

An example of such untraditional thinking is the oxidation of alkenes to glycols with a concommitant production of fructose from glucose, most of the processes being catalyzed neatly by enzymes (1). Unfortunately, the processes do not seem to be economically competitive.

Some of the important challenges seem far from a solution. For example, the industrial use of enzymes depending on cofactors such as NADH and the enzymatic hydrolysis of lignocellulose. In both these instances it appears to be necessary to develop methods not known today to solve the problems.

ISBN 0-12-040306-4

I. PROTEOLYTIC ENZYMES

The detergent proteases have now, for over 15 years, maintained a position as the most important enzyme products. And there is a steady growth in the application of these enzymes (2).

In the dairy industry, microbial rennet also have a solid and growing position, especially since the chemical derivation of the enzymes have improved their properties. Apart from these two large applications, microbial proteases only find minor application, for example in the brewing industry, the tanning industry and in baking (3). An interesting development is the use of proteases for production of protein hydrolysates, which are soluble at isoelectric point and have a bland taste (4,5). These products may be used in beverages or for other food purposes.

A. *Bacillus Protease*

Commercial protease products for detergents produced by *Bacillus licheniformis* were believed to contain only one proteolytic component, subtilopeptidase A (subtilisin Carlsberg) and isoenzymes thereof. It has now been found (6) that small amounts (5%-10%) of another protease is present in traditional commercial protease products. This protease is apparently a new type of Bacillus protease, it is not a serine- or metallo protease. The enzyme has a high allergenic potential and is therefore undesirable. Its presence in the product can be avoided by using mutants deficient in this enzyme. The kinetics of protease production by *B. licheniformis* in defined media has been studied, and it was found that the protease is produced during exponential growth and that the production is not influenced by catabolite repression (7,8). The connection with the sporulation process was not clarified. *Bacillus subtilis* (or better *Bacillus amyloliquefaciens*) protease is of minor economic importance but is still an attractive subject for study, it produces a serine protease and a neutral, metallo-protease. It has been shown that the neutral protease will be inactivated by the serine protease at neutral pH. The process may be reduced by addition of up to 2% of protein hydrolyzates or by lowering the pH to 5.5-6 (9). A closer study of the proteolytic components in a hyper producing strain of *B. subtilis* has revealed that the cell membranes contain the two extracellular enzymes bound to phosphorlipids and small amounts of an additional protease of esterase type (10). Sporulation and protease formation are closely linked in the Bacilli, and it has been believed that proteolysis plays a central role in the initiation of sporulation. In an admirable review Maurizi and Switzer (11) evaluate the current knowledge on the proteolytic apparatus of the Bacillus species and conclude, that the evidence for the involvement of proteolysis in the regulation of gene expression during sporulation is not conclusive although proteases serve processes that are necessary for sporulation. The properties of the thermostable metallo protease, Thermolysin has been compared with the properties of the metallo protease of *B. amyloliquefaciens* (NRRL 2411) (12). Although the C-terminal end of the molecules show extensive homology, this is not so in the N-terminal end, furthermore, a larger helicity of the *B. amyloliquefaciens* enzyme is revealed by circular dichroism tests. There is no immunological cross reaction between the two enzymes. It is believed, that there are similarities in the enzymatically active part of the molecule and differences elsewhere. Mutants of *B. subtilis* producing metallo protease with decreased thermostability have been found (13). The pH optimum of the proteases was changed, it is assumed, that the mutations took place in the active center. A thermostable collagenase has been isolated from a strain of *Bacillus stearothermophilus*-like organism (14). The enzyme is a metallo protease and appears to have properties similar to Termolysin.

Around 1970 allergic symptoms in workers handling proteases for detergents created a strong public reaction. Over the past 10 years safety regulations have been strict and the health of workers handling these products has been monitored carefully. A study (15) on the sensitization of workers handling detergent enzyme Esperase® reveals that few workers were sensitized and no permanent damages to their health occurred. The report concludes that there is no significant risk of consumer sensitization.

B. Proteases from Serratia

The metallo protease from *Serratia marcescens* is, in some countries, used in medicine as an anti-inflammatory agent. The organism produces, just like the *Bacilli*, both serine and metallo proteases (16). An extensive purification of the metallo protease has shown (17) that it is microheterogenous and forms several components with similar activity but variation in molecular weight and isoelectric point.

C. *Proteases from Actinomycetes*

Several proteases derived from Streptomyces have been developed and marketed (e.g., *Streptomyces fradiae, Streptomyces griseus, Streptomyces rectus*) but none of them seem to have found a permanent application. The proteolytic system Thermitase of *Thermoactinomyces vulgaris* contains three components (18) of which the main component is a serine protease (MW 37,000, pI 9) which like the subtilisins show the amino acid sequence Thr•Ser•Met•Ala around the active serine group. The enzyme is however different from the subtilisins in having a cystein group necessary for activity (19). In this respect the enzyme is similar to the proteases from *Bacillus sphaericus, Bacillus thuringiensis* and *S. rectus*. It has been reported that this protease contains about 4% carbohydrate but the majority of this may not be covalently bound but seen as a result of the strong adsorption of carbohydrate by the enzyme. This protease is useful at temperatures up to 70°C and the specificity is low (20).

A protease which is found useful in depilation, is made by a newly isolated strain named *Streptomyces caligosus*. The enzyme has pH optimum 7.5, MW 27,000, pI 3.7 and its is rapidly inactivated at temperatures about 50°C (21).

D. *Fungal Proteases*

Despite the variation in the large number of known fungal proteases the industrial use is still limited to the Aspergillus proteases and their application in digestive aid and various food processes. Both acid, neutral and alkaline proteases are formed and an investigation of a large number of species (22) has not unexpectedly shown that species with strong acid production produce acid protease and little or no alkaline protease, whereas the opposite is found in species with little acid production. In a study of the secretion of acid proteases by *Aspergillus oryzae*, it has been found (23) that the organism has two intracellular acid proteases, which have identical catalytic and immunological properties to the carbohydrate free extracellular acid proteases but which are different from the membrane bound proteases also found in the mycelium. This situation appears to be similar to that of *B. licheniformis*.

The peptidases of the Aspergilli are of potential commercial interest for complete hydrolysis of proteins. An acid carboxypeptidase from *Aspergillus niger var. macrosporus* has been purified (24). *Monascus koaliang,* which is used as a coloring agent in rice wine and other oriental food also produce an acid aspartate protease, which is a glycoprotein (MW 34,000) (25).

Yeast proteases are rately described in literature, but in a screening of 147 strains from 47 different species for milk coagulating activity, a few were found with acceptable enzyme activity (26). The best yield was obtained from *Cryptococcus albidus var. aerius,* which produced approximately 0.25 g per liter of an extracellular acid protease, which coagulate milk to make a firm coagulum. The specificity of the enzyme was not determined.

Proteolytic enzymes are used in several medical areas such as digestive aids, debridement of wounds and treatment of burns and ulcers, in the treatment of chataract and as fibrinolytic and anti-inflammatoric agents. In most of these applications, animal enzymes such as pancreatin, trypsin, plasmin and urokinase are used, but a few microbial enzymes are in regular use, e.g., streptokinase for plasminogen activation. Some fungal proteases have been tested extensively for use as fibrinolytic agents, e.g., Brenolase, the alkaline serine protease from *A. oryzae,* Ocrase, the corresponding protease from *Aspergillus ochraceus* and similar proteases from *Trichothecium roseum* and *Armillaria mellea* (27,28). Although the enzymes have shown some promise and have been tested extensively over the last ten years, they have never been widely accepted for clinical use, perhaps because they are not, as originally believed, specific fibrinolytic agents but generally proteolytic.

E. *Mucor Proteases*

The two acid aspartate proteases from *Mucor miehei* and *Mucor pusillus* are important industrial enzymes, their single use being as milk coagulating agents in the dairy industry (2). *M. pusillus* forms an extracellular, unspecific acid protease, which is less thermostable than the milk coagulating enzyme from which it may be separated by ammonium sulphate fractionation (29). The regulation of enzyme formation by *M. miehei* has been studied by Lasure, who finds that protein is necessary in the medium in order to produce the enzyme and that free amino acids and high levels of glucose will repress the enzyme synthesis (30). The amino acid composition of the *M. miehei* rennet has been determined on different occasions and various results have been found. These differences are often attributed to strain differences (31) but they might more likely be caused by differences in purification methods since this could easily cause varying degrees of autolysis. Strain differences as the cause of variation should only be accepted, if the recovery and purification has bee made in exactly the same way for the enzymes to be compared.

The proteases from *M. pusillus, M. miehei* and chymosin are very closely related enzymes, hence their successful use as milk coagluants. Nevertheless they show small but for the application in the dairy industry important differences. A kinetic study has been made on the hydrolysis of a hexapeptide, similar in composition to the sequence in kappa casein, which is attacked by these enzymes in only one bond. This peptide was thought to be used as a substitute for milk clotting analysis, but there is no direct correlation between milk coagulation and hydrolysis of the peptide. Chymosin has a much higher clotting activity relative to the proteolytic activity toward the peptide than the microbial enzymes (32).

Etoh, et al. (33), have studied the physiochemical and immunochemical properties of the acid proteases of *M. pusillus* and *M. miehei*. They find that the enzymes are similar but also that they show distinct differences. The two enzymes show partial immunological identity, a property which is often found in enzymes of related species (34).

The good heat stability of the *M. miehei* rennet, in relation to calf rennet, has as a result that it is difficult to remove milk coagulating activity in the whey and this is a problem, when the whey is used for other milk products. The problem has been solved by chemical derivation of the rennet, whereby its thermostability is reduced about 5°C, so that it comes close to the thermostability of calf rennet. The destabilization is effected simply by adding oxidizing agents like sodiumhypochlorite, hydrogen peroxide or peracetic acid to the enzyme. The mechanism of the reaction is not known, but it is assumed, that uncharged non aromatic aminoacid residues are attached, e.g., methionine. The carbohydrate part of the molecule is not influenced (35,36).

In the Soviet Union a rennet enzyme has been developed from *Mucor renninus*, a species not generally known. It grows at 30°C and produces a rennet enzyme with pH optimum for proteolysis at 3.5 (MW 34,000, pI 5.1). The enzyme contains 8% carbohydrate, 8 moles of cystine and 9 moles of methionine per mole (37). In addition to the rennet enzyme, this fungus produces two unspecific proteases with low milk coagulating effect, which are similar to acid proteases from other fungi.

In a screening for milk coagulating activity, Fraile, et al., found, that some strains of *Mucor bacilliformis* will form a milk coagulating enzyme with a similar ratio between milk coagulating activity and proteolytic activity as *M. miehei* and *M. pusillus*, but with a much lower thermostability (38).

II. AMYLOLYTIC ENZYMES

In the competition between commercial enzymes on the market inferiority in a single property is often decisive for the use. α-Amylase from *B. amyloliquefaciens* was for many years dominating, but after the development of the more thermostable α-amylase from *B. licheniformis* this enzyme has almost completely taken over the market of the starch industry, whereas the *B. amyloliquefaciens* amylase still is used in desizing and the distilling industry.

A. Bacillus α-Amylase

For the liquefying enzymes used in the starch industry, thermostability is of major importance and therefore the amylase of *B. licheniformis* with its 10-20°C higher operating temperature has almost completely superseded the traditional amylase from *B. amyloliquefaciens* over the ten years, it has been on the market. The enzyme is produced by most strains of *B. licheniformis* (39,40). Its structure is very similar to that of the amylase of *B. amyloliquefaciens*, in the N-terminal part of the molecule 13 out of 15 aminoacids are identical (41). *B. licheniformis* amylase has MW 48,700 by sedimentation equilibrium, 13,500 by gel filtration and 62,000 by polyacrylamid electrophoresis. It has not been proved whether the enzyme is an oligomer. It contains one gram atom Ca per 26,000 Daltons. The molecule contains cystein residues and no carbohydrate. pH optimum is 6-7 and the enzyme is stable between 6 and 10 (42). In a synthetic medium the amylase synthesis takes place in the exponential growth phase, it is subject to catabolite repression and phosphate is needed in the synthesis (43,44).

B. licheniformis usually produces a substantial amount of protease along with the α-amylase. This is undesired in starch hydrolysis, because proteins present as impurities in the starch, may be hydrolyzed and subsequently cause browning reaction. Fortunately this protease may readily be inactivated by heating at 70-90°C. There is only a negligible loss of amylase activity, provided sufficient calcium ion is present (45). The molecular structure of the α-amylase of *B. amyloliquefaciens* has been studied by several authors and it appears, that the molecular weight of the monomer is

48,000 and not 24,000 as previously believed (46). The active center is not known, but evidence points to a tyrosine residue as an important part (47).

The N-terminal sequence of amylase from *B. licheniformis*, *B. amyloliquefaciens*, *B. subtilis Marburg* and *B. subtilis var. amylosaccharitꞏicus* has been compared, and it has been found, that there is a striking homology between *B. licheniformis* and *B. amyloliquefaciens* amylase and a similar homology between the saccharifying enzymes from the *B. subtilis* strains (Table 1) (41).

The regulatory system for α-amylase production in *B. subtilis* (Marburg) has been studied by mutation (48,49). At least five different and identified mutations may act synergestically to increase the yield. If this is combined with certain genetic elements found in cycloserine and ampicillin resistant mutants, it may be possible to increase the yield 1,500-2,000 times over the parent strain. In some of the mutants there is a concomitant increase of amylase and protease production.

A new α-amylase with operational pH between 4 and 5.2 at 80°C has been found in certain strains of *B. stearothermophilus* (53) (MW 96,000). Its thermostability is better than any other amylase known. Unfortunately very low yields are reported for the fermentation process, which must take place at temperatures above 50°C. The acid amylase from *Bacillus acidocaldarius* has a pH optimum at pH 4.5, it is highly dependent on calcium ion and its heat stability is relatively low, its maximum operating temperature is about 60°C, furthermore the enzyme is produced in extremely low yield (54). An alkaline amylase has been found in *Bacillus alcalophilus*, NRRL 3881. The optimum temperature was 50°C at the pH optimum of 9.2. The alkaline amylase was less sensitive to chelating agents such as EDTA than the α-amylase from *B. amyloliquefaciens*. The enzyme is an endo amylase, but a study of the optical rotation of the hydrolysis products indicate, that it is also a β-amylase. It is possible that several different amylolytic components are produced (54).

B. Actinomyces Amylase

α-Amylase production in ther thermophile, *Thermomyces vulgaris* has been known since 1967. In a new screening work it has been found, that the organism apart from the well known neutral amylase also produces an acid amylase. This enzyme has optimum at pH 5.0 and will operate at temperatures up to above 70°C. Ca stabilizes the enzyme, which is of the saccharifying type with a hydrolysis spectrum similar to that

Table 1. N-terminal Sequences of α-Amylase

α-Amylase	Mol wt	Sequence	Reference
B. licheniformis ERY R9 (liquefying, thermostable)	62,000	Ala-Asn-Leu-Asn-Gly-Thr-Leu-Met-Glu-Tyr-Phe-Glu-X[a]-Tyr-Met-Pro-Asn-Asp- (1 ... 5 ... 10 ... 15)	41
B. amyloliquefaciens N (liquefying, thermolabile)	50,000	Val-Asn-Gly-Thr-Leu-Met-Glu-Tyr-Phe-Glu-Trp-Tyr-Thr-Pro-Asn-Asp-	48
B. subtilis Marburg (saccharifying)	68,000	Leu-Thr-Ala-Pro-Ser-Ile-Lys-(Ser)[b]-Gly-(Thr-)-Ile-Leu- (1 ... 5 ... 10)	49
B. subtilis (saccharifying)	42,000	Thr-Ala-Pro-Ser-Ile-Lys-Ala-Gly-Thr-Ile-Leu-	50
Aspergillus oryzae (TAA)	51,000	Ala-Thr-Pro-Ala-Asp-Trp-Arg-Ser-Gln-Ser-Ile-Tyr-Phe-Leu-Leu- (1 ... 5 ... 10 ... 15) Thr-Asp-Arg-Phe-Ala-Arg-Thr-Asp-Gly-Ser-Thr-Thr-Ala-Thr- (20 ... 25 ... 29)	57

[a] X, Not identified.

[b] ( ), Assignment not definite.

of *A. oryzae* amylase. The enzyme is interesting in that it hydrolyzes pullulan, mainly under the formation of panose. Only one active center appears to be present in the enzyme (55,56).

C. *Fungal Amylase*

The most important fungal α-amylase is derived from *A. oryzae.* The sequencing of its 470 amino acid groups has been going on for several years and now 66 residues from the C-terminal serine and 29 residues from the N-terminal chains are known. By comparison (Table 1) it is seen, that there is not much homology with the N-terminal of the other known α-amylases (57). It will be interesting to see whether these amylases have a similar active site sequence as one would expect.

In a study of the kinetics of maltooligosaccharide hydrolysis, it has been found (58) that this amylase will hydrolyze more than one glucosidic bond during some enzyme-substrate encounters. It has furthermore been shown, that at higher concentration of the lower oligosaccharides, transglucosylation plays a significant role in the hydrolysis (59).

D. *Yeast Amylases*

With the growing interest in the production of single cell mass and alcohol from starch and the utilization of starchy waste products has followed a desire for yeast strains, which may syntesize their own amylolytic enzyme system. A number of yeast species have been screened in order to find a strain, which would grow rapidly on a starch substrate (60). In all cases pH fell below 3 and growth rate dropped. If the pH was maintained by buffering, the growth rate did not increase. *Schwanniomyces castellii* produces an α-amylase and two glucoamylase enzymes (61), they have pH optimum at 6.0 and temperature optimum at 60°C. The enzymes are normally adsorbed to the cell wall, but they can be excreted by addition to the growth medium of mineral salts or surfactants, this is necessary for their action in starch hydrolysis. In all circumstances, the yields of these enzymes are so low, that they limit the amount of biomass produced on starchy media. In *Lipomyces starkeyi* the amylolytic system is also associated with the cell walls (62). It was not possible to separate the activity in α-amylase and α-glucosidase. *Lipomyces kononenkoae* is a promising candidate for the direct conversion of starch into biomass (63). It produces an α-amylase, a glucoamylase and a debranching transferase. The α-amylase has MW 38,000, pH optimum at 5.5 and temperature optimum at 40°C (15 min.), calcium ion is apparently not required and EDTA does not inhibit the enzyme. The glucoamylase has MW 81,000, pH optimum 4.5 and temperature optimum at 50°C (20 min.).

α-Amylase hyperproducing mutants were found by selecting for growth in a 2-dioxyglucose medium (64). It is assumed, that the mutants are derepressed in the catabolite repression, normally shown.

E. Glucoamylase

Glucoamylase is one of the most important of the industrial enzymes and calculated on enzyme protein, it is also one of the cheapest enzymes (2). Highly saccharified starch hydrolyzates and dextrose is almost exclusively made with this enzyme. Commercial production is based on *Rhizopus*, *Endomyces* or *A. niger*, the latter being the most important source. The properties of the glucoamylases have been studied over many years (65). The enzyme is a glucoprotein and usually occurs as at least two components with different properties. In a recent study (66) of a commercial glucoamylase preparation from *A. niger*, the two components I and II of the enzyme were purified and it was found, that the amino acid sequence appears to be identical, but that the sequence of component I extends over the C-terminal of component II. The sequence of 32 amino acids of the N-terminal has been determined and is identical for the two components (Ala-thre-leu-asp-). The C-terminal amino acids is *thr* for component I and *ser* for component II. The carbohydrate content is 17%-19% for component I and 22% for component II. The carbohydrate is bound to serine and threonine as previously found, but the results indicate, that there are about 30 glucosidically linked groups per molecule, corresponding to an average chain length of about 2. The carbohydrate composition of component II from another commercial *A. niger* preparation has MW 70,000 and contains per mole 45 mole mannose, 3 mole galactose, 4 mole glucose and 3 mole glucosamine (67). The carbohydrates are attached to serine and threonine and the carbohydrate chains have the structure man - man - x, where x can be glucose, galactose, mannose or N acetylgalactosamin. A structure of this type has also been reported in Rhizopus glucoamylase.

A purified glucoamylase from the commercial preparation, Molsin, prepared from *Aspergillus saitoi (Aspergillus phoenicis)* has similar properties to the enzyme isolated from *A. niger*. Component I has MW 90,000, a carbohydrate content of 18% and the N-terminal ala (68). Component II has a similar sequence, N-terminal (ala - val - ile - val) and kinetic properties; furthermore, lower affinity to branched substrates like amylopectin, MW 70,000 (69,70). The results suggest, that component II is formed from component I by proteolytic hydrolysis in the C-terminal end of the molecule during growth of the organism. The glucoamylase of *Aspergillus awamori var. kawachi* consists of at least three components, I has MW 90,000 and 7.6% carbohydrate, N-terminal ala and C-terminal arg, it digests raw starch and it may be converted to component I' by the action of proteolytic enzymes. This component is assumed to be an aggregate of two amino acid chains, on having N-terminal ala and C-terminal arg and the other having N-terminal ile and C-terminal thr. By further proteolytic and glycolytic digestion component II is formed. It has MW 57,000, N-terminal ile, C-terminal ser and a carbohydrate content of 1% (71,72).

The proteolytic conversion could be performed both with alkaline proteases such as subtilisin and with acid proteases such as the protease formed by the glucoamylase producing strain. With glycolytic enzymes, such as mannosidase, several intermediate components could be formed. A mutant free of acid protease produced a high concentration of Component I, whereas a mutant high in protease produced a high proportion of component I' (73). Based on this information, it may be concluded, that one major glucoamylase component is formed by the *A. niger* group and that this enzyme after its formation is proteolytically and glycolytically degraded to other active enzymes. It is interesting that the specificity apparently changes since component I in all instances has higher affinity for branched substrates than the other components.

It has some interest for the enzyme industry to develop more heat resistant glucoamylase preparations. The main purpose has been to permit an economical use of immobilized preparations for starch hydrolysis. Consequently, screening programs, searching especially the thermophilic fungi are justified. An enzyme from *Talaromyces dupontii* has pH optimum at 4.0 and temperature optimum at 75°C (10 min.), MW 31,000 (74). Another thermophile *Thermomyces lanuginosus* forms a glucoamylase with pH optimum 6 and temperature optimum at 70°C (15 min.). This enzyme is reported not to have any activity towards isomaltose (75). It is not disclosed if these enzymes will operate better in starch hydrolysis than the *A. niger* glucoamylase.

F. Debranching Enzymes

The α 1,6-bonds in amylopectin are much more resistant to attack by glucoamylase than the α 1,4-bonds and they will therefore be decisive for the final rate of hydrolysis and thus for the dosage of glucoamylase. This can be counteracted by using a debranching amylase early in the process, whereby the large, branched molecules are converted to large linear molecules, which are readily attacked by glucoamylase. This process has been known for several years and the pullulanase from *Klebsiella* was used. Unfortunately the pH activity of this enzyme meant that the reaction had to be performed at a pH too high for optimal glucoamylase application. If a debranching enzyme with optimum pH close to that of glucoamylase is used, it is possible to reduce the glucoamylase level considerably and to obtain a higher final degree of hydrolysis (76). Another useful application of debranching amylase is in the preparation of high maltose syrups together with β-amylase. Debranching enzymes are found in many species of microorganisms. In a screening experiment including 297 strains of 26 species it was found, that half of the strains hydrolyze pullulan (77).

The isoamylase from *Cytophaga (Flavobacterium)* which is commercially available has been purified by affinity chromatography on con A sepharose saturated with glucogen (78). The enzyme has pH optimum at 5-6, temperature optimum at 45°C (20 min.). It is of interest to note, that contrary to what was previously believed, it reacts slowly with pullulan and it will hydrolyze branched oligosaccharides with branches as short as maltose. A similar enzyme has been found in an unidentified *Flavobacterium sp.* (79). Their high pH optimum and low temperature stability make these enzymes little suited for industrial processes.

G. β-Amylases

Many sources of microbial β-amylases are known and since no enzyme with completely satisfactory properties for industrial applications has yet been found, screening for new sources is important. A strain of *Bacillus polymyxa,* a well known source of β-amylase, has been found, which produces the enzyme free from α-amylase and debranching enzyme (80). Murao, et al., made a screening for β-amylase producing bacteria by using agar plates with a medium containing starch and an inhibitor for amylase and glucoamylase (produced by *Streptomyces diastaticus*). In this way they found a strain of *B. polymyxa* which produces β-amylase in high yield (81).

The organism produces two β-amylases with a small difference in isoelectric point. They are SH enzymes, pH optimum 7.5, temperature optimum 45°C (10 min.), MW 44,000. Amylose is hydrolyzed 100% to maltose.

H. α-Glucosidases

α-Glucosidases from bacteria do not play any commercial role at the moment, but they are nevertheless of interest due to their importance for the metabolism of enzyme producing organisms and for their potential industrial use. *Bacillus amylolyticus*, which forms α-amylase and pullulanase, will practically only produce α-glucosidase, when grown on a starch containing medium. The enzyme is cell bound and has pH optimum at 7 and temperature optimum at 40°C. When incubated with maltose, it produces transglucosylation products (82). *Arthrobacter globiformis* produces an exo 1,6-α-glucosidase (an exodextranase), which has an activity ratio dextran:starch of 100:0.5. The activity ratios on isomaltose: maltose:nigerose and kojibiose were 100:3.9:0.03:0.006 (83).

III. OTHER CARBOHYDRASES

A. β-glucanases

Cellulases will not be described here. They were recently reviewed by Mandels (84) and by Halliwell (85) who coveres all β-glucanases.

In the brewing industry glucanases have been used for over 20 years to reduce viscosity of wort and beer by hydrolyzing the barley gum. The major commercial enzymes are prepared either from *B. amyloliquefaciens* or from *A. niger*. The latter enzyme has been purified (86). The enzyme is a single polypeptide chain without carbohydrates, MW 45,000, pI 4.5, it is insensitive to SH reagents and EDTA.

Several enzyme complexes, which hydrolyze yeast cell walls have been described. *Physarium polycephalum* produces a conglomerate of β-glucanases, α-glucanases, citinase and other glucodidases. The enzyme hydrolyzes cell walls of several fungi and may be used in the treatment of mycoses in combination with antimycotic agents (87). Another preparation, which is very powerful in hydrolyzing yeast cell walls, is Novozym 234, which is prepared from *Trichoderma harzianum* (88). This enzyme is widely ised in the preparation of protoplasts from yeast and fungi. The commercial preparation 'Kitalase' is prepared from a *Basidiomycete* belonging to the Aphyllophoratis, it is marketed as a yeast cell wall lytic enzyme. It contains both endo and exo β-1,3-glucanase activity, the exo enzyme has MW 73,000 and is stable between pH 5 and 9.5. Its optimum temperature is 55°C (10 min.) with laminarin as substrate (89). An enzyme mixture, Lyticase from *Oerskovia xanthineolytica* also hydrolyzes yeast cell walls. It contains an endo β-1,3-glucanase MW 59,000, pH optimum 6.0 on laminarin. In order to effect yeast cell wall lysis, 2 mercaptoethanol or an alkaline protease present in the culture liquid was necessary (90).

B. Lactase

The use of lactase in the dairy industry is slowly growing, although the production of dairy products with hydrolyzed lactose is much smaller than one would expect based on the large percentage of the world population, which suffers from maldigestion of lactose. The commercial products are primarily made from yeast (*Saccharomyces fragilis* and *Saccharomyces lactis*) and from *A. niger*. The yeast enzymes have pH optimum at 6.5-7, close to the normal pH of milk and the *Aspergillus* lactase has pH optimum at 4.8. In hydrolysis of lactose in milk with yeast lactase, the formation of oligosaccharides is small, but if higher concentrations of lactose is used, substantial amounts of oligosaccharides are formed. These carbohydrates are preferably β-1,6-linked and they are slowly digested in the gut, where they may cause problems (91,92). The development of an industrial type production process using *A. niger* has been described (93). Mutants with 3-4 fold yield improvement were produced, they were characterized by slow growth and biomass formation. A simple medium without lactose but other wise of undisclosed composition was used (94). A commercial product of *A. niger* lactase could be separated in three lactase components (95) with MW 124,000, 150,000 and 173,000 and carbohydrate contents of 12.5%, 20.5% and 29% respectively. It is assumed, that this multiple form pattern is dependent on the culture conditions (95). *A. oryzae* lactase may similarly be separated in two forms with MW 110,000 and carbohydrate contents of 7% and 4.6% (96).

Lactases have been found in *Penicillium multicolor* (97) and in *Fusarium moniliforme* (98). The properties appear to be similar to those of the *Aspergillus* lactase. Bacterial lactases are usually intra cellular and have pH optimum around neutrality. Especially thermotolerant or thermophilic strains have been of interest in the search for thermostable lactases. The operating temperatures for the enzymes appear to be a little over the 50°C usually mentioned for the *Aspergillus* enzymes (99,100).

C. *Hemicellulases*

Hemicelluloses are as abundant in nature as lignin and second only to cellulose. They are present in plant cell walls and are homo- or heteropolysaccharides containing primarily arabinose, xylose, mannose and galactose. Industrial use of hemicellulases is limited to digestive aids and side activities in pectic enzymes for removal of haze (101).

The recent interest in enzymatic hydrolysis of wood has also resulted in an increasing interest in the study of hemicellulases, especially xylanase. *A. niger* is a good source of both endoxylanase and xylosidase. Up to five different xylanases have been found (102,103) and were shown to be able to hydrolyze arabinoxylan completely to pentose sugars. The pH optimum is between 4 and 5 but the operating temperature is below 40°C. More thermostable enzymes have been found in the thermophiles *Talaromyces byssoclamoides* (opt. 70°C at 1 hour) and in *Humicola insolens* (104). Other fungi reported to produce xylanase are *Fusarium avenaceum* (105), *Sclerotium rolfsii* (106) and *Trametes hirsuta* (107). The enzymes are usually accompanied by a strong cellulolytic action.

An acidophilic and thermophilic *Bacillus sp.* produce an extracellular endo xylanase during growth on xylan, MW 56,000, pH optimum 4 and temperature optimum 80°C (10 min.). The enzyme hydrolyzes both xylan and CMC (β-1,4 bonds) (108).

Xylanase activity of *Streptomyces sp.* also capable of glucose isomerase production has been of industrial interest, because it meant that xylose necessary for induction of the isomerase could be replaced by inexpensive plant raw materials. In most industrial processes constitutive strains have now been developed so xylanase activity no longer is important in the isomerase producing organisms (109).

IV. OTHER HYDROLYTIC ENZYMES

A. *Pectinases*

Commercial pectinase products are all produced by members of *A. niger* group, despite the fact that pectinolytic enzymes are present in many different microorganisms. A significant part of the industrial production still takes place in semisolid culture on wheat bran (110). The commercial preparations are complex mixtures of several pectinolytic enzymes. A separation in single components has been achieved by HPLC chromatography (111). With the increasing knowledge of the pectic enzymes and with the improved methods of separation available, special products with single activities are developed. An example is pectinesterase, which is purified from a complex pectinase mixture produced by *Aspergillus japonicus* (112). This enzyme may be used to demethylate high methoxy pectin so that it may form strong gels in the presence of calcium ion. A similar esterase has abeen prepared in pure form from *A. niger* (113). Since the pectin formed in the demethylation process is homogenous, a multichain reaction is proposed.

A so called proto pectinase is produced by the yeast *Trichosporon penicellatum* (114). The enzyme will convert insoluble pectin to soluble pectin and is useful in the production of pectin from citrus peel.

B. *Lipases*

Over the last five years there has been a growing interest in lipolytic enzymes and several new applications have been described, but this has not yet materialized in important industrial utilization of lipases. One of the most interesting possibilities is the use of lipolytic enzymes for interesterification of triglycerides in order to upgrade the properties of cheap fats. This process may take place in a system with low water activity and with the enzyme cell bound or simply immobilized by being precipitated on a suitable carrier such as celite (115,116). Enzymes from *Aspergillus*, *Rhisopus* or *Mucor* have been used for this purpose and the position specificity of these enzymes mean, that only the 1,3 groups are exchanged. The reaction mechanisms is these multiphase systems with low water activity is little understood. Apparently some of the enzymes have a larger capacity for interesterification than hydrolysis, for example are the *Rhizopus* enzymes superior to the *Aspergillus* enzymes in interesterification based on the same degree of hydrolytic activity (117).

Lipases from *Aspergillus*, *Rhizopus*, *Geotrichum* and *Penicillium* will synthesize esters from oleic acid and various primary alcohols, but only the *Geotrichum* lipase will synthesize esters of secondary alcohols. None of these lipases will synthesize esters of tertiary alcohols, phenols or sugar alcohols. Esters of glycols are synthesized with large concentrations of alcohol (118). Lipase from *Aspergillus* and *Rhizopus* will synthesize oleyl- esters of various fatty acids, but the lipases from *Geotrichum* and *Penicillium* synthesize esters only from medium or long fatty acids. Similarly terpene alcohol esters used in fragrance formulations may be synthesized by these lipases (119). The lack of reaction with sugar esters may be utilized in purifying chemically formed sugar esters with impurities of other, hydrolysable esters (120).

The molecular structure of *Aspergillus* and *Rhizopus* lipase has been studied by Tombs and Blake (121) and it was found, that the specific activity of the *Rhizopus* enzyme was four times that of the *Aspergillus* enzyme. Both lipases were glycoproteins and both consisted of two subunits with a molecular weight of about 25,000. The enzymes are not particularly hydrophobic as might be expected. It is concluded, that these fungal lipases are globular proteins containing two subunits, interacting probably through small hydrophobic patches with triglyceride in a way that causes relatively small conformation changes.

The lipase from *Candida cylindraceae* hydrolyses triolein completely without position specificity. The reaction conditions are important, over 40 mM calcium ion is necessary to avoid product inhibition, pH optimum is 7-7.5. The enzyme has little thermostability and is rapidly inactivated at temperatures above 45°C (122).

Bacterial lipases are not used in industry despite extensive research in their properties. The cause may be the low yield in which these enzymes usually are formed. Some bacterial lipases are extremely heat resistant. A lipase from a psychrophilic *Pseudomonas sp.* resists 3.6 min. at 140°C before 90% of its activity is destroyed (123). This creates problems when these lipases are present in food products, but the good thermostability might be useful in other connections.

An enzyme similar to pancreas lipase has been found in a strain of *Alcaligenes* after an extensive screening work. The enzyme has pH optimum 9.0, temperature optimum 40°C, it is activated by bile salts, has cholesterol esterase activity and MW 30-40,000. It is furthermore produced in high yield. An abvious application is in the preparation of digestive aids (124).

C. *Penicillinase*

This enzyme is found in both gram positive and gram negative bacteria. There appear to be several different types of the enzyme with distinctly different properties and active site. *Bacillus cereus* produces two different types, A has serine in the active center, the B is requiring metal ions for its activity. Some β-lactamases of gram negative organisms are inhibited by thiol reagents. The similarity to the different types of proteases is striking. The type A. enzyme is best known and there is a striking homology between these enzymes derived from as widely separated species as *B. cereus*, *B. licheniformis*, *E. coli* and *S. aureus* (125,126). An extensive review on β-lactamases has been made in 1979 (127).

D. Penicillin Acylase

The enzymatic hydrolysis of *Penicillin* G or V to 6 amino penicillanic acid is a well established commercial process with the enzyme from *E. coli* being the most important source for hydrolysis of Pen G. In the last few years there has been an increasing interest in the production of 6 APA from Pen V and a number of organisms which can make a suitable enzyme have been found. One example is an enzyme from *Pseudomonad acidovorans*, but the most promising source described in the literature is *Bovista plumbea*, which gives a very stable enzyme with good kinetic properties so that a yield of up to 97% in 8% Pen V may be obtained (128).

E. Dextranase

The dextran like polysaccharide consisting of primarily α-1,3 linked glucose, which is produced by *Streptococcus mutans*, is usually regarded as one of the major cariogenic factors. Therefore over the years a number of attempts have been made to produce enzymes, which may decompose this dextran and thus be useful to prevent caries. A new enzyme of this type has been produced by *Streptomyces chartreusis*. The crude enzyme preparation will hydrolyze both α- and β- 1,3 glucans, but the purified enzyme is specific for the α-1,3-glucanase. NW 61,000, pH optimum, 5.5-6 and temperature optimum 55°C (129).

V. OTHER ENZYMES

A. Glucose Isomerase

The enzyme has large industrial importance in the production of fructose containing syrups from starch hydrolyzates. The enzyme is only used in immobilized form and the most important sources for industrial use are *Bacillus coagulans*, *Actinoplanes missouriensis* and *Streptomyces sp*. The enzymes from different organisms are remarkably similar. They are all xylose isomerases, stabilized by Co and activated by Mn, and they have a remarkable heat stability, which allow operation at 60°C for 4 weeks in immobilized form. Despite its large economical interest, this enzyme has attracted little interest from the people who study the basic enzyme chemistry. Several review articles have appeared laterly (130-133).

B. *Ligninase*

The position of lignin as the second most abundant organic compound and the industrial production of large quantities of waste lignin, naturally creates a strong interest in search for methods for its utilization. Thus the microbial enzyme systems which degrade lignin in nature are intensively studies (134). There are several microorganisms which in culture will degrade lignin at an appreciable rate, first of all the traditional wood rotting fungi, but also microfungi like *Aspergillus* and *Fusarium,* bacteria like *Bacillus* and *Pseudomonas sp.* and *Streptomyces sp.* (135,136,137,138).

Unfortunately the phenolic structure of lignin means, that no simple hydrolytic enzyme system will degrade it. The degradation must involve a number of oxidases, which presents difficulties in an industrial application since such enzymes are difficult to isolate and apply. One might hope, that a single enzyme would give a partial degradation, which would result in an industrial useful process for the production of a particular component, but so far no such developments have materialized. An interesting possibility is the partial delignification of kraft pulp by incubation with *Phanerochaete chrysosporium* which will decrease the consumption of chlorine for bleaching considerably, however the incubation time was uneconomically long and there was also a substantial loss of cellulose (139).

VI. GENE TECHNOLOGY

No attempt will be made here to cover this rapidly moving area in depth. So far, no industrial enzyme product has been marketed, as a result of gene splicing, but the possibilities are in legion and from the published patents and papers it is to be expected, that these methods soon will be used in practice.

One of the most interesting possibilities is transfer of genes between species. This has also been achieved by more traditional methods. In 1979, the α-amylase from *B. amyloliquefaciens* was transferred to *B. subtilis Marburg* with a transducing phage (140) and the same year, α-amylase from a thermophilic bacterium was transferred to a strain of *B. subtilis* lacking restriction and modification enzymes by transformation (141). Later a similar gene transfer was accomplished using a phage as vector (142). Patents have been applied for an elaborate system of cloning and transferring amylolytic enzymes (143). The fast development of cloning systems in *B. subtilis*, will in the near future allow production by this organism of a number of enzymes, otherwise difficult to produce (144).

Cloning of the important animal enzyme rennin is well under way (145). Another possibility is yield improvement. This has been described for glucose isomerase by cloning in *E. coli* (146).

VII. REVIEW ARTICLES AND BOOKS

A comprehensive overview over enzyme production and application is given in *Applied Biochemistry and Bioengineering Vol. 2: Enzyme Technology* (147). In the new edition of *Microbial Technology*, a chapter is devoted to the technical aspects of enzyme production (148). Volume 5 of the series *Economic Microbiology* has chapters dealing with proteinases, amylolytic enzymes, glucose oxidase, glucose isomerase, lactase and invertase, pectic enzymes, cellulase and penicillin acylases and β-lactamases (149). Hirokoshi in his monograph on alkalophilic microorganisms also gives a review of enzymes of industrial interest (150).

REFERENCES

1. Neidleman, S.L., Amon, W.F., and Geigert, J. European Patent Application 7176 (1980).
2. Aunstrup, K. (L.B. Wingard, Jr., E. Katchalski-Katzir, and L. Goldstein, eds.), *Appl. Biochem. and Bioeng. 2*, 27 (1979).
3. Aunstrup, K. (A.H. Rose, ed.), *Economic Microbiol 5*, 49 (1980).
4. Olsen, H.S. and Adler-Nissen, J. *Process Biochem. 14* (1979).
5. Adler-Nissen, J. U.S. Patent 4,100,024 (1978).
6. Tang, P., Nielsen, G.L., Gibson, K., and Aunstrup, K., and Schiff, H. U.S. Patent 4,266,031 (1981).
7. Hanlon, G.W. and Hodges, N.A. *FEMS Microbiol Lett. 11*, 51 (1981).
8. Hanlon, G.W. and Hodges, N.A. *J. Bact. 147*, 427 (1981).
9. Feder, J., Kochavi, D., and Anderson, R.G. *Biotech. Bioeng. 20*, 1865 (1978).
10. Mäntsälä, P. and Zalkin, H. *J. Bact. 141*, 493 (1980).

11. Maurizi, M.R. and Switzer, R.L. *Current Topics in Cellular Regulation 16,* 163 (1980).
12. Grandi, C., Vita, C., Dalzoppo, D., and Fontana, A. *Int. J. Peptide Protein Res. 16,* 327 (1980).
13. Uehara, H., Yamane, K., and Maruo, B. *J. Bact. 139,* 583 (1979).
14. Miwa, N., Masauda, Y., Kawarabuki, S., Sai, T., and Saito, T. U.S. Patent 4,315,988 (1982).
15. Zachariae, H., Høegh-Thomsen, J., Witmeur, O., and Wide, L. *Allergy 36,* 513 (1981).
16. Kaska, M., Lysenko, O., and Chaloupka, J. *Folia Microbiol. 21,* 465 (1976).
17. Lyerly, D. and Kreger, A. *Infection and Immunity 24,* 411 (1979).
18. Hausdorf, G., Krüger, K., and Höhne, W.E. *Int. J. Peptide Protein Res. 15,* 420 (1980).
19. Stepanov, V.M., Chestukhina, G.G., Rudenskaya, G.N., Epremyan, A.S., Osterman, A.L., Khodova, O.M., and Belyanova, L.P. *Biochem. Biophis. Res. Comm. 100,* 1680 (1981).
20. Behnke, U., Kleine, R., Ludewig, M., and Ruttloff, H. *Acta. Biol. Med. Germ. 37,* 1205 (1978).
21. Belloc, A., Florent, J., Lunet, J., Palla, J.C., and Mancy, D. U.S. Patent 4,288,556 (1981).
22. Matsushima, K-I, Hayakawa, K., Ito, M., and Shimada, K. *J. Gen. Appl. Microbiol. 27,* 423 (1981).
23. Tsujita, Y. and Endo, A. *Appl. Environ. Microbiol. 36,* 237 (1978).
24. Kumagai, I., Yamasaki, M., and Ui, N. *Biochim. Biophys. Acta 659,* 334 (1981).
25. Tsai, M.S., Hseu, T.H., and Shen, Y.S. *Int. J. Peptide Protein Res. 12,* 293 (1978).
26. Alessandro, M. and Federico, F. *J. Dairy Sci. 63,* 1397 (1980).
27. Roschlau, W.H.E. *Handb. Exp. Pharmako. 46,* 337 (1978).
28. Christie, R.B. (Alan Wiseman, ed.), *Topics on Enzyme and Fermentation Biotechnology 4,* 25 (1980).
29. Khan, M.R., Blain, J.A., and Patterson, J.D.E. *Appl. Env. Microbiol. 37,* 719 (1979).
30. Lasure, L.L. *Mycologia 72,* 483 (1980).
31. Lagrange, A., Paquet, D., and Alais, C. *Int. J. Biochem. 11,* 347 (1980).
32. Martin, P., Raymond, M-N., Bricas, E., and Dumas, B.R. *Biochim. Biophys. Acta 612,* 410 (1980).
33. Etoh, Y., Shonn, H., Beppu, T., and Arima, K. *Agr. Biol. Chem. 43,* 209 (1979).
34. Salton, M.R.J., ed. Immunochemistry of Enzymes and Their Antibodies, New York (1977).

35. Branner-Jørgensen, S., Eigtved, P., and Schneider, P. *Neth. Milk Dairy Sci. 35,* 361 (1981).
36. Cornelius, D.A. U.K. Patent 2,024,828 (1980).
37. Belyauskaite, I.P., Palubinskas, V.J., Anchenko, O.E., Vesa, V.S., and Glemzha, A.A. *Enzyme Microb. Technol. 2,* 37 (1980).
38. Fraile, E.R., Muse, J.O., and Bernandinelli, S.E. *Eur. J. Appl. Microbiol. Biotechnol. 13,* 191 (1981).
39. Morgan, F.J. and Priest, F.G. *J. Appl. Bact. 50,* 107 (1981).
40. Medda, S. and Chandra, A.K. *J. Appl. Bact. 48,* 47 (1980).
41. Kuhn, H., Fietzek, P.P., and Lampen, J.O. *J. Bact. 149,* 372 (1982).
42. Chiang, J.P., Alter, J.E., and Sternberg, M. *Starch 31,* 87 (1979).
43. Chandra, A.K., Medda, S., and Bhadra, A.K. *J. Ferment. Technol. 58,* 1 (1980).
44. Thirunavukkarasu, M. and Priest, F.G. *FEMS Microbiol. Lett. 7,* 315 (1980).
45. Leach, H.W. and Hebeda, C. U.S. Patent 4,235,9970 (1980).
46. Detera, D.S. and Friedberg, F. *Int. J. Peptide Protein Res. 14,* 364 (1979).
47. Detera, D.S. and Friedberg, F. *Int. J. Peptide Protein Res. 17,* 93 (1981).
48. Chung, H. and Friedberg, F. *Biochem. J. 185,* 387 (1980).
49. Mäntsälä, P. and Zalkin, H. *J. B. C. 254,* 8540 (1979).
50. Nagata, Y., Suga, S., Kado, O., and Marko, B. *Agr. Biol. Chem. 44,* 215 (1980).
51. Hitotsuganagi, K., Yamane, K., and Maruo, B. *Agr. Biol. Chem. 43,* 2343 (1979).
52. Yoneda, Y. *Appl. Env. Microbiol. 39,* 274 (1980).
53. Tamuri, M., Kanno, M., and Ishii, Y. U.S. Patent 4,284,722 (1981).
54. Boyer, E.W., Jugle, M.B., and Mercer, G.D. *Starch 31,* 166 (1979).
55. Shimizo, M., Kanno, M., Tamura, M., and Suekane, M. *Agr. Biol. Chem. 42,* 1681 (1978).
56. Sakano, Y., Hiraiwa, S., Fukashima, J., and Kobayashi, T. *Agr. Biol. Chem. 42,* 1121 (1982).
57. Toda, H., Kondo, K., and Narita, K. *Agr. Biol. Chem. 44,* 1945 (1980).
58. Allen, J.D. and Thoma, J.A. *Carbohydrate Research 61,* 377 (1978).
59. Suganuma, T., Matsuno, R., Ohnishi, M., and Hiromi, K. *J. Biochem. 84,* 293 (1978).
60. Oteng-Gyang, K., Moulin, G., and Galzy, P. *Acta. Microbiol. Acad. Sci. Hung. 27,* 155 (1980).
61. Oteng-Gyang, K., Moulin, G., and Galzy, P. *Eur. J. Appl. Microbiol. 9,* 129 (1980).

62. Moulin, G.-J. and Gaizy, P. *Agr. Biol. Chem. 43,* 1165 (1979).
63. Spencer-Martins, I. and van Uden, N. *Eur. J. Appl. Microbiol. Biotechnol. 6,* 241 (1979).
64. van Uden, N., Cabeca-Silva, C., Madeira-Lopes, A., and Spencer-Martins, I. *Biotech. Bioeng. 22,* 651 (1980).
65. Fogarty, W.M. and Kelly, C.T. *Progress in Industrial Microbiol. 15,* 87 (1979).
66. Svensson, B., Pedersen, T.G., Svendsen, I., Sakai, T., and Ottesen, M. *Carlsberg Res. Commun. 47,* 55 (1982).
67. Manjunath, P. and Rao, M.R.R. *J. Biosci. 3,* 343 (1981).
68. Takahashi, T., Inokuchi, N., and Iric, M. *J. Biochem. 89,* 125 (1981).
69. Inokuchi, N., Takahashi, T., and Iric, M. *J. Biochem. 90,* 1055 (1981).
70. Inokuchi, N., Iwawa, M., Takahashi, T., and Iric, M. *J. Biochem. 91,* 125 (1982).
71. Hayashida, S. and Yoshino, E. *Agr. Biol. Chem. 42,* 927 (1978).
72. Yoshino, E. and Hayashida, S. *J. Ferm. Technol. 56,* 289 (1978).
73. Hayashida, S. and Flor, P.G. *Agr. Biol. Chem. 45,* 2675 (1981).
74. Tamura, M., Shimizo, M., and Togo, M. U.S. Patent 4,247,637 (1981).
75. Rao, V.B., Sastri, N.V.S., and Rao, P.V.S. *Biochem. J. 193,* 379 (1981).
76. Norman, B.E. U.S. Patent 4,335,208 (1982).
77. Morgan, F.J., Adams, K.R., and Priest, F.G. *J. Appl. Bact. 46,* 291 (1979).
78. Evans, R.M., Manners, D.J., and Stark, J.R. *Carbohydrate Res. 76,* 203 (1979).
79. Sato, H.H. and Park, Y.K. *Starch 32,* 352 (1980).
80. Hensley, D.E., Smiley, K.L., Boundy, J.A., and Lagoda, A.A. *Apl. Env. Microbiol. 39,* 678 (1980).
81. Murao, S., Ohyama, K., and Arai, M. *Agr. Biol. Chem. 43,* 719 (1979).
82. Kelly, C.T., Heffernan, M.E., and Fogarty, W.M. *Biotech. Letters 2,* 351 (1980).
83. Onya, T., Suzuki, H., and Sawai, T. *Agr. Biol. Chem. 44,* 203 (1980).
84. Mandels, M. *Annual Report Ferm. Process 5,* (1980).
85. Halliwell, G. *Progress Ind. Microbiol. 15,* 1 (1979).
86. Svensson, B. *Carlsberg Res. Commun. 43,* 103 (1978).
87. Davies, D.A.L. and Pope, A.M.S. U.K. Patent 1,576,891 (1980).

88. Hamly, P.F., Bradshaw, R.E., Mellon, F.M., Santiago, C.M., Wilson, J.B., and Peberdy, J.F. *Enzyme Microb. Technol.* 3, 321 (1981).
89. Tsujisaka, Y., Hamada, N., and Kobayashi, R. *Agric. Biol. Chem.* 45, 1201 (1981).
90. Scott, J.H. and Schekman, R. *J. Bact.* 142, 414 (1980).
91. Burvall, A., Asp, N.G., and Dahlqvist, A. *Food Chemistry* 4, 243 (1979).
92. Asp., N.G., Burvall, A., Dahlqvist, A., Hallgren, P., and Lundblad, A. *Food Chemistry* 5, 140 (1980).
93. Nevalainen, K.M.H. *Appl. Env. Microbiol.* 41, 593 (1981).
94. Mustranta, A., Karvonen, E., Ojamo, H., and Linko, M. *Biotech. Letters* 3, 333 (1981).
95. Widmer, F. and Leuba, J.L. *Eur. J. Biochem.* 100, 559 (1979).
96. Ogushi, S., Yoshimoto, T., and Tsuru, D. *J. Ferm. Tech.* 58, 115 (1980).
97. Miwa, T., Kobayashi, R., and Takita, K. U.S. Patent 4,229,539 (1980).
98. Macris, B.J. *Eur. J. Appl. Microbiol. Biotech.* 13, 161 (1981).
99. Bungard, S.J. and Byrom, D. U.S. Patent 4,234,687 (1980).
100. Levin, R.E. and Mahoney, R.R. *Ant. Leuwenhoek* 47, 53 (1981).
101. Dekker, R.F.H. *Easter Sch. Agric. Sci.* 27, 93 (1979).
102. Conrad, D. *Biotech. Letters* 3, 345 (1981).
103. John, M., Schmidt, B., and Schmidt, J. *Can. J. Biochem.* 57, 125 (1979).
104. Yoshioka, H., Chavanich, S., Nilubol, N., and Hayashida, S. *Agric. Biol. Chem.* 45, 579 (1981).
105. Zalewski-Sobczak, J. and Urbanek, H. *Arch. Microbiol.* 129, 247 (1981).
106. Sadana, J.C., Shewale, J.G., and Deshpande,. *Appl. Env. Microbiol.* 39, 935 (1980).
107. Kubachova, M., Karacsonyi, S., Bilisics, L., and Toman, R. *Carbohydrate Research* 76, 177 (1979).
108. Uchino, F. and Nakane, T. *Agric. Biol. Chem.* 45, 1121 (1981).
109. Ishaque, M. and Kluepfel, D. *Biotech. Letters* 3, 481 (1981).
110. Ghildyal, N.P., Ramakrishna, S.V., Devi, P.N., Lonsane, B.K., and Asthana, H.N. *J. Food Scil Tech.* 18, 248 (1981).
111. Mikes, O. and Sedlackova, J. *J. Chromat.* 207, 99 (1981).
112. Ishii, S., Kiho, K., Sugiyama, S., and Sugimoto, H. *J. Food Sci.* 44, 611 (1979).
113. Baron, A., Rombouts, F., Drilleau, J.F., and Pilnik, W. *Lebensm. Wiss. u Technol.* 13, 330 (1980).

114. Sakai, T. and Okushima, M. *Appl. Env. Microbiol. 39,* 908 (1980).
115. Coleman, M.H., Deir, A., and Macrae, R. U.K. Patent 1,577,933 (1980).
116. Takaharu, M., Norio, S., Yukio, H., and Wataru, H. Eur. Pat. Appl. 813,009,388 (1981).
117. Okumura, S., Iwai, M., and Tsujisaka, Y. *Agri. Biol. Chem. 45,* 185 (1981).
118. Okumura, S., Iwai, M., and Tsujisaka, Y. *Biochim. Biophys. Acta 575,* 156 (1979).
119. Iwai, M., Okumura, S., and Tsujisaka, Y. *Agr. Biol. Chem. 44,* 2731 (1980).
120. Mitsui Toatsu, Chem. Inc. Jap. Pat. Appl. 5,651,496 (1981).
121. Tombs, M.P. and Blake, G.G. *Biochim. Biophys. Acta 700,* 81 (1982).
122. Schifree, R.S. and Carr, P.W. *Anal. Lett. 12,* 47 (1979).
123. Andersson, R.E., Hedlund, C.B., and Jonsson, U. *J. Dairy Sci. 62,* 361 (1979).
124. Kokusho, Y., Machida, H., and Iwasaki, S. U.S. Patent 4,283,494 (1981).
125. Ambler, R.P. *Phil. trans. R. Soc. Lond. B 289,* 321 (1980).
126. Hill, H.A.O., Sammes, P.G., and Waley, S.G. *Phil. Trans. Soc. Lond. B 289,* 333 (1980).
127. Hamilton-Miller, J.M.T. and Smith, J.T. Beta Lactamase, Academic Press, London (1979).
128. Lowe, D.A., Romancik, G., and Elander, R.P. *Development in Industrial Microbiol. 22,* 163 (1980).
129. Takehara, T., Ihoue, M., Morioka, T., and Yokogawa, K. *J. Bact. 145,* 729 (1981).
130. Antrim, R.L., Colilla, W., and Schnyder, B.J. In "Applied Biochemistry and Bioengineering 2, Enzyme Technology, (L.B. Wingard, E. Kachalski-Katzir, and L. Goldstein, eds.), New York (1979).
131. Hemmingsen, S.H. In "Applied Biochemistry and Bioengineering 2, Enzyme Technology, (L.B. Wingard, E. Katchalski-Katzir, and L. Goldstein, eds.), New York (1979).
132. Wai-Pin, C. *Process Biochemistry 15*(5), 30 (1980).
133. Wen-Pin, C. *Process Biochemistry 15*(6), 36 (1980).
134. Crawford, D.L. and Crawford, R.L. *Enz. Microb. Technol. 2,* 11 (1980).
135. Deschamps, A.M., Mahondeau, G., and Lebeault, J.M. *Eur. J. Appl. Microbiol. Biotech. 9,* 45 (1980).
136. Keyser, P., Kirk, T.K., and Zeikus, J.G. *J. Bact. 135,* 790 (1978).

137. Janshekar, H. and Fiechter, A. *Eur. J. Appl. Microbiol. Biotech. 14,* 47 (1982).
138. Janshekar, H., Haltmeier, T., and Brown, C. *Eur. J. Appl. Microbiol. Biotech. 14,* 174 (1982).
139. Kirk, T.K. and Yang, H.H. *Biotech. Lett. 79,* 347 (1979).
140. Yoneda, Y. *Appl. Env. Microbiol. 39,* 274 (1980).
141. Shinomiya, S., Yamane, K., and Oshima, T. *Biochem. Biophys. Res. Comm. 96,* 175 (1980).
142. Shinomiya, S., Yamane, K., Mizukami, T., Kawamura, F., and Saito, H. *Agric. Biol. Chem. 45,* 1733 (1981).
143. Antoine, C. and Emile, P. U.K. Patent 2,069,503 (1981).
144. Young, F.E. *J. Gen. Microbiol. 119,* 1 (1980).
145. Ychiyama, H., Uozumi, T., Beppu, T., and Arima, K. *Agr. Biol. Chem. 44,* 1373 (1980).
146. Wovcha, M.G. and Brooks, K.E. U.K. Patent 2,031,905 (1980).
147. Wingard, L.B., Katchalski-Katzir, E., and Goldstein, eds. *Appl. Biochem. Bioeng. 2 Enzyme Technology* (1979).
148. Aunstrup, K., Andresen, O., Falch, E.A., and Nielsen, T.K. In "Microbial Technology," Vol. 1, 2nd edition, (H.J. Peppler and D. Perlman, eds.), New York (1979).
149. Rose, A.H., ed. "Economic Microbiology Microbial Enzymes and Bioconversions," Vol. 5, London (1980).
150. Horikoshi, K. and Akiba, T. *Alaklophilic Microorganisms,* Berlin (1982).

CHAPTER 8

APPLICATION OF BIOSENSOR IN FERMENTATION PROCESSES

Isao Karube

Tenabe Seiyaku Co., Ltd.,
Osaka, Japan

Shuichi Suzuki

Tokyo Institute of Technology
Yokohama, Japan

Recently, considerable worldwide interest has arisen in the development of biotechnology and bioengineering. The determination processes play important roles in biotechnology. The determination of organic compounds such as whole cells, nutrients, and products in fermentation broths is important for control of fermentation processes. Most of these compounds in broths can be determined by spectrophotometric methods. However, the fermentation broths are not optically clear.

Electrochemical monitoring of these compounds has definite advantages. Many biosensors have utilized enzyme-catalyzed reactions because of their specificity. Many reports on enzyme electrodes have been published (1-5). However, enzymes are generally expensive and unstable. Recently many methods have been developed for immobilization of whole cells (6,7). Sensors using immobilized whole cells and electrochemical devices have been developed by the authors. These microbial sensors can be classified into three categories: 1) Microbial sensors consisting of immobilized whole cells and an oxygen electrode (the sensors based on amperometric determination of microbial respiration); 2) Microbial sensors consisting of immobilized whole cells and a fuel cell type electrode (the sensors based on amperometric determination of electroactive metabolites); 3) Microbial sensors consisting of immobilized whole cells and membrane electrodes (pH, CO_2 electrode),(the sensors based on potentiometric determination of electroactive metabolites).

Bacterial electrodes using whole cell suspensions and an ammonia gas electrode also have been reported by Rechnitz, et al (8-11).

Application of microbial sensors to fermentation processes are described in this chapter.

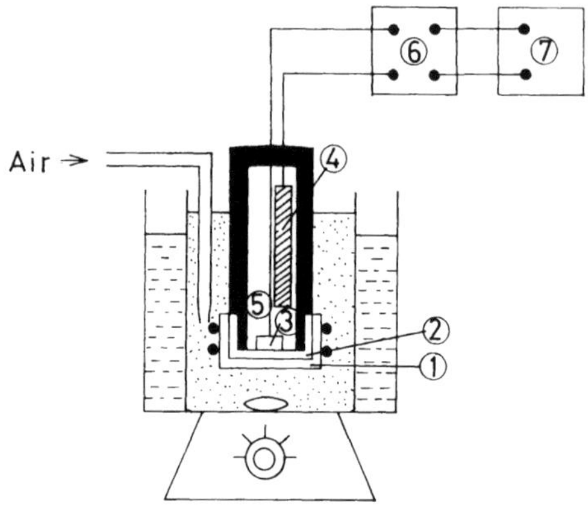

Figure 1. Scheme of the microbial electrode sensor for glucose. 1) bacteria-collagen membrane; 2) Teflon membrane; 3) platinum cathode; 4) lead anode; 5) electrolyte (KOH); 6) ammeter; 7) recorder

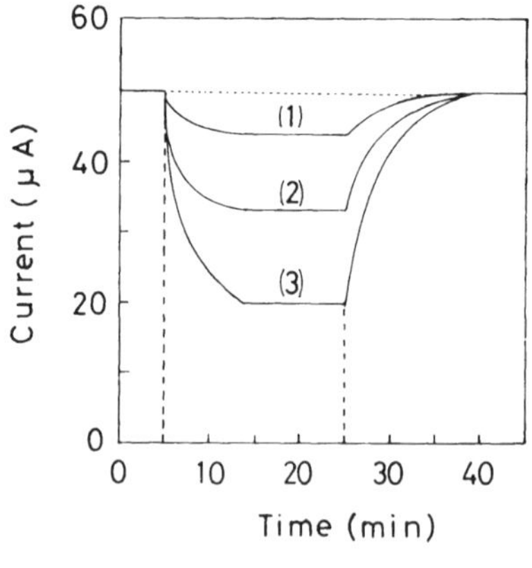

Figure 2. Response curves of the microbial sensor. Glucose concentrations were 1) 3.6 mg•l^{-1}; 2) 10 mg•l^{-1}; 3) 16 mg•l^{-1}

I. MICROBIAL SENSORS FOR FERMENTATION PROCESSES

A. *Glucose Sensors*

Many useful materials such as drugs and foods are produced in fermentation industries. Control of the fermentation process is needed for effective production of these useful materials. Therefore, determination of raw materials is required for process control. Generally molasses is used as a carbon source for the cultivation of microorganisms. Molasses contains glucose, sucrose, and other sugars.

The determination of glucose is important for the process control. Preliminary experiments showed that assimilation of glucose by microorganisms could be determined by an oxygen electrode. Therefore, it is possible to construct a microbial sensor for glucose using immobilized whole cells which utilize mainly glucose and an oxygen electrode (12).

The bacteria used in this study was *Pseudomonas fluorescens* which utilized mainly glucose. Whole cells of *P. fluorescens* were immobilized in collagen membrane. The bacteria-collagen membrane was then treated with a 0.1% glutaraldehyde solution for 1 min and dried again at 4°C.

A schematic diagram of the microbial sensor is illustrated in Fig. 1. The sensor consisted of double membranes of which one layer was the bacteria-collagen membrane (thickness 40 μm), the other an oxygen permeable Teflon membrane (thickness 47 μm), an alkaline electrolyte, a platinum cathode, and a lead anode. The double membranes are in direct contact with the platinum cathode and is tightly secured with rubber rings. The microbial sensor was inserted into a sample solution saturated with dissolved oxygen. The temperature of a sample solution was maintained at 30 ± 0.1°C. The current was measured by a milliammeter and the signal was displayed on a recorder.

Figure 2 shows typical response curves of the sensor. The current at time zero was that obtained in a sample solution saturated with dissolved oxygen. The bacteria began to utilize glucose in a sample solution when the sensor was placed in it. Then, consumption of oxygen by the bacteria in collagen membrane began. Consumption of oxygen by bacteria caused a decrease in dissolved oxygen on the membrane. As a result, the current of the electrode markedly decreased with time until a steady state was reached. The steady state indicated that the consumption of oxygen by the bacteria and the diffusion of oxygen from the solution to the membrane were in equilibrium. The steady state current was attained within 10 min at 30°C. The steady state depended on the concentration of glucose.

When the electrode was removed from the sample and placed in a solution free of glucose, the current of the microbial sensor gradually increased and returned to the initial level. ('Current' means the steady state current hereafter.)

This sensor responded slightly to fructose, galactose, mannose and saccharose. However, no response was observed in the case of amino acids. Therefore, the selectivity of the microbial sensor for glucose was satisfactory.

A linear relationship was observed between the current and the concentration of glucose below 20 $mg \cdot l^{-1}$. The minimum concentration for determination was 2 $mg \cdot glucose \cdot l^{-1}$. The reproducibility of the current was examined using the same sample solution. The current was reproducible with ± 6% of the relative error when a sample solution containing 10 $mg \cdot l^{-1}$ of glucose was employed. The standard deviation was 0.6 $mg \cdot l^{-1}$ in 20 experiments.

The microbial sensor for glucose was applied to molasses. The concentration of glucose was determined both by the microbial sensor and by the enzymatic method (13). Satisfactory comparative results were obtained. Glucose in molasses was determined on an average within 10% of relative error by the microbial sensor.

The reusability of the microbial sensor was examined. A glucose solution (10 $mg \cdot l^{-1}$) was used for the long term stability test of the sensor. The current of the sensor was determined 10 min after it was immersed in a glucose solution. The above procedure was repeated more than ten times per day. No decrease in current output was observed over a two-week period and 150 assays. The storage stability of the microbial sensor was examined. The sensor was suspended in 0.1 M phosphate buffer at 5°C. The current output was almost constant for 30 days. Therefore, the bacteria in collagen membrane were metabolically active for a long time and the microbial sensor could be used over a long period for the glucose assay.

The concentration range of glucose determination was from 10^{-4} to 10^{-5} M by the microbial electrode, while that from 10^{-2} to 10^{-4} M was by ordinary enzyme electrodes (14-16). The sensitivity of the microbial sensor was, therefore, better than that of ordinary enzyme electrodes. Furthermore, an expensive enzyme (glucose oxidase) must be employed for an enzyme electrode. The microbial electrode sensor was more stable than enzyme electrodes when they were applied to determination of glucose in molasses (a microbial electrode sensor was stable for 3 weeks and an enzyme electrode for 2 weeks).

B. Total Assimilable Sugar Sensor

In the cultivation of microorganisms in cane molasses, which contains various sugars, determination of the total assimilable sugars in a broth is important for the control of the fermentation process. Reduced sugars and sucrose in culture broth are determined by the ferricyanide method (17). However, this method is not completely reliable because unassimilable substances are also included.

As described above, assimilation of organic compounds by microorganisms can be determined from the respiratory activity of the microorganisms, which can be directly measured by an oxygen electrode.

A microbial sensor consisting of immobilized living whole cells of *Brevibacterium lactofermentum* and an oxygen electrode was prepared for continuous determination of total assimilable sugars (glucose, fructose, and sucrose) in a fermentation broth for glutamic acid production (18). *Brevibacterium lactofermentum* was immobilized in a strip of nylon net (1 cm x 1 cm, 20 mesh) and fixed on the oxygen electrode. Total assimilable sugars were estimated from oxygen consumption by the immobilized microorganisms. When a sample solution containing glucose was applied to the sensor system, increase in consumption of oxygen by the microorganisms caused a decrease in the dissolved oxygen around the Teflon membrane of the oxygen electrode and the current of the electrode decreased markedly with time until steady state was reached. The response time was 10 min by the steady state method and 1 min by the pulse method. A linear relationship was found between the decrease in current and the concentration of glucose (1 mM), fructose (1 mM), and sucrose (0.8 mM) respectively. The ratio of the sensitivity of the microbial sensor to glucose, fructose, and sucrose was 1.00: 0.80:0.92. The decrease in current was reproducible within 2% of the relative standard deviation when a sample solution containing glucose (0.8 mM) was employed for experiments. The selectivity of the microbial sensor for assimilable sugars was satisfactory for use in the fermentation process. The additivity of the response of the microbial sensor for glucose, fructose, and sucrose was examined. The difference between the observed and calculated values was within 8%. The microbial sensor was applied to a fermentation broth for glutamic acid production. Total assimilable sugars can be determined by the microbial sensor which can be used for more than 10 days and 960 assay.

C. *Acetic Acid Sensor*

On-line measurement of acetic acid concentrations is required in fermentation processes. In the cultivation of microorganisms with acetic acid as the carbon source, acetic acid inhibits growth above a certain concentration, so that the optimal concentration should be maintained by on-line measurements. Gas chromatography can be used but is unsuitable for on-line measurements.

As described above, assimilation of organic compounds by microorganisms can be determined from the respiration activity of the microorganisms, which can be measured directly with an oxygen electrode.

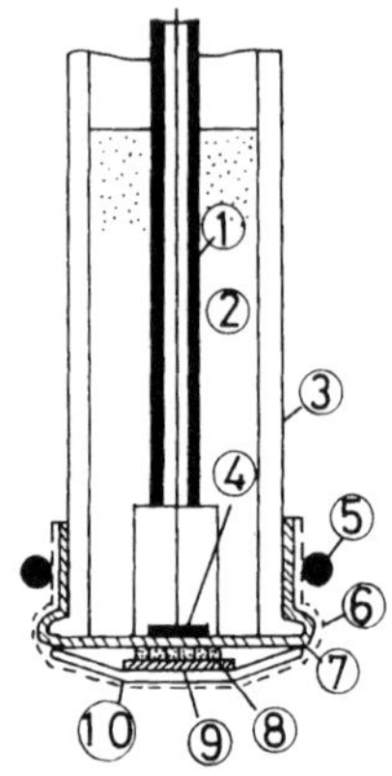

Figure 3(A). The microbial sensor for acetic acid: 1) aluminum anode; 2) electrolyte; 3) insulator; 4) platinum cathode; 5) rubber ring; 6) nylon net; 7) Teflon membrane; 8) microorganisms; 9) acetylcellulose membrane; 10) porous Teflon membrane

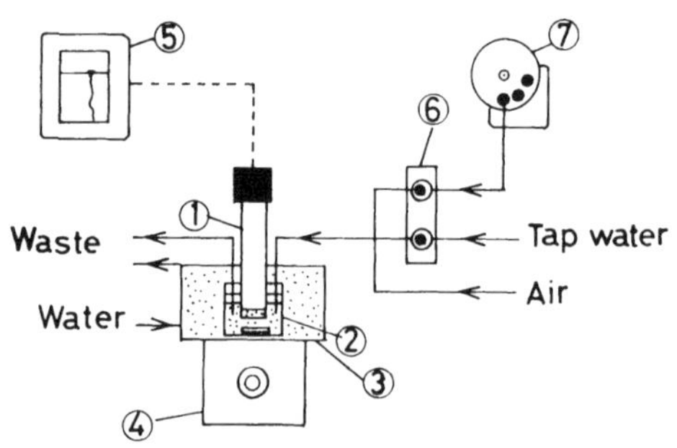

(B). The sensor system: 1) microbial electrode; 2) flow cell; 3) jacket; 4) magnetic stirrer; 5) recorder; 6) peristaltic pump; 7) sampler

A microbial sensor consisting of immobilized yeasts, *Trichosporon brassicae*, a gas-permeable Teflon membrane and an oxygen electrode is described for the determination of acetic acid (19). The microorganisms were immobilized onto a porous acetylcellulose membrane. The microbial sensor is illustrated in Fig. 3A. The electrode consisted of a Teflon membrane (50 µm thick), a platinum cathode, an aluminum anode and a saturated potassium chloride electrolyte. The porous membrane with the immobilized microorganisms was cut into a circle (1.4-cm diameter) and fixed on the surface of this Teflon membrane. This was covered with a gas-permeable Teflon membrane (0.5 µm size) or a silicone rubber membrane, the whole arrangement being held together with a nylon net. Thus, the microorganisms were trapped between the two porous membranes.

The system (Fig. 3B) consisted of a jacketed flow cell (3 cm diameter, 2.7 cm high, 19 ml capacity) a magnetic stirrer (1000 rpm), a peristaltic pump, an automatic sampler, a current recorder.

The temperature of the flow cell was maintained at 30 ± 0.1°C by passing warm water through the jacket. Tap water was adjusted to pH 3 with 0.05 M sulfuric acid and saturated with air. Then, it was transferred to the flow cell at a rate of 1.4 ml•min^{-1} together with air at a flow rate of 200 ml•min^{-1}. When the output current of the sensor became constant, sample was passed into the system at a rate of 0.8 ml•min^{-1} for 3 min.

When an acetic acid solution was pumped through the flow system, the current decreased to a steady state with a response time of 8 min.

The calibration graphs obtained showed linear relationships between the current decrease and the concentration of acetic acid up to 72 mg•l^{-1}. The minimum concentration for determination was 5 mg of acetic acid per liter. The reproducibility of the current difference was examined using the same sample. The current difference was reproducible within ± 6% for an acetic acid sample containing 54 mg•l^{-1}. The standard deviation was 1.6 mg•l^{-1} in 20 experiments.

The selectivity of the microbial electrode for acetic acid was examined. The sensor did not respond to volatile compounds such as formic acid and methanol or to involatile nutrients such as glucose and phosphate ions. As the microbial sensor was covered with a gas-permeable membrane, only volatile compounds could penetrate through the membrane. The response to organic compounds depends on the assimilability by the immobilized microorganisms. *Trichosporon brassicae*, which utilizes acetic acid, did not assimilate formic acid or methanol, but utilized propionic acid, n-butyric acid and ethanol. Fortunately, the latter three compounds are not generally present in fermentation broths.

The microbial sensor for acetic acid was applied to a fermentation broth of glutamic acid. The concentration of acetic acid was determined by the microbial sensor and by a gas chromatographic method. Good agreement was obtained; the regression coefficient was 1.04 for 26 experiments.

The long-term stability of the microbial sensor was examined for acetic acid solutions (72 $mg \cdot l^{-1}$). The current output (0.29 - 0.25 μA) of the sensor was constant (within ± 10% of the original values) for more than 3 weeks and 1500 assays.

D. *Alcohol Sensor*

On-line measurements of methyl and ethyl alcohol concentrations in culture broth are required in fermentation industries. In the cultivation of yeasts using sugar as a carbon source, it is well known that ethyl alcohol as a by-product, decreases the sugar basis yield of whole cells (20). In the cultivation of microorganisms using methyl alcohol as a carbon source, the concentration of methyl alcohol must be maintained at the optimal level to avoid substrate inhibition (21).

Many microorganisms utilize alcohols as carbon. As described above, assimilation of organic compounds by microorganisms can be determined from the respiration activity of microorganisms. The respiration activity is directly measured by an oxygen electrode. Therefore, it is possible to construct a microbial sensor for alcohols using immobilized microorganisms. The respiration activity is directly measured by an oxygen electrode. Therefore, it is possible to construct a microbial sensor for alcohols using immobilized microorganisms and an oxygen electrode.

A microbial electrode consisting of immobilized yeasts or bacteria, a gas permeable Teflon membrane, and an oxygen electrode is prepared for the determination of methyl and ethyl alcohols. Furthermore, the microbial sensor is applied to continuous determination of alcohols in a fermentation broth (22).

Unidentified bacterium (AJ 3993) and *Trichosporon brassicae* were used for the methyl and ethyl alcohol sensor, respectively. Alcohol sensors were assembled with the same method as that described in the acetic acid sensor.

When the sample solution containing alcohol was injected into the system, ethyl alcohol permeated through the gas-permeable membrane and was assimilated by the immobilized microorganisms. Oxygen consumption by immobilized microorganisms began and caused a decrease in dissolved oxygen around the membranes. As a result, the electrode current decreased markedly with time until a steady state was reached.

The time required for the determination of ethyl alcohol is long by the steady-state method. Therefore, the pulse method was employed for the determination. The assay can be done within 6 min by the pulse method. The total time required for the assay of ethyl alcohol was 30 min by the steady-state method and 15 min by the pulse method.

The pulse method was, therefore, employed for further work. A linear relationship was observed between the current decrease and the concentration of ethyl alcohol below 22.5 $mg \cdot l^{-1}$ by the steady-state and pulse methods. The minimum concentration for the determination was 2 mg ethyl alcohol l^{-1}. The reproducibility of the current difference was examined using the same sample. The current difference was reproducible within ± 6% of the relative error when a sample solution containing 16.5 $mg \cdot l^{-1}$ of ethyl alcohol was employed. The standard deviation was 0.5 $mg \cdot l^{-1}$ in 40 experiments. The selectivity of the microbial sensor for ethyl alcohol was satisfactory.

The microbial sensor for ethyl alcohol was applied to fermentation broths of yeasts. The concentration of ethyl alcohol was determined by the microbial sensor and by gas chromatography. Satisfactory comparative results were obtained between them. The correlation coefficient was 0.98 with 20 experiments. The reusability of the microbial sensor for ethyl alcohol was examined. Ethyl alcohol solutions (from 5.5 to 22.3 $mg \cdot l^{-1}$) were used for long-term stability testing of the sensor. The current output of the electrode was almost constant for more than three weeks and 2100 assays. The microbial sensor can be used for a long time for assay of ethyl alcohol.

A microbial sensor consisting of immobilized bacteria, a gas permeable membrane, and an oxygen electrode was applied to the determination of methyl alcohol. A methyl alcohol utilizing bacterium (AJ 3993) was employed for the sensor. The determination was performed under the same conditions described in the ethyl alcohol sensor by the steady-state method. A linear relationship was also observed between the current decrease and the concentration of methyl alcohol. Therefore, the sensor can also be applied to the determination of methyl alcohol.

E. Sensors for Cell Populations

The determination of microbial populations in fermentors is important for the control of fermentation processes. Several methods have been used for the determination of cell numbers (23,24,25). However, most of these methods are time consuming and are not suitable for the continuous determination of cell numbers in fermentors. Turbidimetry is a simple method for monitoring the number of cells in a fermentor. But sampling is required, and a colored or suspended broth is not suitable for this method.

Simple and continuous methods for the direct determination of cell populations are still required for fermentation control. It was found that bacteria were oxidized directly on the surface of the anode and a current was generated. This electrochemical system can be applicable to the determination of microbial populations (26,27).

The electrode system for the determination of the microbial population is shown in Fig. 4. The apparatus contains a similar system for determination and for reference (the surface of the anode is covered with cellulose dialysis membrane). Each system has a working electrode (platinum, diameter: 0.3 cm), the counter electrode (platinum, 0.6 cm x 3 cm) and the saturated calomel electrode (S.C.E.). Saturated KCl was used as an electrolyte. Three electrodes were connected through a potentiostat. The current was measured by a millivolt ammeter and the signal obtained was displayed on a recorder.

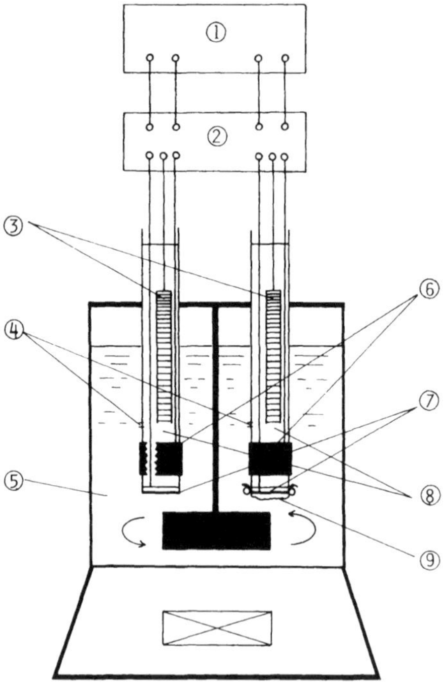

Figure 4. Scheme of systems for electrochemical determination of cell populations. 1) recorder; 2) potentiostat; 3) saturated calomel electrodes; 4) junctions (ceramic); 5) fermentation broth; 6) counter electrodes (Pt); 7) working electrodes (Pt); 8) electrolyte (saturated KCl); 9) cellulose dialysis membrane

The effect of potential on the current difference between both systems was examined. The currents of both systems increased with increasing potential. The current difference also increased, and became constant above + 0.2 V. Therefore, the experiments were performed at the anode potential of + 0.2 V relative to the SCE.

Two systems were immersed in a culture broth containing 1.4×10^9 cells ml^{-1} of *Bacillus subtilis*. The potentials of both systems were at + 0.2 V relative to the SCE. High currents were obtained initially and then the steady state current was obtained. Therefore, the diffusion of micro-organisms and electroactive substances to the electrode seemed to become the rate determining factor. The response time (the time required for the current to reach a steady state) of both systems was 3-5 min. The steady state current was measured 5 min after the insertion of the system in a medium.

A culture broth containing 1.4×10^9 cells ml^{-1} of *B. subtilis* was sterilized for 10 min at 110°C. Then, both systems were immersed in the broth. The current difference of both systems was very small. Therefore, the current difference between both systems might result from reaction of the living bacteria with the anode.

A linear relationship was obtained between the current difference and the cell population of *B. subtilis* below 2.0×10^9 cells ml^{-1}. The current difference was reproducible with an average relative error of 4% when 30 samples of culture broths containing 1.4×10^9 cells ml^{-1} were employed.

The continuous determination of microbial population in a fermentor was performed by the system. The cell population plots measured by the conventional method are located near the line obtained by the electrochemical methods. The continuous monitoring of cell population in the actual cultivation was repeated twenty times. The same growth curves were obtained. The current difference was reproducible with an average relative error of 8%. This shows that the systems can be used for more than 400 h.

The current obtained from the determination system results from oxidation of microorganisms and electroactive substances. The current from the reference system is attributed to oxi-dation of electroactive substances, because microorganisms cannot penetrate through the cellulose dialysis membrane. Therefore, the current difference between the two systems is proportional to the number of microbial cells in a culture broth. However, the current difference decreased markedly, when the culture broth of *B. subtilis* was sterilized for 10 min at 110°C. This fact shows that only living microorgan-isms contribute to the current generation. Non-viable cells and non-microbial particles were also detected by the conven-tional methods such as turbidimetry and electronic particle counts (28). On the other hand, the number of living cells is determined by this electrochemical system. The mechanism of the current generation of microorganisms may be the same as previously reported (29).

F. Formic Acid Sensor

Formic acid is a commonly occurring intermediate of cellular metabolism. It is found in culture media, urine, blood, and gastric juices, and as a product of many chemical reactions. Formic acid is now attracting attention as an intermediate of biomass conversion which is easily converted to hydrogen. It has been determined by gas and ion-exchange chromatography and spectrophotometry (30). Recently, selective spectrophotometric enzymatic methods utilizing formate dehydrogenase, malate dehydrogenase and tetrahydrofolic acid synthetase have been reported (31). However, colored sample cannot be determined by spectrophotometric methods and also therefore enzymatic methods. Moreover, these conventional methods are not suitable for on-line measurement.

The fuel cell type electrode consisting of a platinum anode and a silver peroxide cathode is available for measuring the hydrogen produced from glucose by hydrogen producing bacteria (32). Some anaerobic bacteria such as *Escherichia coli, Clostridium butyricum,* and *Rhodospirillum rubrum* produce hydrogen from formic acid (33). The reactions involved are:

$$\text{Formic acid} \longrightarrow \text{Ferredoxin}_{\text{reduced}} + CO_2$$

$$\text{Ferreodoxin}_{\text{reduced}} \xrightarrow{\text{hydrogenase}} \text{Ferredoxin}_{\text{oxidized}} + H_2$$

$$\text{Formic acid} \xrightarrow{\text{formate dehydrogenase}} \text{Cytochrome C}_{\text{reduced}} + CO_2$$

$$\text{Cytochrome C}_{\text{reduced}} \xrightarrow{\text{hydrogenase}} \text{Cytochrome C}_{\text{reduced}} + H_2$$

Therefore, determination of formic acid is possible by using *Clostridium butyricum* and a fuel cell type electrode (34).

The fuel cell type electrode consisted of a Teflon membrane (50 μm thick), a platinum anode (1 cm diameter), Ag_2O_2 cathode (2 cm x 4 cm) and an electrolyte (0.1 M phosphate buffer, pH 7.0). Bacteria were immobilized on an acetylcellulose filter with agar. The bacteria membrane was attached on the surface of this Teflon membrane. This was covered with a porous Teflon membrane.

The system for the determination of formic acid was composed of the microbial sensor, a cell (2.8 cm diameter, 3.8 cm height), an incubator, a millivolt-ammeter and a recorder.

When the sensor was inserted into a sample solution containing formic acid, formic acid permeated through the porous Teflon membrane. Hydrogen, produced from formic acid by *C. butyricum,* penetrated through the Teflon membrane, and was oxidized on the platinum anode. As a result, the current increased until it reached a steady state, which indicated that the production of hydrogen by microorganisms and the diffusion of formic acid from the sample solution to the immobilized cells were in equilibrium. The steady state was attained in all cases within 20 min. The steady state current depended on the concentration of formic acid. When the sensor was removed from the sample solution and placed in the buffer solution, the current of the sensor gradually decreased. As shown, the current returned to the initial level within 20 min.

A linear relationship was obtained between the steady state current and the formic acid concentration below 1,000 $mg \cdot l^{-1}$. The minimum concentration for determination was 10 $mg \cdot l^{-1}$. The currents were reproducible with an average relative error of 5% when a medium containing 200 $mg \cdot l^{-1}$ of formic acid was used. The standard deviation was 3.4 $mg \cdot l^{-1}$ in 30 experiments.

The sensor did not respond to nonvolatile nutrients such as glucose, pyruvic acid and phosphate ions. Volatile compounds such as acetic acid, propionic acid, n-butyric acid, methyl alcohol and ethyl alcohol can permeate through the porous Teflon membrane. However, no current was obtained from these compounds, because *C. butyricum* did not produce hydrogen from them.

The microbial sensor was applied to the determination of formic acid in the cultivation medium of *Aeromonas formicans.* The formic acid concentration was measured by the gas chromatography and by the microbial sensor. Good agreement was obtained between both methods; the regression coefficient was 0.98 for 10 experiments.

To study the stability of the immobilized *Clostridium butyricum* in the sensor, it was stored in 0.1 M phosphate buffer at 5°C. Formic acid (200 $mg \cdot l^{-1}$) was determined at 5-day intervals with these stored microorganisms. The current output obtained from each experiment was constant for 20 days. The bacteria in the sensor are, therefore, active for 20 days.

G. Cephalosporin Sensor

For control of an antibiotic fermentation, antibiotics are usually determined by microbioassay based on turbidimetric

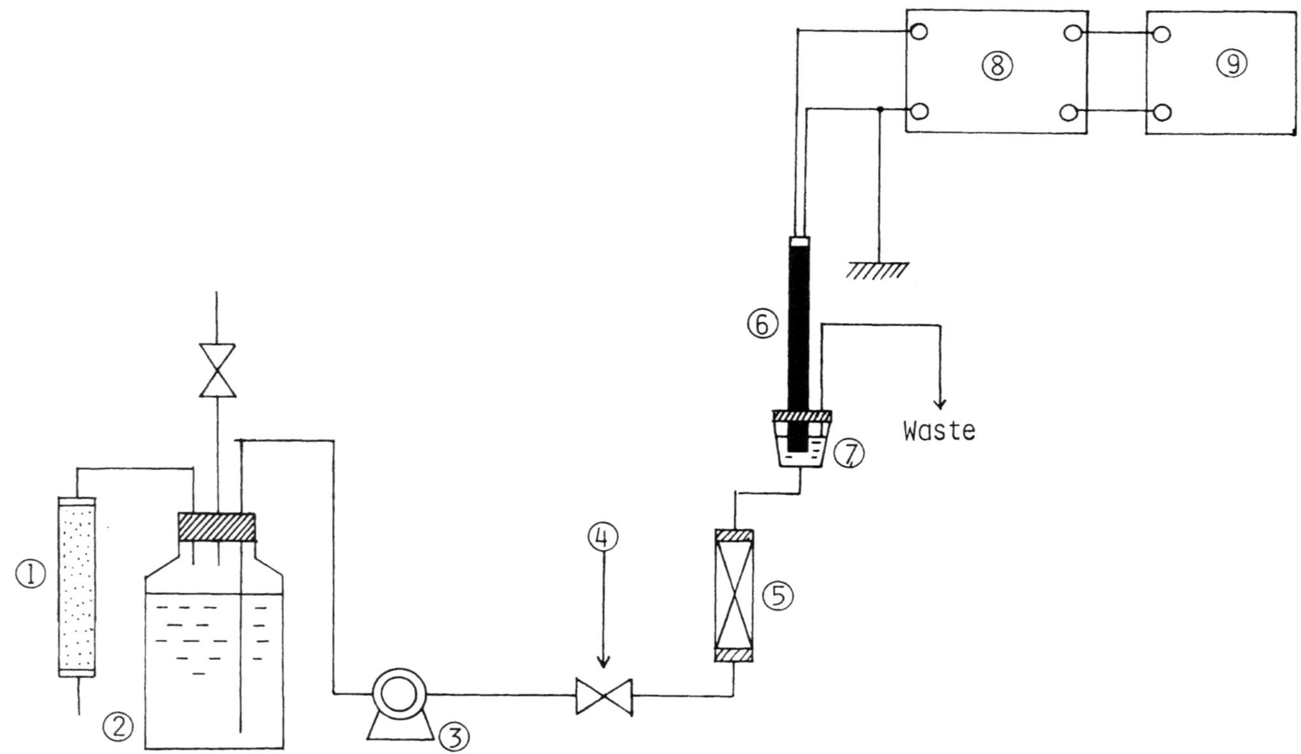

Figure 5. Immobilized whole cell-based flow-type sensor for cephalosporins. 1) soda lime; 2) buffer reservoir; 3) peristaltic pump; 4) sample inlet; 5) immobilized whole cell reactor; 6) combined glass electrode; 7) sensing chamber; 8) amplifier; 9) recorder

or titrimetric methods. However, these methods required a long time for cultivation of bacteria. Simple continuous methods for antibiotic determination are required for fermentation control.

It was found that *Citrobacter freundii* produced cephalosporinase, which catalyzes the following reaction of cephalosporin, which liberated hydrogen ions:

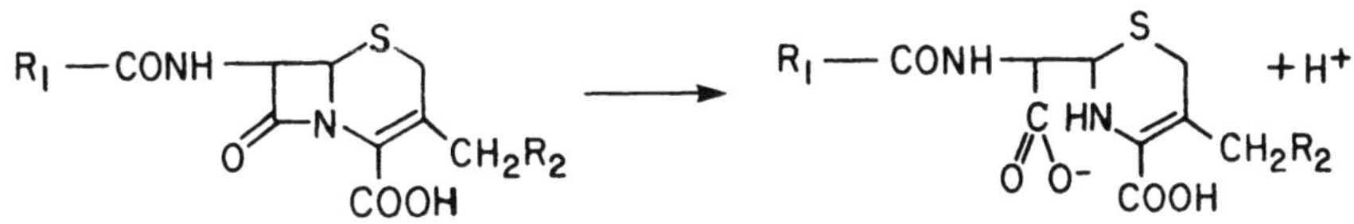

Cephalosporin may, therefore, be determined from the proton concentration generated in a medium by using immobilized cephalosporinase. Immobilization of cephalosporinase was difficult because the molecular weight of the enzyme is only 30,000 and the enzyme is unstable. Therefore, whole cells of *Citrobacter freundii* were immobilized in a collagen membrane. A microbial sensor composed of a bacteria-collagen membrane reactor and a combined glass electrode can be applied to the determination of cephalosporins in fermentation media (35).

The system used for continuous determination of cephalosporins is illustrated in Fig. 5. The reactor was biocatalytic type (acrylic plastic, 1.8 cm diam., 5.2 cm long) with a spacer (glass rod, 1.4 cm diam., 5.0 cm long) located in the center. The inner volume of the reactor was 4.1 ml. The bacteria-collagen membrane (10 x 5.5 cm^2; 53,276 units) was rolled up with a plastic net (5 x 20 cm^2, 20 mesh) and inserted into the reactor. The pH of the sample solution was measured by a combined glass electrode and displayed on a recorder.

Sample solutions (10 ml) containing various amounts of cephalosporins were transferred to the reactor. The electrode potential difference in the sensing chamber increased with time until a maximum was reached. The time required for the maximum to be reached (the response time) depended on the flow rate and activity of the bacteria-collagen membrane. The maximum potential difference was attained in 10 min at a sample flow rate of 2 $ml \cdot min^{-1}$.

A linear relationship was obtained between the logarithm of the cephalosporin concentration and the potential difference. 7-Phenyl-acetylamidodesacetoxy-sporanic acid (phenyl-acetyl-7ADCA), cephaloridine, cephalothin and cephalosporin c were determined by the cephalosporin sensor. Each determination took about 10 min.

The reusability of the microbial sensor was examined with a solution containing 125 $\mu g\ ml^{-1}$ of phenylacetyl-7ADCA. The cephalosporin determination was carried out several times a day, and no change in the potential difference response was observed for a week.

The system was applied to the determination of cephalosporin c in a broth of *Cephalospolium acremonium*, and was compared with a method based on high-pressure liquid chromatography (h.p.l.c.) (36). The relative error of the determination by the microbial system was 8%. Accordingly, the method is suitable for continuous analysis of cephalosporins in fermentation broths.

H. Glutamic Acid Sensor

Large quantities of glutamic acid are produced by fermentation, so that rapid automatic measurements of glutamic acid in fermentation media are required. Autoanalyzer-based on enzymatic reaction (37) can be used, but the method has disadvantages such as consumption of expensive enzyme.

Glutamate decarboxylase catalyzes the decarboxylation of glutamic acid, which produces carbon dioxide and amine, but the enzyme is expensive and unstable. However, certain microorganisms contain glutamate decarboxylase. Consequently, it has been possible to construct a microbial sensor for glutamic acid by using immobilized microorganisms having glutamate decarboxylase activity, in conjunction with a carbon dioxide gas-sensing electrode (38).

A gas-sensing electrode for carbon dioxide was used, consisting of a silicone rubber membrane, a combined glass electrode and a sodium hydrogen carbonate and sodium chloride buffer solution. The freeze-dried cells of *Escherichia coli* was mixed with one drop of water and coated on both sides of a nylon mesh (60-mesh, 7 mm diameter) which was placed on the surface of the silicone rubber membrane of the electrode and covered with a cellophane membrane to entrap the microorganisms between the two membranes. The cellophane membrane was fastened with a rubber ring.

The microbial sensor system includes a 0.5-ml flow cell, a peristaltic pump, an automatic sampler, a potentiometer and a recorder.

Preliminary experiments showed that the bacteria *(E. coli)* did not evolve carbon dioxide under anaerobic conditions, in the absence of glutamic acid. Therefore the normal respiration activity of the bacteria was depressed under anaerobic conditions, and any carbon dioxide produced by these bacteria under such conditions would result from the glutamate decarboxylase reaction. Nitrogen gas was passed through the flow cell at 500 $ml \cdot min^{-1}$ in order to remove dissolved oxygen in buffer and sample solution.

The initial constant potential (E) was obtained with the buffer solution only. When the sample solution containing glutamic acid was injected into the system, glutamic acid permeated through the cellophane membrane and was metabolized by the microorganisms to produce carbon dioxide:

$$HOOC\text{-}CH_2\text{-}CH_2\text{-}\underset{}{\overset{NH_2}{\overset{|}{C}}}H\text{-}COOH \xrightarrow[\text{decarboxylase}]{\text{Glutamate}} HOOC\text{-}CH_2\text{-}CH_2\text{-}CH_2NH_2 + CO_2$$

The enzyme reaction was carried out at pH 4.4, which was sufficiently below the pK_a value (6.34 at 25°C) of carbon dioxide to allow the carbon dioxide around the membranes to increase. As a result, the potential of the carbon dioxide gas-sensing electrode increased with time. The assay can be done by using a 1-3 min injection period and measuring the maximum potential with little loss of sensitivity.

The plot of the maximum potential vs. the logarithm of the glutamic acid concentration was linear over the range shown in Fig. 4; the slope over this range was approximately Nernstian. When a glutamic acid solution (400 $mg \cdot l^{-1}$) was measured in replicate, the standard deviation was 1.2 $mg \cdot l^{-1}$ (20 experiments).

The sensor responded to glutamic acid and glutamine and very slightly to some other amino acids. The response to glutamine can be decreased, if necessary, using acetone-treated *E. coli*. The selectivity of the microbial sensor was compared with that of the Warburg method using microorganisms prepared by the same method. The selectivity of the sensor was better than that of Warburg's method. The microbial sensor did not respond under the anaerobic conditions to organic substances such as glucose (7900 $mg \cdot l^{-1}$) and acetic acid (200 $mg \cdot l^{-1}$). The influence of inorganic ions on the response was negligible.

The microbial sensor was applied to the determination of glutamic acid in fermentation broth. Known amounts of glutamic acid were added to a broth and the concentration of glutamic acid was determined. Satisfactory recovery data (99-103%) were obtained. The concentrations of glutamic acid in some fermentation broths were determined by the microbial sensor and by the Autoanalyzer method. For the microbial sensor method, the broths were diluted (1 + 99) with deionized water before injection. The results were in good agreement. Dissolved carbon dioxide in the culture borth at pH 4.4 was removed by passing nitrogen gas, and did not influence the determination.

Glutamic acid solutions (240-800 $mg \cdot l^{-1}$) and fermentation broths of glutamic acid were used to test the long-term stability of the sensor. The response of the electrode was constant for more than 3 weeks and 1500 assays. Thus, the microbial sensor appears to be very attractive for the determination of glutamic acid.

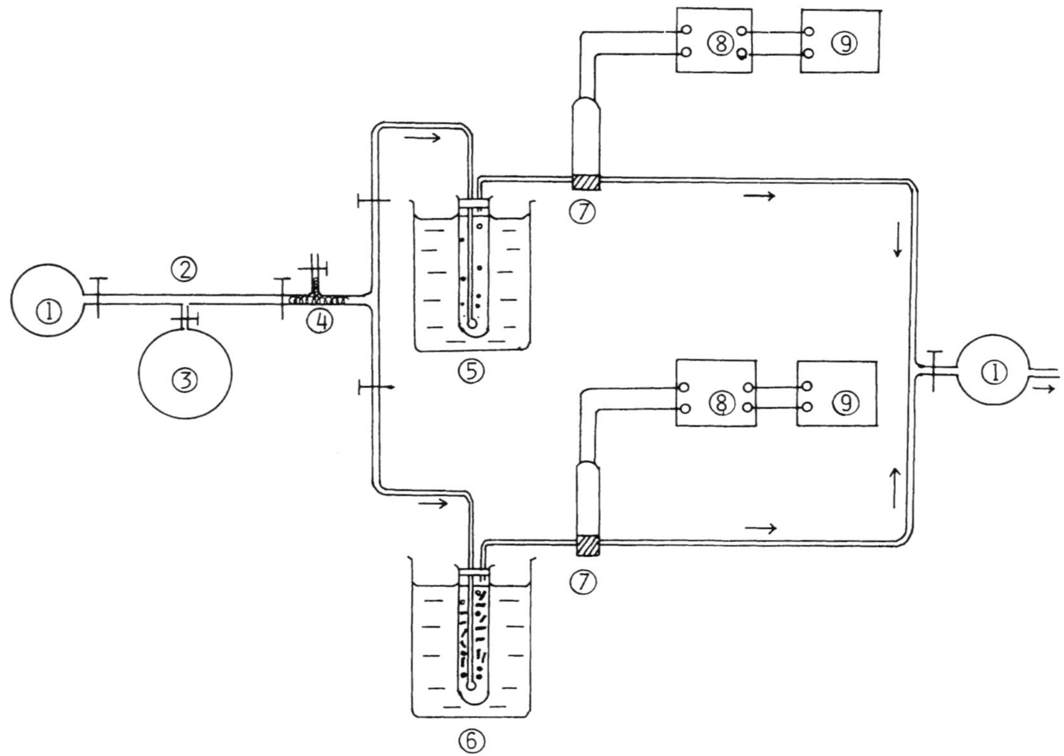

Figure 6. Schematic diagram of the methane gas sensor. 1) pump; 2) gas sampler; 3) sample gas; 4) cotton filter; 5) reference reactor; 6) methane oxidizing bacterial reactor; 7) oxygen electrode; 8) amplifier; 9) recorder

I. Methane Sensor

Methane is an attractive energy source and a main component of natural gases. It is used as gaseous fuel in many fields, and it is well known that methane forms explosive mixtures with air (5-14%). Therefore, rapid methods for the determination of methan in air are required in various fields such as coal mining and gasification processes. Methane is also produced by methanogenic bacteria. The biological formation of methane is the result of a specific type of bacteria energy-yielding metabolism. World-wide interest has arisen in the production of methane by fermentation of biomass, renewable resources. Control of the fermentation process is needed for effective production of methane. The determination of methane in the reactor is required for process control.

Whittenbury, et al., (39) isolated more than 100 gram-negative, strictly anaerobic, methane-utilizing bacteria. More recently, an easily-grown methane-oxidizing bacterium was isolated in a pure culture from a natural source through studying the production of single cell protein (40). This bacterium was identified as a new species, *Methylomonas flagellata*, capable of growing by oxidizing methane as the main carbon and energy source.

In general, methane-oxidizing bacteria utilize methane, and oxygen is consumed by the respiration as follows:

$$CH_4 + O_2 \xrightarrow{\text{methane-oxidizing bacteria}} CH_3OH + H_2O$$

Therefore, the methane concentration may be determined with a microbial sensor containing immobilized *M. flagellata*, and an oxygen electrode.

The microbial sensor system is schematically illustrated in Fig. 6. The system is composed of an immobilized micro-organism reactor, a control reactor and two oxygen electrodes. Methane gas was introduced into both reactors by a pump at a controlled flow rate. The flow rate of each was carefully adjusted with control valves. The partial pressure of oxygen in each line was monitored with an oxygen electrode. The difference between the output currents of the two electrodes is related to the amount of methane in the flow lines. The oxygen electrode consisted of a Teflon membrane (50 μm thick), a platinum cathode, a lead anode and sodium hydroxide (30%) as the electrolyte. The system consisted of jacketed flow reactors containing immobilized bacteria (1.9 cm diameter; 18 cm thick; 51 ml volume; 44 ml liquid volume, 300 mg immobilized cells).

The temperature of the reactors was maintained at $30 + 0.1^\circ C$ in a water bath. When the output current of the sensor became constant, sample gas was transferred to the flow cell together with air at a flow rate of 80 $ml \cdot min^{-1}$.

When sample gas containing methane was transferred to the immobilized bacteria cells, methane was assimilated by the microorganisms. Oxygen was then consumed by the microorganisms so that the concentration of dissolved oxygen in the reactor decreased. The current decreased until it reached a steady state. The steady state current depended on the concentration of methane. When air passed through the flow reactor, the current of the sensor returned to its initial level within 1 min. The response time required for the determination of methane gas was 1 min. The total time required for an assay of methane gas by this steady state method was 2 min.

A linear relationship was observed between the current difference between the electrodes and the concentration of methane (below 6.6 mM). The minimum concentration for the determination was 13.1 μM. The reproducibility of the current difference was examined by repeated assay of one sample. The current decrease was reproducible within ± 5% in 25 experiments with sample gas containing 0.66 mM methane. The long-term stability of the microbial sensor was examined with a sample gas containing 0.66 mM of methane.

The current output of the sansor system was almost constant for more than 20 days and 500 assays. The microbial sensor can, therefore, be used to assay methane over a long period of time. In the same experiment the concentration of methane was determined by both the electrochemical sensor and the conventional method (gas chromatography). A good correlation was obtained between the methane concentrations determined by the two methods (correlation coefficient 0.97), which also indicates that the microbial sensor is stable for a long time.

J. *Assimilation Test Sensor*

Assimilation characteristics of microorganisms are very important for classification and utilization of microorganisms. A number of methods for the assimilation test have been developed, but they are based on detection of growth in a basic medium containing a specific carbon compound which serves as the sole source of carbon. These methods (growth tests) require a long cultivation time (27-72 hr) and special techniques such as aseptic precautions. Assimilation of substrate by microorganisms can be detected from their respiration activity. The respiration activity is directly measured by an oxygen electrode. Therefore, it is possible to detect assimilation characteristics of microorganisms by using a microbial sensor consisting of the immobilized whole cells and an oxygen electrode (41).

Microorganisms (ca. 0.015 g) were coated on the surface of an acetylcellulose membrane (diameter, 14 mm: thickness, 150 μm). The membrane retaining the microorganisms was placed on the Teflon membrane of the oxygen electrode so that the microorganisms were trapped between the two membranes. The membranes were covered with a nylon net and fastened with rubber rings. The system consisted of a flow cell (diameter, 3 cm: height, 4 cm; volume, 30 ml; liquid volume, 5 ml) with the microbial electrode, a water bath, a magnetic stirrer (1000 rpm), a peristaltic pump, an automatic sampler and a recorder.

The temperature of the flow cell was maintained at 40 ± 0.2°C (for molds) or at 30 ± 0.2°C by passing warm water through the water bath. Phosphate buffer solution (0.01 M, pH 7.0) saturated with dissolved oxygen was transferred to the flow cell at flow rate of 3.9 $ml \cdot min^{-1}$ with 2.0 $ml \cdot min^{-1}$ of air. When the output current of the microbial electrode reached a steady state value, a sample was transferred to the flow cell at a flow rate of 0.1 $ml \cdot min^{-1}$ during a period of 5 min. A sample was injected at 30 min intervals. In this study, concentration of substrate means that in the flow cell.

When a sample solution containing an assimilable substrate was injected into the system, the substrate permeated through the porous acetylcellulose membrane and was assimilated by the microorganisms. This respiration of the microorganisms was activated and dissolved oxygen around the membranes was consumed. As a result, the current of the electrode decreased markedly with time until a steady state was reached. The steady state current depends on many factors such as the concentration of the stubstrate, the assimilation activity of the immobilized microorganisms and the rate of metabolism (oxidation) of the substrate. When the injection of the sample solution was stopped, the current of the microbial electrode returned to the initial level. It is, therefore, possible that assimilation of a substrate by the microorganisms can be tested by the current of the electrode.

Figures 7A and B show typical response curves of the microbial sensor to various substrates (29 substrates shown in Table 1). The pulse method was employed for the assimilation test. In the case of the pulse method, the maximum current decrease (the current difference between the initial and the minimum) for the glucose was 70% of that obtained by the steady state method. A linear relationship was observed between the maximum current decrease and the concentration of a substrate when the concentration of substrate was below 100 $mg \cdot l^{-1}$. In this case, the rate of metabolism (oxidation of a substrate) of the immobilized microorganisms does not seem to be a rate-determining step and the current decrease obtained from the microbial sensor for a definite concentration

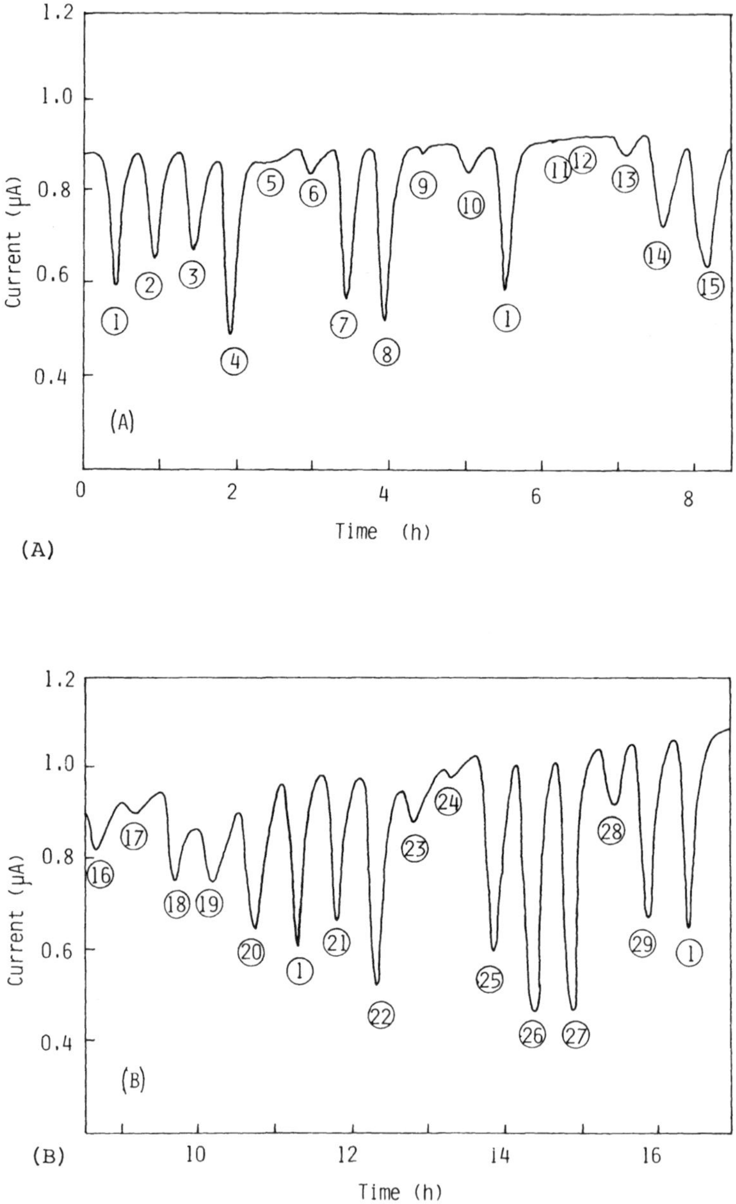

Figure 7 (A) and (B). Typical response curves of the microbial electrode for various substrates. Immobilized Aspergillus sojae *was employed for the electrode. A sample solution (0.5 ml) containing 50 mg$\cdot$l^{-1} of glucose and 100 mg$\cdot$l^{-1} of various substrates (shown in Table 3) were injected during a period of 5 min into the system.*

of a substrate is, therefore, proportional to the apparent rate of metabolism (oxidation) of the immobilized microorganisms. Therefore a concentration of 50 $mg \cdot l^{-1}$ or 100 $mg \cdot l^{-1}$ for each substrate tested was employed in this study.

Various microbial sensors were prepared and substrates were applied to the sensor. In order to generalize the results, glucose was used as a standard substrate and maximum current decrease for a substrate was normalized.

$$\frac{\text{maximum current decrease for a substrate } (\mu A) \times \text{concentration of glucose } (mg \cdot l^{-1})}{\text{maximum current decrease for glucose } (\mu A) \times \text{concentration of a substrate } (mg \cdot l^{-1})} \times 100(\%) \qquad (1)$$

The normalized maximum current decrease for a substrate gives the apparent rate of metabolism (oxidation) of the substrate by the microorganisms being examined. As shown in Fig. 7 (Number 1 in Figs. 7A, B), the maximum current decrease for glucose increased gradually with time, since the activity of the microbial gradually increased. Therefore, the normalization of maximum current decrease for a substrate was carried out by injecting a glucose standard solution for every 10 samples. Table 1 shows the results of assimilation tests for molds *(Aspergillus sojae* and *A. niger)*.

As shown in Table 1, molds can assimilate a wider range substrates is well known. In the case of molds, the test must be carried out as soon as possible after cultivation. Otherwise, molds form spores which show little or no respiration activity.

The assimilation characteristics of various microorganisms could be tested with the microbial sensor system. There is a fundamental difference between this method and conventional growth tests. In the case of the electrochemical method, substrate assimilation by microorganisms is determined by detecting an increase in oxygen consumption of the microorganisms. Therefore, a substrate which is gradually utilized by microorganisms such as soluble starch is difficult to test since the increase in oxygen consumption by immobilized microorganisms is too small to be detected. Similarly, a substrate which requires the induction of an enzyme is also difficult to test because the time for measurement is too short to permit enzyme induction. In spite of these problems, the results of this method showed fairly good agreement with those of the conventional growth tests. This study provides a rapid, simple and automatic method for an assimilation test. This system is also applicable to the following test:

- simple identification of microorganisms
- selection of culture medium for microorganisms

- measurement of the enzymatic activity of microorganisms
- estimation of the biodegradability of substances in waste waters
- selection of microorganisms for waste water treatment
- assimilation test of activated sludges
- estimation of biological degradation of materials

II. MICROBIAL SENSORS FOR WASTE WATERS CONTROL

A. *BOD Sensors*

The biochemical oxygen demand (BOD) test is one of the most widely used and important tests in the measurement of organic pollution. The five day BOD test has remained a standard pollution monitoring tool since 1936. In practice, the BOD test requires a five day incubation period at 20°C and demands skill in determination. A number of papers concerning methods for rapid estimation of the five day BOD test have been published in recent years. However, simple and reproducible methods for estimation of five day BOD are still required for pollution control.

Bacteria isolated from soil were immobilized in collagen membrane and a microbial electrode using the bacteria-collagen membrane and oxygen electrode was prepared. Here, we describe the use of the microbial sensor in estimating the BOD of waste waters (42,43,44).

Trichosporon cutaneum was used for the BOD sensor. Microorganisms were adsorbed on the membrane by suction. The membrane retaining microorganisms was dried and stored at room temperature. The porous membrane (diameter 14 mm) retaining the microorganisms was placed on the Teflon membrane of the oxygen probe so that the microorganisms were trapped between the two membranes.

The system consisted of a jacketed flow cell (diameter 1.7 cm, height 0.6 cm, volume 1.4 ml) contained a microbial sensor, a peristaltic pump, an automatic sampler and a recorder.

The temperature of the flow cell was maintained at 30 ± 0.2°C by passing warm water through the jacket. Phosphate buffer solution (0.01 M, pH 7) saturated with dissolved oxygen was transferred to the flow cell at a flow rate of 1 ml • min^{-1}. When the current reached a steady state value, a sample was injected into the flow cell at a rate of 0.2 ml • min^{-1} for 20 min. The sample was injected with 60 min intervals.

When the sample solution containing glucose and glutamic acid was injected into the system, organic compounds permeated through the porous membrane and were assimilated by the immobilized microorganisms. Consumption of oxygen by the immobilized

Table 1. Results of the assimilation test for molds

No. Substrate	Aspergillus sojae IFO 4200 Normalized maximum Current decrease* (%) Run 1	Run 2	Aspergillus niger ATCC 6275 Normalized maximum Current decrease* (%) Run 1
1. D-Glucose	100	100	100
2. D-Galactose	31	41	36
3. Sucrose	55	37	41
4. D-Maltose	61	67	60
5. Lactose	6	2	0
6. L-Sorbose	9	10	4
7. D-Cellobiose	64	57	58
8. Trehalose	68	66	44
9. Melibiose	5	3	18
10. Raffinose	2	11	9
11. Melezitose	0	0	0
12. Inulin	2	0	0
13. Soluble starch	3	8	0
14. D-Xylose	21	31	20
15. L-Arabinose	19	40	8
16. D-Arabinose	25	15	21
17. D-Ribose	29	6	4
18. Rhamnose	41	27	20
19. Erythritol	29	21	8
20. D-Fructose	50	41	49
21. Dulcitol	9	43	0
22. D-Mannitol	22	64	24
23. D-Sorbitol	17	14	12
24. α-Me-Glucoside	5	8	0
25. Salicin	57	55	28
26. DL-Lactic acid	60	71	8
27. Succinic acid	19	71	28
28. Citric acid	11	16	0
29. Arbutin	43	48	40

*Calculated from Eq. (1)

microorganisms began and caused a decrease in dissolved oxygen around the membranes. As a result, the current of the electrode decreased markedly with time until a steady state was reached within 18 min.

The steady state current depended on BOD of the sample solution. Then the current of the microbial sensor finally returned to the initial level. The response time of the microbial sensor (time required for the current to reach steady state) depended on the kind of sample solution. The response time was 8 min for acetic acid solution, and 18 min for glucose solution. Therefore, injection time of 20 min was employed for further work.

A linear relationship was observed between the current difference (between initial and steady state currents) and the 5-day BOD of the standard solution below 60 mg$\cdot$l^{-1}. The minimum measurable BOD was 3 mg$\cdot$l^{-1}. The current was reproducible within ± 6% of the relative error when BOD 40 mg$\cdot$l^{-1} in 10 experiments. The current means the current differences hereinafter.

The microbial sensor was applied to estimation of 5-day BOD for untreated waste waters from a fermentation factory. The 5-day BOD of the waste waters was determined by JIS method (Japanese Industrial Standard Committee)(45). As shown in Fig. 8, good comparative results were obtained between BOD estimated by the microbial sensor and those determined by JIS method. Regression coefficient was 1.2 in 17 experiments and the ratios (BOD estimated by the microbial sensor/5-day BOD determined by JIS method) were in the range from 0.85 to 1.36. This variation might have been caused from change in composition of organic waste water compounds. Furthermore, the BOD of various kinds of untreated industrial waste waters were estimated by the sensor. The BOD values estimated by the sensor depended on compounds in waste waters.

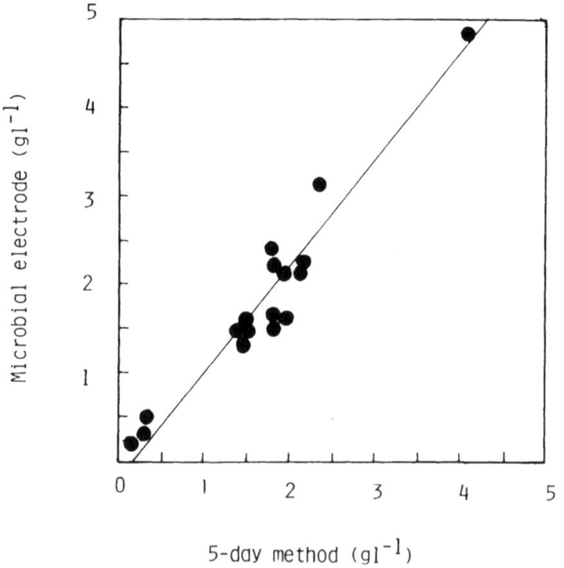

Figure 8. Comparison between 5-day BOD obtained by JIS method and BOD estimated by the microbial sensor. Waste waters from a fermentation factory were employed for experiments. The line was fitted by the least Square method.

Stable responses to the standard solution (BOD 20 $mg \cdot l^{-1}$) were observed for more than 17 days (400 tests). Fluctuations of the current and the base line (endogeneous level) were within 20% and 15% respectively for 17 days. The microbial sensor could be used for a long time for the estimation of BOD. However, it will be difficult to use this system to estimate the BOD of waste waters which contain organic compounds which cannot be degraded or which can be assimilated rapidly by the microorganisms.

B. *Ammonia Sensor*

The determination of ammonia is important in clinical, environmental, and industrial process analyses. The determination of ammonia is presently performed by spectrophotometric methods. However, these methods require a long reaction time and additional reagents (46). On the other hand, electrochemical monitoring of ammonia may have definite advantages. An ammonia gas electrode consisting of a combined glass electrode and a gas permeable membrane is usually used for this purpose. In this case, the determination must be performed under strong alkaline conditions (above pH 11). The ammonia electrode is based on potentiometric detection of ammonia. However, volatile compounds such as amines often interfere with the determination of ammonia. Therefore, an ammonia sensor based on amperometry is desirable for the electrochemical determination of ammonia.

Nitrifying bacteria contains two genera of bacteria. One genus (i.e., *Nitrosomonas* sp.) of bacteria utilizes ammonia as the sole source of energy and oxygen is consumed by the respiration as follows:

$$NH_3 + 1.5O_2 \xrightarrow{\textit{Nitrosomonas}\ \text{sp.}} NO_2^- + H_2O + H^+$$

The other genus (i.e., *Nitrobacter* sp.) of bacteria oxidizes nitrite to nitrate as follows:

$$NO_2^- + 0.5O_2 \xrightarrow{\textit{Nitrobacter}\ \text{sp.}} NO_3^-$$

The oxidation of both substrates (NH_3, NO_2^-) proceeds at a high rate, and oxygen uptake by the bacteria can be directly determined by the oxygen electrode attached to the immobilized bacteria. Therefore, ammonia is determined by the microbial sensor using immobilized nitrifying bacteria and an oxygen electrode (47,48).

The pH of a sample solution had to be kept sufficiently above the pK value for ammonia (9.1 at 30°C) because ammonium ions cannot pass through the gas permeable membrane.

When the ammonia solution was injected into the buffer solution, ammonium ion changed to ammonia gas. Ammonia gas permeated through the gas permeable membrane and was assimilated by the immobilized bacteria. Oxygen was then consumed by the bacteria so that the concentration of dissolved oxygen around the membrane decreased. The current decreased until it reached a steady state. The steady state current depended on the concentration of ammonia. When the sensor was inserted in tap water, the current of the sensor returned to its initial level within 5 min.

A microbial sensor consisting of immobilized *Nitrosomonas europaea* and an oxygen electrode has been prepared for the determination of ammonia (47). However, the reusability of the sensor was very poor. This result was caused by the inactivation of bacterium with nitrite formed. Therefore, stable nitrifying bacteria are required for the microbial sensor. The nitrifying bacteria which contained *Nitrosomonas* sp. and *Nitrobacter* sp., were isolated from a waste water treatment facility.

A linear relationship was observed between the current decrease (the current difference between the initial and the steady state) and the ammonia concentration below 42 $mg \cdot l^{-1}$ (current decrease 4.7 μA). The minimum concentration for the determination of ammonia was 0.1 $mg \cdot l^{-1}$ (signal to noise, 20; reproducibility, ± 5%). The reproducibility of the current decrease was examined by using the same sample. The current decrease was reproducible within ± 4% of the relative error when a sample solution containing 21 $mg \cdot l^{-1}$ of ammonium hydroxide was employed. The standard deviation was 0.7 $mg \cdot l^{-1}$ in 20 experiments.

Thus the amperometric determination of ammonia became possible by the microbial sensor. The sensitivity of the microbial sensor was almost at the same level as that of a glass electrode, and its minimum measurable concentration was 0.1 $mg \cdot l^{-1}$.

The sensor did not respond to volatile compounds such as acetic acid, ethyl alcohol, and amines (diethylamine, propylamine, and butylamine) or to involatile nutrients such as glucose, amino acids, and metal ions (potassium ion, calcium ion, and zinc ion).

Nitrifying bacteria, which could utilize ammonia, did not assimilate acetic acid, ethyl alcohol, volatile amines, glucose, amino acids, and metal ions. Therefore, these substances did not affect the determination of ammonia by the sensor. Therefore, selectivity of this microbial sensor became satisfactory.

The long-term stability of the microbial sensor was examined with a sample solution containing 33 mg$\cdot$l^{-1} of ammonia. The current output of the sensor was almost constant for more than two weeks and 1500 assays. Therefore, the microbial sensor can be used for a long time for the assay of ammonia.

The microbial sensor was applied to the determination of ammonia in waste waters of a fermentation factory. The concentration of ammonia of the sample solutions were determined by the distillation-acidimetry. A good agreement was obtained between the ammonia concentration determined by the conventional method and that obtained by the microbial sensor. The relative difference between the two methods was less than 6%.

C. *Nitrite Sensor*

The principal compounds of nitrogen and oxygen that form gases of interest in air pollution sampling and analysis are nitric oxide (NO), and nitrogen dioxide (NO_2). Nitrogen dioxide is the most reactive of the gaseous oxides of nitrogen. It exists as nitrite in an aqueous solution. Therefore, the determination of nitrite is important in industrial waste waters. The determination of nitrite is presently performed by spectrophotometric method (49). However, these methods require a long reaction time and additional reagents. The concentration of organic compounds could be determined from the respiration activity of immobilized whole cells with an oxygen electrode. *Nitrobacter* sp. utilize nitrite as the sole source of energy and oxygen is consumed by the respiration as follows:

$$2\ NO_2^- + O_2 \xrightarrow{\textit{Nitrobacter}\ \text{sp.}} 2\ NO_3^-$$

Oxygen uptake by the bacteria can be directly determined by the oxygen electrode attached to the immobilized bacteria. Therefore, nitrite in the buffer (pH 2.0) can be determined by the microbial sensor using immobilized *Nitrobacter* sp. and an oxygen electrode (50).

The porous membrane retaining immobilized *Nitrobacter* sp. was cut into a circle (0.8 cm diameter) and carefully attached on the surface of a Teflon membrane of the oxygen electrode, and covered with a gas-permeable Teflon membrane (0.5 μm pore size) and fastened with rubber rings.

The system consisted of a jacketed flow cell (23 mm diam. 10 mm height, and 1.0 ml liquid volume) with microbial sensor, a peristaltic pump, an amplified, and a recorder.

The temperature of the flow cell was maintained at 30 ± 0.1°C by a thermostated bath. A buffer solution (pH 2.0) saturated with dissolved oxygen was transferred to the flow cell at a flow rate of 1.6 ml$\cdot$min^{-1}. When the current of the sensor reached a steady state value, a sample was injected into the flow cell at a flow rate of 0.4 ml$\cdot$min^{-1} for 2 min.

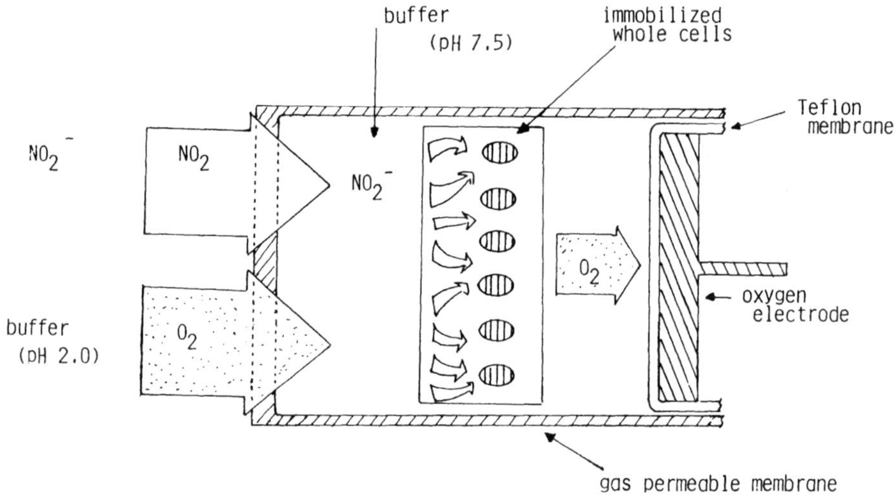

Figure 9. Principle of sodium nitrite sensor

Figure 9 shows the principle of the sensor for sodium nitrite. When the sample solution (sodium nitrite solution) is transferred into the flow cell, nitrous ion changes to nitrogen dioxide gas at the pH 2.0. Then nitrogen dioxide passes through the gas-permeable membrane. Nitrogen dioxide gas changes to nitrous ion in the bacterial layer and it is utilized by *Nitrobacter* sp. as a sole source of energy. The consumption of oxygen around the membrane is determined by the oxygen electrode. Therefore, the concentration of sodium nitrite can be indirectly determined from the current decrease of the oxygen electrode.

The current of the sensor decreased markedly with time until a steady state was reached. The steady state current is obtained within 10 min. When sufficient quantity of the bacteria is immobilized in the sensor, the current of the sensor for a sodium nitrite solution depends mainly on the rate of diffusion of nitrite from the sample solution to the immobilized bacteria. Therefore, steady state current depended on the concentration of sodium nitrite. The current difference between initial and steady state currents were directly proportional to the concentration of sodium nitrite.

A linear relationship was observed between the current decrease (the current difference between the initial and the steady state) and the sodium nitrite concentration below 0.59 mM (current decrease 0.63 μA). The minimum concentration for the determination of sodium nitrite was 0.01 mM (signal to noise, 20: reproducibility, ± 5%). The current decrease was reproducible within ± 4% of the relative error and the standard deviation was 0.01 mM in 25 experiments when a sample solution containing 0.25 mM of sodium nitrite was employed. Thus the amperometric determination of sodium nitrite became possible by the microbial sensor.

Selectivity of this microbial sensor was satisfactory in presence of difference various substances. The current output of the sensor was almost constant for more than 21 days and 400 assays. The microbial sensor can be used to assay sodium nitrite for a long period. In the same experiments the concentration of sodium nitrite was determined by both the sensor proposed and the conventional method (dimethyl-α-naphtylamine method). A good correlation was obtained between the sodium nitrite concentrations determined by both methods (correlation coefficient 0.99).

III. CONCLUSION

Many microbial sensors have been developed as described above. These microbial sensors appears to be quite promising for the control of fermentation processes. The advantages of these sensors are as follows: 1) organic compounds in samples can be directly measured without any pretreatment, such as filtration and dialysis because microbial cells and colored compounds in a fermentation broth do not affect the determination, 2) no reagent except for buffer solution is required, 3) microbial sensors are stable for a long time, 4) the microbial sensor can be readily prepared according to the method proposed.

The microbial sensors for acetic acid, ethyl alcohol, glutamic acid, cell populations, ammonia, and BOD are actually used in industrial fermentation processes in Japan.

ACKNOWLEDGMENTS

The authors are grateful to Dr. M. Hikuma and Mr. T. Yasuda, Central Research Laboratories, Ajinomoto Co., Inc., Dr. T. Matusnaga, Tokyo University of Agriculture and Technology, for their help during this study.

REFERENCES

1. Chang, T.M.S., Ed. "Biomedical Applications of Immobilized Enzymes and Proteins," Vol. *2*, Plenum Publishing Co., New York, 1977.
2. Guilbault, G. G., "Handbook of Enzymatic Methods of Analysis," Marcel Dekker, New York, 1976.
3. Aizawa,M., I. Karube, and S. Suzuki, *Anal. Chim. Acta, 69,* 431 (1974).
4. Satoh, I., I. Karube, and S. Suzuki, *J. Solid-Phase Biochem., 2,* 1 (1977).
5. Satoh, I., I. Karube, and S. Suzuki, *Biotechnol. Bioeng., 19,* 1095 (1977).
6. Vandamme, E. J., *Chem. Eng., 1070* (1976).
7. Karube, I., T. Matsunaga, T. Tsuru, and S. Suzuki, *Biochem. Biophys. Acta, 444,* 338 (1976).
8. Rechnitz, G. A., P. K. Kobos, S. J. Riechel, and G. R. Gebauer, *Anal. Chim. Acta, 94,* 357 (1977).
9. Kobos, R. K. and G. A. Rechnitz, *Anal. Lett., 10,* 751 (1977).
10. Rechnitz, G. A., M. A. Arnold, and M. E. Meyerhoff, *Nature, 278,* 466 (1979).
11. Arnold, M. A. and G. A. Rechnitz, *Anal. Chem., 52,* 1170 (1980).
12. Karube, I., S. Mitsuda, and S. Suzuki, *Europ. J. Appl. Microbiol. Biotechnol., 7,* 343 (1979).
13. Washka, M. E. and E. W. Rice, *Clin. Chim. Acta, 7,* 542 (1961).
14. Nagy, G., C. H. von Strop, and G. G. Guilbault, *Anal. Chim. Acta, 66,* 443 (1973).

15. Weibel, M. K., W. Dritschilo, H. Bright, and A. Humphrey, *Anal. Biochem.*, *52*, 402 (1973).
16. Hirose, S., M. Hayashi, N. Tamura, S. Suzuki, and I. Karube, *J. Molec. Catal.*, *6*, 251 (1979).
17. Technicon Industrial Systems, No. *142-71A*, (1972)
18. Hikuma, M., H. Obana, T. Yasuda, I. Karube, S. Suzuki, *Enzyme Microb. Technol.*, *2*, 234 (1980).
19. Hikuma, M., T. Kubo, T. Yasuda, I. Karube, and S. Suzuki, *Anal. Chim. Acta*, *109*, 33 (1979).
20. Wang, H. Y., C. I. Cooney, and D.I.C. Wang, *Biotechnol. Bioeng.*, *19*, 69 (1977).
21. Reuss, M., J. Gnieser, H.G. Reng, and F. Wagner, *Europ. J. Appl. Microbiol.*, *1*, 295 (1975).
22. Hikuma, M., T. Kubo, T. Yasuda, I. Karube, and S. Suzuki, *Biotechnol. Bioeng.*, *21*, 1845 (1979).
23. Pringle, J. R., in "Methods in Cell Biology" (P.M. Prescott ed.), Vol. *11*, 131, Academic Press, New York, 1975.
24. Hadley, W. K. and G. Senyk, in "Microbiology" (D. Schlessinger ed.), *12*, American Society of Microbiology, Washington, DC, 1975.
25. Postgate, J. R., in "Method in Microbiology" (J.R. Norris and D.W. Ribbons eds.), Vol. *1*, 661, Academic Press, New York, 1969.
26. Matsunaga, T., I. Karube, and S. Suzuki, *Appl. Environ. Microbiol.*, *37*, 117 (1979).
27. Matsunaga, T., I. Karube, and S. Suzuki, *Europ. J. Appl. Microbiol. Biotechnol.*, *10*, 125 (1980).
28. Kubitschek, H. E. in "Methods in Microbiology" (J.R. Norris and D.W. Ribbons eds.), Vol. *1*, 593, Academic Press, New York, 1969.
29. Matsunaga, T., I. Karube, and S. Suzuki, *Anal. Chim. Acta*, *98*, 25 (1978).
30. Heftmann, E., (ed.) in "Chromatography", Van Nostrand Reinhold, New York, 1975.
31. Karube, I., T. Matsunaga , and S. Suzuki, *J. Solid-Phase Biochem.*, *2*, 97 (1977).
32. Matsunaga, T., I. Karube, and S. Suzuki, *Biotechnol. Bioeng.*, *22*, 2607 (1980).
33. Gray, C. T. and H. Gest, *Science*, *148*, 186 (1965).
34. Matsunaga, T., I. Karube, and S. Suzuki, *Europ. J. Appl. Microbiol. Biotechnol.*, *10*, 235 (1980).
35. Matsumoto, K., H. Seijo, T. Watanabe, I. Karube, I. Satoh, and S. Suzuki, *Anal. Chim. Acta*, *105*, 429 (1975).
36. Miller R. D. and R. Neuss, *J. Antibiot.*, *29*, 902 (1976).
37. Industrial Method No. *210-72A*, Technicon Industrial Systems, New York, (1973).
38. Hikuma, M., H. Obana, T. Yasuda, I. Karube, and S. Suzuki, *Anal. Chim. Acta*, *116*, 61 (1980).

39. Whittenbury, R., K. C. Phillips, and J. F. Wilkinson, *J. Gen. Microbiol., 61,* 205 (1970).
40. Y. Morinaga, S. Yamanaka, S. Otsuka, and Y. Hirose, *Agric. Biol. Chem., 40,* 1539 (1970).
41. Hikuma, M., H. Suzuki, T. Yasuda, I. Karube, and S. Suzuki, *Europ. J. Appl. Microbiol. Biotechnol., 9,* 305 (1980).
42. Karube, I., T. Matsunaga, S. Mitsuda, and S. Suzuki, *Biotechnol. Bioeng., 19,* 1535 (1977).
43. Karube, I., T. Matsunaga, and S. Suzuki, *J. Solid-Phase Biochem., 2,* 97 (1977).
44. Hikuma, M., H. Suzuki, T. Yasuda, I. Karube, and S. Suzuki, *Europ. J. Appl. Microbiol. Biotechnol., 8,* 289 (1979).
45. Japanese Industrial Standard Committee, in "Testing Methods for Industrial Waste Water," p. 33, *JIS K 0102,* Japan (1974).
46. Japanese Industrial Standard Committee, in "Testing Methods for Ammonium Ion, p. 116, *JIS K 0101,* Japan (1979).
47. Hikuma, M., T. Kubo, T. Yasuda, I. Karube, and S. Suzuki, *Anal. Chem., 52,* 1020 (1980).
48. Karube, I., T. Okada, and S. Suzuki, *Anal. Chem., 53,* 1852 (1981).
49. Japanese Industrial Standard Committee, in "NO and NO_2 Standard Samples," p. 1, *JIS K 0516,* Japan (1976).
50. Karube, I., T. Okada, H. Suzuki, M. Hikuma, and T. Yasuda, *Europ. J. Appl. Microbiol. Biotechnol., 15,* 127 (1982).

CHAPTER 9

VENTURES IN YEAST UTILIZATION[1]

Henry J. Peppler

Universal Foods Corporation
Whitefish Bay, WI 53201

Dr. David Perlman, in whose memory we are assembled, distinguished himself in the development of the fermentation industries during the period of unprecedented research and growth that began in the 1940s. Superimposed on his prolific scientific activities was a life-long devotion to timely reporting and interpretation of the current and future states of the art and science of fermentation technology.

His initial review, prepared in 1949 with Asger Langlykke and C. V. Smythe (1), focused on the production of enzymes for industrial use, and some of the practical problems in the fermentation industry. In 1952, Dave (2) joined the team of William Brown and Sylvan Lee to prepare the fermentation review series started in 1948 by Lee and Herbert Silcox (3), and published annually in *Industrial and Engineering Chemistry.* This series of reports (4,5,6) covered current fermentations, newer products which could be produced on a commercial scale by microorganisms, and factors concerned with fermentations as a unit process. World directories of major manufacturers of microbes, and their products, appeared as early as the reviews of the mid-1950s, and were revised as needed in subsequent reviews. Following the industry surveys of 1953 (7) and 1954 (8), co-authored by Perlman, Brown, Charles Kroll, and Arthur Tempel, the burden of continuing the fermentation

[1] *The third David Perlman Memorial Lecture sponsored by the Microbial and Biochemical Technology Division of the American Chemical Society, and BioChem Technology, Inc., Malvern, Pennsylvania. Presented in Kansas City, Missouri, on September 14, 1982.*

ISBN 0-12-040306-4

reports, for the next three years, was assumed by Gilbert Shull and Samuel Beesch (8,9,10). After a five-year lull, Perlman and Ralph Anderson assembled the Fermentation Report for *Chemical Week* in 1962 (11,12).

The rising challenge of global competition for markets, and in the development of new products, highlighted the 1967 Perlman assessment (13), also for *Chemical Week*. After moving in 1967 from the Squibb Institute for Medical Research to the Madison campus of the University of Wisconsin, his alma mater, the reports (14,15,16,17,18,19), critiques (20,21) and editorial activities continued at regular intervals throughout his years as dean of the School of Pharmacy, and professor of pharmaceutical biochemistry.

Following Perlman's 5th biennial report (19), the fermentation reviews became an annual series in 1977 as the *Annual Reports On Fermentation Processes* (22). It was edited by Perlman and George Tsao, and published by Academic Press. Now in its 6th volume, the contributed series is edited by George Tsao, and associates Michael Flickinger and Robert Finn.

David Perlman also edited the long-standing series *Advances in Applied Microbiology*, from Volume 10 through Volume 27. The current editor is Allen Laskin.

Early in 1976, Dave promoted the need for a revised and expanded edition of the book *Microbial Technology*, first published in 1967. Together, with Dave as co-editor and 57 able and willing contributors, the two volume Second Edition of *Microbial Technology* was published by Academic Press in late 1979. It came in time for Dave's final review.

As activities in biotechnology expand and stimulate fresh growth in microbial and biochemical technology, let us hope there will be enough dedicated reporters and editors to cover it. In the new, emerging ventures, some kinds of yeasts may well be the microbes of choice over *Escherichia coli*.

The topic I have chosen, "Ventures in Yeast Utilization," covers a brief review of the major yeast-dependent industries in the United States--their production, problems, and prospects. It was prompted by the market battles and technical troubles all segments of the yeast business encountered in the past year or two.

Some giants have fallen; a few are ailing. Some of the startling events include, for example: Fleischmann yeast--the leading baker's yeast producer for one hundred twelve years (23), has curtailed production drastically (24). The beer capitol of the world moved to Los Angeles (25,26). A major distillery closed down. *Candida utilis* food and feed yeast production may be in jeopardy.

Before discussing the details of these and other recent events in the yeast business, I will briefly review the major yeast-dependent processes and products.

Yeast products and yeast-based industries can be grouped, in general, into seven categories (27,28):

- Baker's yeast and bread-type products
- Food and feed yeast biomass
- Products derived from yeast biomass; e.g., autolysates, and biochemicals
- Beer yeast and brewing
- Wine yeast and winemaking
- Distiller's yeast and distilled spirits
- Industrial grade ethanol (solvents) and fuel alcohol

This classification differs somewhat from the divisions used by the Bureau of Industrial Economics (SIC categories) for sales and market data compiled on major industries (29). The main segments of industry in this system are:

- Bread and related products (SIC 2051)
- Alcoholic beverages (SIC 2081)
- Malt beverages (SIC 2082)
- Wine and brandy (SIC 2084)
- Distilled liquor (SIC 2085)

The major industrial ventures in yeast processes and yeast utilization usually involve one or more of four species of yeast (27,28): Two species of *Saccharomyces* for the production of bread and bread-type products, alcoholic beverages, power and industrial-grade ethanol, and yeast biomass for food and feed use, autolysates and as a source of enzymes and biochemicals. And two others, *Candida utilis* and *Kluyveromyces fragilis*, grown on ethanol, cheese whey, pulp mill liquor and other process wastes. They are also used for food, feed and derived products. Also included are those processes which alter, enrich, or extract components of living yeast cells. Yeast-derived products of major importance are yeast extracts, 5' nucleotides, and the enzyme invertase and lactase (31,32,33).

Yeasts, and the products they make, are produced worldwide (Table 1). The economic importance of yeast in the food and beverage industries far outweighs its other applications. And many governments exploit the popularity of fermented beverages by levying heavy excise taxes. In the United States in 1981, the alcoholic beverage industries paid more than $12 billion in excise taxes (35,39)--second only to federal and state income taxes.

Table 1. Director of Baker's Yeast Plants [(a)]

	Companies	Plants
United States of America	6	14
Canada	2	3
Mexico	4	4
Central America	6	9
South America	13	14
England	3	3
Europe	24	28
The Orient	8	8
Asia	11	12
Africa	7	7

Source: In part, ref. 19, and Universal Foods Corp., International Division, Milwaukee, Wisconsin, U.S.A.

[(a)] Soviet Union countries not included.

According to the Bureau of Industrial Economics, the major United States industries in which yeast plays a decisive role, had a combined product and service value greater than \$45 billion--second highest in the food field, led by the red meat industry valued at \$56 billion. Dairy products, on the same basis, were valued at \$37 billion (29)(Table 2).

Table 2. Production and Sales of Major Yeast-Based Industries

	United States			World
	Consumer Expenditures ($ billion)	Units (millions)	Excise Tax ($ billion)	Production (million)
Malt beverages (SIC 2052)	23	184 bbl.	2.4	470 bbl.
Bread and related Products (SIC 2051)	14	8 tons	0	ND[a]
Distilled Liquor (SIC 2085)	4	450 gal.	9.0	ND
Wine and Brandy (SIC 2084)	4	500 gal.	0.8	9,300 gal.
Fermentation Fuel and Industrial Alcohol	ND	150	ND	530 gal.

Source: refs. 29, 30, 35, 41.

[a] ND = not determined.

Malt beverages (beer and ale) lead the fermentation industries with products and service valued at more than $23 billion. The 1981 production of 184 million barrels (5.7 billion gal.) amounted to 40% of the world output. In 1982 the brewing industry is expected to outperform the entire alcoholic beverage business with a 6.5% increase in sales (30,75).

Bread and related yeast-leavened goods shipped in 1981 were valued at $14 billion on production of approximately 8 million tons, which required around 300 million pounds of compressed yeast. Although white pan bread consumption, per capita, has decreased 34% during the past 15 years, sales in this segment of bakery foods have shown progressive increases in the demand for variety breads, hearth breads, rolls, muffins, and buns (36). The annual growth rate of this industry is 2.5%.

American is in a wine boom, in quality as well as in quantity. In 1980, for the first time, we consumed as much wine as hard liquor, and 80% of the wine was made in the United States (37,41,42).

Domestic wine and brandy production increased 5.4% last yeast, reaching a record volume of 507 million gallons, valued at about $4 billion (29). Compared with world production of 9.2 billion gallons, U.S. wine making, mainly in California, is an infant industry. However, economic forecasters predict that domestic volume will double to one billion gallons within 10 years (37). Last year Gallo led the field with 26% of national sales, on more than 100 million gallons of wine--twice the output of the nearest competitor, United Vintners. Eighty percent of the world's wine is made in Europe. Italy, the largest producer, has a 30% share of the European output; France follows with 25% of the business (40).

Distilled liquor, excluding brandy, is a $4 billion enterprise in the U.S., and the industry is growing at a compounded annual rate of 5%. American distillers in 1981 produced about 450 million wine gallons (29), Kentucky, traditionally the major producer of beverage alcohol, in 1981 made 54 million proof gallons--about 75% of the United States manufacture (35,39). Of 20 licensed distilleries, only four did not operate.

Industrial-grade alcohol made by fermentation last year amounted to about 150 million gallons, or approximately one-third of the world output. Included in the domestic volume is about one million gallons anhydrous ethanol for use as a partial substitute for gasoline. However, the gasohol program is faltering as motor fuel usage continues to decline (6.8% drop in the last 12 months). Among the reasons for the lower consumption are more efficient vehicles, motorist's conservation and the persistence of the economic recession. Of greater concern to the fuel alcohol fermentation projects is the absence of a clear-cut public policy favoring renewable fuels (43).

One independent study estimated that if all the fuel alcohol ventures that were planned in 1981 were to be built and go on stream this year, their capacity would exceed 900 million gallons, and it could go to 2 billion gallons by 1986 (44). ADM, based in Decatur, Illinois, is the largest producer of fuel alcohol, and the most successful today.

Food and feed yeasts are heat-killed, inactive (nonfermentative) yeasts, generally classed as "dried yeast." They are obtained either by recovering surplus brewer's yeast, of by harvesting yeast grown primarily for the purpose of supplementing human diets and animal rations (34).

Primary yeasts include molasses-grown baker's yeast (*Saccharomyces cerevisiae*), whey grown *Kluyveromyces fragilis* and *Saccharomyces lactis* (in Europe), and *Candida utilis* propagated on wood sugars and ethanol. Production estimates for this segment of the yeast business are highly variable, and are often confused with baker's yeast data.

Based on limited information, a reasonable estimate would be one million tons of dried food and feed yeast, which is about double the world baker's yeast output, on the same basis. USSR is believed to have an output close to one million tons annually, all for livestock supplements (45). The main usage of dried yeasts is in the formulation of feed. Brewer's yeast and *C. utilis*, as well as other species, grown on a variety of industrial and agricultural wastes, are the main feed yeasts.

Food yeasts are also custom-grown, with enhanced B-vitamin contents and increased nutritional complexes of the essential trace elements chromium and selenium (27,46,47). Under special growth conditions, baker's yeast converts inorganic chromium to a biologically active factor called glucose tolerance factor (GTF), which promotes sugar transport in mammals. Yeast with high levels of GTF is under clinical evaluation to determine the effective dosages for alleviating age-onset diabetes in humans.

Selenium taken up during yeast growth is bound to sulfur containing amino acids, and also is a constituent of glutathione peroxidase. Such complexes of selenium are involved in the prevention of cell degeneration. Both new yeast products are produced in significant tonnage.

The major products derived from self-digestion of yeast are yeast extracts, 5' nucleotides, and the enzymes invertase and lactase. Yeast extracts are prepared primarily from brewer's yeast, and secondarily, in smaller amounts, from food yeast grown on molasses, cheese whey, wood sugars, and ethanol (32). Yeast extracts are used as condiments in the preparation of meat products, sauces, soups, gravies, cheese formulations, bakery products, vegetable products, seafoods, and seasonings. They are a source of free amino acids, vitamins of the B-complex and other nutrients desired as additives in the formulation of baby foods, health foods, and feed supplements. Yeast autolysates also serve as feed supplements, enrichments of laboratory and production media for microorganisms and other biological culture systems. Annual United States production is estimated to be around 9 million pounds (33).

5' nucleotides are produced from yeast RNA. They are used widely as flavor enhancers, usually in combination with monosodium glutamate and yeast extracts. Japan is the principal producer of yeast RNA and its derivatives.

Liquid concentrates of invertase are obtained from baker's yeast after autolysis under mild process conditions. Its main use is the candy industry to liquify the fondant in chocolates.

Lactase is a dried extract recovered from autolysed *S. lactis* grown in Europe. It is used mainly in milk products to increase sweetness, prevent lactose crystallization in whey concentrates, and to alleviate lactose intolerance in humans, poultry, and cattle.

Fermentation industries have always been in a state of flux, more or less. In the yeast business the action is livelier than elsewhere. In the past year or two, corporate, marketing and technical changes have caused some startling realignments, especially in the brewing industry, and in the baker's yeast business.

Brewers have been battling for brewing capacity and greater market penetration ever since beer became legal in 1933. Alcoholic beverage production had been restricted by the Volstead Act, enacted in 1919 to conserve cereal grains for food use. This fourteen-year "noble experiment" is known as the Prohibition Era. In 1933, 750 brewers were in business; there are 45 today. While acquisitions, mergers, and bankruptcy diminshed the number, Anheuser-Busch grew bigger and stronger, and has dominated the industry for the past 25 years. Anheuser-Busch is the world's largest brewer, and getting more of the market with sharp merchandising and expanded capacity, which, today is about 60 million barrels, or 2 billion gallons per year. In 1981, Anheuser-Busch's growth rate was four times that of the entire brewing industry (38). Anheuser-Busch and Miller Brewing Company share more than half of the beer business--30% for Anheuser-Busch, 22% for Miller.

Of the many changes in the brewing industry, the most startling ones are happening in Milwaukee, the former beer capitol of the world. Just three months ago, Schlitz, "the beer that made Milwaukee famous," closed its local brewery and headquarters, and merged its remaining five plants with The Stroh Brewing Company, based in Detroit. Thus ended 133 years of operations. The decline of Schlitz is attributed to technical changes in its top brand (Schlitz), and unwise market strategy made in the early 1970s (26). Pabst Brewing Company, also a major Milwaukee brewer, and struggling for survival, is seeking merger partners. As Schlitz and Pabst faltered, Miller Brewing Company moved up to challenge Anheuser-Busch for market supremacy with the successful introduction and promotion of light beer, the low-calorie beverage. About 15% of the beer brewed in the U.S. is light beer containing from 70 to 135 calories per 12 ounces, as compared with around 156 calories in regular beer (48). No new technologies are needed to make light beer. Brewing formulas are adjusted to achieve the "less filling" characteristic and variable flavor of light beers. However, Anheuser-Busch, the beer king, uses a "beechwood aging" step to develop and control the flavor profile for all its beers, including the light beers.

The baker's yeast industry is continuing its recovery from a sluggish period resulting from excessive production capacity, variance of plant manufacturing efficiency, depressed yeast prices, and the escalated price of molasses, the principal raw material in yeast manufacture. Currently the U.S. baker's yeast market is estimated at $225 million, and it is growing at a rate of about 2%.

Until about eighteen months ago, Fleischmann, a former division of Standard Brands, dominated the baker's yeast and supply market, with more than 50% of the business, for many decades. Competitors readjusted market share every now and then, and acquisitions eventually reduced the dozen or so yeast companies to five today. They operate twelve plants scattered across the nation: Milwaukee, St. Louis, Old Bridge and Belleville, NJ, Baltimore (2 plants), Belle Chasse, LA, Gastonia, NC, Oakland (2 plants), Bakersfield, CA, and Sumner, WA.

In December, 1980, the industry was shocked by the announcement by Fleischmann, now a subsidiary of Nabisco Brands, that it would close its three plants east of the Rocky Mountains (that meant Dallas and Chicago--Peekskill, NY had closed earlier), *and* abandon the sale of compressed yeast in the same area. The plants did close, early in 1981, but production continues in Oakland, CA and Sumner, WA, to manufacture Fleischmann's new direct instant dry yeast (24,49), a new trend in active dry yeast production (50).

With that change, a market bonanza opened up for five manufacturers in the eastern and midwestern states. These were: Red Star (Universal Foods Corp.), Budweiser (Anheuser-Busch, Ind.), Federal, and two small companies, American in Baltimore; and Dixie Yeast Co., a new venture begun in 1980 in Gastonia, NC. The next major change in the baker's yeast field occurred in April, 1981. Red Star Yeast, a division of Universal Foods Corp., based in Milwaukee, WI, acquired Federal Yeast Co. (Baltimore, MD), thereby increasing Red Star Yeast capacity about 50%, and boosting Red Star to sales Leadership (51). In a counter move, Anheuser-Busch began enlargement of its Bakersfield, CA, yeast plant (built in 1976) to more than double production capacity (52).

Soon to enter the realigned baker's yeast business is the joint venture of Kroger/Corning Glass Works. The plant under construction in Winchester, KY, will utilize Corning immobilized enzyme technology to hydrolyze lactose in cheese whey for baker's yeast production (53). The operation is expected to supply 60% of the yeast needed for nine Kroger bakeries.

More than one hundred baker's yeast plants are located around the world (Table 3). Some may produce a variety of yeast products. Among these are: Food yeasts--as nutritional food additives; autolyzed yeast and yeast extracts; wine, distiller's and brewer's yeasts which may be grown similar to baker's yeast in molasses media under highly aerobic, sugar-limiting conditions (27,28).

Distilleries have had their economic problems too. For several years they have been joining the booming wine business. Joseph E. Seagram & Sons, the world's largest producer and marketer of distilled spirits and wines, has acquired four vintners, including Paul Masson, Gold Seal, and two others. In 1981, Seagram sales of food and beverages were reported at $2.8 billion, up 7% from the previous year.

Many of the major distillers have substantial interests in the U.S. wine business (54): National Distillers markets Almaden; Brown-Forman owns Korbel; Glenmore Distillers recently acquired the new Lawrence winery established in 1979; Hiram Walker and Sons owns Callahan.

In a recent surprise move, the Peoria plant of Hiram Walker was closed. The high cost of modernization needed to meet standards for pollution control, and greater labor costs than current in the industry, were given as the reasons for moving Peoria production to Walkerville, Ontario (39).

Table 3. Baker's Yeast Production 1980 Estimates (58)

	Fresh Yeast[(a)] *Tons*
Africa	*52,000*
Asia	*145,000*
Australia	*11,000*
Eastern Europe (b)	*433,000*
Western Europe (c)	*487,000*
North and Central America	*363,000*
South America	*83,000*

(a) 28-32% dry matter

(b) Includes 297,000 tons estimated for the USSR

(c) Includes Turkey

Treatment of process waste and control of air pollution are problems for all industries. Yeast-based processes have found novel systems for meeting local and federal standards; for example: in the production of baker's yeast at one plant, the spent molasses medium, formerly spilled into the Mississippi River, is evaporated to form MYCS--molasses--yeast condensed solubles--which goes to cattle feeders. The condensate from the evaporators is combined with all other waste streams from the plant and treated in a conventional aerobic basin/trickling filter/clarifier system to complete BOD reduction.

The Bakersfield yeast plant of Anheuser-Busch, now being expanded, will dispose its high nitrogen fermentor beer on a turf farm (400 acres) for sod nutrition (52).

A new brewery also condenses plant residues to form BCS--brewers condensed solubles--used for supplementing cattle feed (55).

A soft drink bottler, faced with non-renewal of a waste water discharge permit, installed an alcohol fermentation and distillation system to convert high sugar (about 6%) waste to ethanol, which is blended with gasoline for use in the soda bottler's vehicles (56). Among the nearly four hundred known species of yeast, are a few likely candidates for commercial development and exploitation. Table 4 lists a few of the species with promise.

Table 4. Yeasts With Commercial Potential

	Reference
I. For industrial and fuel ethanol.	
A. From starch *Schwanniomyces alluvius*	
Schwanniomyces castellii	
.Both possess extracellular alpha-amylase and glucoamylase	
B. From inulin (Jerusalem artichock):	
...*Kluyveromyces fragilis*	
.Converts inulin to fructose	63,64
.Small plant at Wabuska, Nevada	
C. Direct fermentation of xylose:	
...*Pachysolen tannophilus*	65,66
.Ferments both xylose and glucose to ethanol	
...*Candida tropicalis*	67
...*Schizosaccharomyces pombe*	68
.Also deacidifies grape musts and table wines	69
D. From whey:	
...*Kluyveromyces fragilis*	70
...*Candida psudotropicalis*	71
II. Other candidates for exploitation:	
...*Candida quilliermondii*	72
.Citric acid production	
...*Phaffia rhodozyma*	73
.Accumulates astaxanthin, for poultry feed to impart skin color	
...*Saccharomyces cerevisiae* mutants	
.To increase proportion of lysine and/or methionine in yeast protein	74,76
.To produce human interferon	77

I conclude this brief perspective of the yeast business--its practices, problems and prospects--with a short account of a unique yeast-based process and product that transforms rose petals to gragrant beads (57). Dried rose petals of a dozen roses are pulverized, mixed with water and sugar to a peanut butter consistency, and allowed to ferment for two to three weeks. The mash develops the aroma of a fine wine. after fermentation, the dark brown mash is spread on platters to dry. After grinding the dried mash to fineness, the powder is moistened with enough saline solution to form a cooky-dough texture, and rolled to proper thickness on a marble slab. The beads are pressed out on a special mold, drilled for stringing on thin wire, dried and polished. The finished, shiny beads, always black regardless of the original color of the rose petals, are assembled in the proper sequence, a crucifix is added, and the rose rosary is complete.

This art was described to me by Sister Parcelaus at St. Joseph's Convent, a retirement home in Campbellsport, Wisconsin. For thirty years Sister Parcelaus and two assistants have manufactured around 1,000 rose rosaries annually, and shipped many of them to all parts of the world.

That venture is the one high-demand, yeast-essential enterprise practiced "not for profit," but for faith.

I thank the members of the Microbial and Biochemical Technology Division of ACS, and the associate sponsor, BioChem Technology, Inc., for the invitation to give the third David Perlman Memorial Lecture in commemoration of David Perlman's dedication to service shared by all of us.

REFERENCES

1. Langlykke, A.F., Smythe, C. V., and Perlman, D. In "The Enzymes" (Sumner, J. B. and Myrback, K., eds.), Vol. 2, Part 2, p. 1180, Academic Press, New York (1952).
2. Perlman, D., Brown, W.E., and Lee, S.B. *Ind. Eng. Chem. 44*, 1996 (1952).
3. Silcox, H.E. and Lee, S.B. *Ind. Eng. Chem. 40*, 1608 (1948).
4. Lee, S.B. *Ind. Eng. Chem. 41*, 1868 (1949).
5. Lee, S.B. *Ind. Eng. Chem. 42*, 1672 (1950).
6. Lee, S.B. *Ind. Eng. Chem. 43*, 1948 (1951).
7. Perlman, D., Tempel, A.E., and Brown, W.E. *Ind. Eng. Chem. 45*, 1944 (1953).
8. Perlman, D. and Kroll, C.L. *Ind. Eng. Chem. 46*, 1809 (1954).
9. Beesch, S.C. and Shull, G.M. *Ind. Eng. Chem. 47*, 1857 (1955).

10. Beesch, S.C. and Shull, G.M. *J. Commerce, Nov.*, 12 (1956).
11. Beesch, S.C. and Shull, G.M. *Ind. Eng. Chem. 49,* 1491 (1957).
12. Anderson, R.F. *Chemical Week, June 16,* 105 (1962).
13. Perlman, D. *Chemical Week 101*(25), 44 (1967).
14. Perlman, D. *Process Biochem. 4*(6), 29 (1969).
15. Perlman, D. *ASM News 39*(10), 648 (1973).
16. Perlman, D. *Chemtech 4*(4), 210 (1974).
17. Perlman, D. *Process Biochem. 10*(9), 23 (1975).
18. Perlman, D. *Chemtech 7,* 434 (1977).
19. Perlman, D. *ASM News 43*(2), 82 (1977).
20. Perlman, D. *Process Biochem. 13*(5), 3 (1978).
21. Perlman, D. *Devel. Ind. Microbiol. 21,* XV (1980).
22. Perlman, D. *Ann. Rpts. Ferm. Proc. 1,* 1 (1977).
23. Frey, C.N. *Ind. Eng. Chem. 22,* 1154 (1930).
24. *Milling and Baking News, Jan. 20,* 7 (1981).
25. Verespej, M.A. *Industry Week, May 17,* 78 (1982).
26. Taylor, A.L., Koepp, S., and Witteman, P.A. *Time, August 16,* 48 (1982).
27. Reed, G. *Food Technol. 35,* 89 (1981).
28. Reed, G. and Peppler, H.J. "yeast Technology," AVI Press, Westport, Conn. (1973).
29. Seligsohn, M. *Food Eng. 54*(2), 19 (1982).
30. *Brewers Almanac,* U.S. Brewers Assoc., Inc. Washington, DC (1981).
31. Peppler, H.J. *Ann. Rpts. Ferm. Ind. 2,* 191 (1978).
32. Peppler, H.J. In "Microbial Technology," (Peppler, H.J. and Perlman, D., eds.), Second Edition, Vol. 1 (1979).
33. Peppler, H.J. In "Economic Microbiology," (Rose, A.H., ed.) 7, 293, Academic Press, London (1982).
34. Peppler, H.J. In "The Yeasts," (Rose, A.H. and Harrison, J.S., eds.), Vol. 3, Academic Press, London (1976).
35. Distillers Spirits Council of the U.S. (DISCUS), Washington, DC (1981).
36. Chaney, D. *Bakery Production and Marketing 17*(5), 83 (1982); *16*(5), 120 (1981); *15*(6), 95 (1980).
37. *Wines and Vines 83*(6), 18 (1982).
38. *Food Eng. 54*(8), 72 (1982).
39. "The Liquor Handbook," Gavin-Jobson Assoc., Inc., New York (1981).
40. Bulletin de L'office Inter. Vigne Vin, p. 740 (1981).
41. "The Wine Marketing Handbook," Gavin-Jobson Assoc., Inc., New York (1981).
42. Lockwood, C. *Passages, Sept.,* 47 '1982).
43. *Fuel Alcohol 4*(5), 4 (1982).
44. Jenkins, D.M. *Cereal Foods World 26,* 612 (1981).
45. *Food Eng. 48*(2), 36 (1976).
46. Schwarz, K. *Med. Clin. North Am. 60*(4), 745 (1976).
47. Mertz, W. *Science 213,* 1332 (1981).

48. Byers, S. *The Milwaukee J., Insight, June 6,* 31 (1982).
49. *Prepared Foods 151*(8), 104 (1982).
50. Saussele, H. *Food Eng. 54*(5), 129 (1982).
51. Kenney, R. *The Milwaukee J., April 10* (1981).
52. *Food Processing 43*(7), 50 (1982).
53. *Food Development, Jan.,* 34 (1982).
54. *Wines and Vines 82*(12), 21 (1981).
55. Watner, J.N. *Food Eng. 54*(7), 88 (1982).
56. Rainwater, C. and Swientek, R.J. *Food Processing 43*(10), 74 (1982).
57. Stoffel, R. *Catholic Digest 37*(12), 77 (1973).
58. Oura, E., Suomalainen, H., and Viskari, R. In "Economic Microbiology," (Rose, A.H., ed.), *7*, 87; Academic Press, London (1982).
59. Spencer-Martins, I. and Van Uden, N. *European J. Appl. Microbiol. 4,* 29 (1977).
60. Oteng-Gyand, K., Moulin, G., and Galzy, P. *European J. Appl. Microbiol. Biotechnol. 9,* 129 (1980).
61. Wilson, J.J. and Ingledew, W.M. *Appl. Environ. Microbiol. 44,* 301 (1982).
62. Wilson, J.J., Khachatourians, G.G., and Ingledew, W.M. *Biotechnol. Letters 4*(5), 333 (1982).
63. *Fuel Alcohol U.S.A. 4*(3), 21 (1982).
64. Kim, W.Y. and Byun, S.M. *Enzyme Microbiol. Technol. 4,* 239 (1982).
65. Maleszka, R., Veliky, I.A., and Schneider, H. *Biotechnol. Letters 3,* 89, 415 (1981).
66. Mayberry, D. *Agric. Res. 31*(1-2), 8 (1982).
67. Jeffries, T.W. *Biotechnol. Letters 3,* 213 (1981).
68. Ueng, P.P., Hunter, C.A., Gong, C.S., and Tsao, G.T. *Biotechnol. Letters 3,* 315 (1981).
69. Yang, H.Y. *Amer. J. Enol. Viticult. 24*(1), 1 (1973).
70. *Chem. & Eng. News 60*(37), 7 (1982); *Food Eng. 54*(10), 105 (1982).
71. Bailey, R.B., Benitez, T., and Woodward, A. *Appl. Environ. Microbiol. 44,* 631 (1982).
72. *Federal Register,* March 5 (1976).
73. Johnson, E.A. and Lewis, M.J. *J. Gen. Microbiol. 115,* 173 (1979).
74. Momose, H. and Gregory, K.F. *Appl. Environ. Microbiol. 35,* 641 (1978).
75. Taradejna, D. and Leinen, N.J. *Food Processing 42*(12), 28 (1981).
76. Tanner, R.D., Souki, N.T., and Russell, R.M. *Biotechnol. Bioeng. 19,* 27 (1977).

CHAPTER 10

RECENT ADVANCES IN D-XYLOSE CONVERSION BY YEASTS

Cheng S. Gong

Purdue University
West Lafayette, Indiana

I. INTRODUCTION

Next to D-glucose, D-xylose is the most abundant carbohydrate derived from woods, agricultural residues, and other plant materials. D-Xylose along with L-arabinose and other neutral carbohydrates reside in the hemicellulose portion of plant cell walls. Hemicellulose often exists as an integral part of branched-chain heteropolysaccharides consisting of mixed hexosans and pentosans. The pentosan portion is comprised of D-xylose and L-arabinose and the hexosan constituents are glucose, mannose and galactose (1-5).

Unlike the orderly crystalline structure of cellulose, hemicellulose exhibits variability in both structure and in sugar constituents. The degree of polymerization of hemicellulose is usually less than 200 (1,2) whereas native cellulose has a degree of polymerization greater than 1,000 (6). Since hemicellulose and cellulose exist in nature in polymeric form, hydrolysis yielding constituent sugars is often required before they can be utilized by a biological system.

Cellulose is more difficult to convert to its monomeric form (glucose) than hemicellulose. Its physical structure and configuration renders its resistance to many hydrolysis treatments (6). Hemicellulose is more easily converted to its constituents. The hydrolysis of hemicellulose can be accomplished either physically, chemically or enzymatically. The predominant components derived from hemicellulose upon hydrolysis are D-xylose, L-arabinose and D-glucose. Other minor carbohydrate products include mannose, galactose, and glucouronic acid. Since hemicellulose comprises up to 40%

ISBN 0-12-040306-4

of the cell wall material, it is estimated that between 10 to 30 percent of the carbohydrate derived from renewable biomass is xylose (7,8). Thus, xylose along with glucose are potential substrates for conversion to useful products such as ethanol.

Ethanol can be produced from carbohydrates by a wide variety of prokaryotic and eukaryotic microorganisms. Yeasts, in general, are a better choice than other microorganisms for fermentation of carbohydrates to ethanol due to their high fermentation rates, ethanol tolerance and high ethanol yields. In the design of a simple industrial process to produce ethanol from biomass-derived carbohydrates, a biological system that could ferment both pentoses and hexoses simultaneously to produce high yields of ethanol with high rates of conversion is essential. As a general rule, yeasts do not convert pentoses to ethanol as efficiently as they do hexoses. This is primarily due to the enzyme systems in yeasts. Normally pentoses are assimilated into pathways which lead to the production of pentitols.

There have been relatively few studies concerning the utilization of pentoses to useful products by yeasts. This lack of interest may be ascribed to the fact that for many years xylose has not been considered a good fermentation substrate. The recent global energy situation has stimulated a great deal of interest by research organizations to examine more closely the potential of conversion of xylose to ethanol.

In this review, the recent developments in the area of xylose conversion to ethanol by yeasts is described.

II. PENTOSE METABOLISM

Although the metabolism of pentoses in bacteria is well documented, little information is presently available concerning the utilization of these sugars in eukaryotic microorganisms.

In bacteria, such as *Aerobacter aerogenes* (9,10), *Escherichia coli*, (11), and *Salmonella typhimurium* (12,13), the early steps of xylose metabolism involved inducible enzymes (i.e., xylose transport enzymes, xylose isomerase, and xylulokinase). The direct isomerization of xylose to xylulose is the first step in xylose metabolism. Xylulose thus formed is then phosphorylated to xylulose-5-phosphate, the key metabolic intermediate (Fig. 1). Further metabolism of xylulose-5-phosphate involves different metabolic routes that vary from one type of bacterium to another resulting in the production of various types of metabolic products (for recent reviews see Rosenberg (14) and Zeikus (15)). In some bacteria, xylulose-5-phosphate is cleaved to form a 2-carbon

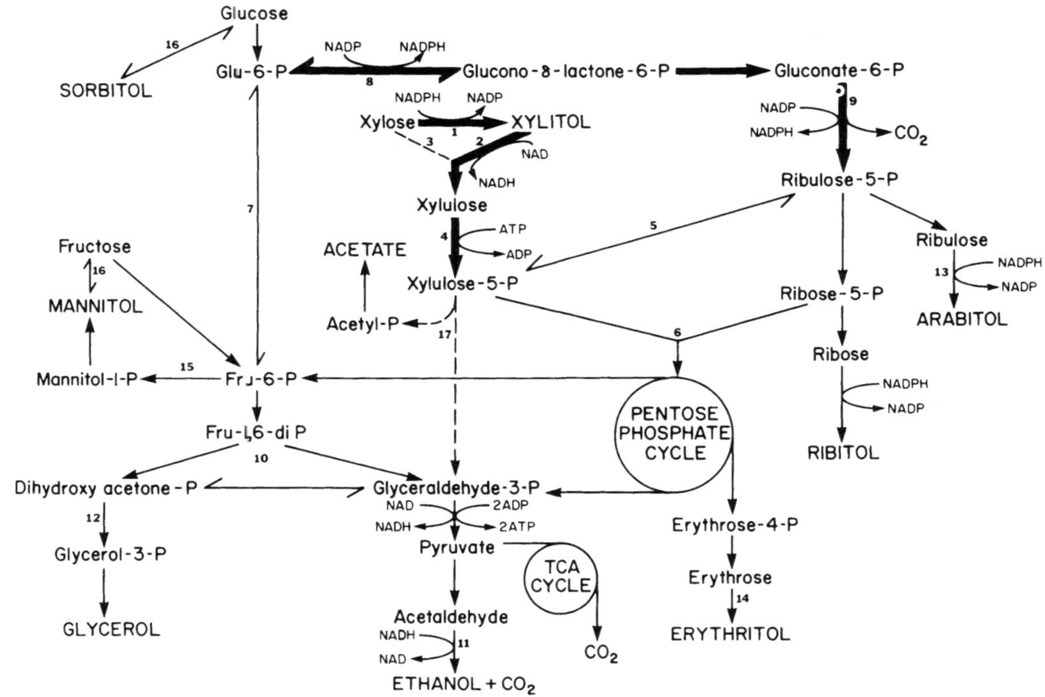

Figure 1. D-Xylose and D-glucose metabolism by yeasts. 1. xylose reductase (aldoreductase); 2. xylitol dehydrogenase (xylulose reductase); 3. xylose isomerase; 4. xylulokinase; 5. phosphoketopentoepimerase; 6. transaldolase and transketolas; 7. phosphohexoisomerase; 8. glucose-6-phosphate dehydrogenase; 9. phosphogluconate dehydrogenase; 10. aldolase; 11. alcohol dehydrogenase; 12. glycerol-3-phosphate dehydrogenase; 13. arabitol dehydrogenase; 14. erythritol dehydrogenase; 15. mannitol-1-phosphate oxidoreductase; 16. polyol dehydrogenase; 17. xylulose-5-phosphate phosphoketolase.

and a 3-carbon intermediate that are the precursors of a 2-carbon (ethanol) and a 3-carbon (lactate) product. In other bacteria, the pentose phosphate and Embden-Meyerhof pathways are employed to convert xylulose-5-phosphate to pyruvate.

In the study of xylose metabolism by *S. typhimurium*, Shamanna and Sanderson (12) observed the sequential appearance of xylose transport enzyme, xylose isomerase and xylulokinase. In addition, xylose negative mutants lacking one or the other of these enzymes were isolated. These mutants fall into five categories (13), those lacking all three enzymes; those lacking xylose isomerase; those lacking xylulokinase; those lacking xylose isomerase and xylulokinase; and one partially deficient in xylose transport enzyme. These studies confirmed that in *S. typhimurium*, xylose catabolism proceeds via xylose to xylulose to xylulose-5-phosphate and that at least four genes govern the early stages of xylose catabolism.

The metabolism of other pentoses in most bacteria also involves the isomerazation step, L-arabinose is isomerized to L-ribulose, L-xylose to L-xylulose and D-arabinose to D-ribulose (9). D-Ribose is metabolized differently; however, the direct phosphorylation of D-ribose to D-ribose-5-phosphate is believed to be first step in the metabolism of this sugar by *E. coli* (16).

In some bacteria, such as *Enterobacter, Corynebacterium* and *Brevibacterium*, the reduction of xylose to xylitol catalyzed by NADPH-dependent xylose reductase has been reported to be the initial step in xylose metabolism (17,18). Reduced pyridine nucleotides required for the reduction of xylose are believed to be generated from the oxidation of 6-phosphogluconate by NADP-dependent 6-phosphogluconate dehydrogenase (18). Further metabolism involves oxidation of xylitol to xylulose followed by phosphorylation of xylulose to xylulose-5-phosphate (Fig. 1).

For the metabolism of pentitols in *Aerobacter aerogenes* and many other bacteria, a family of pentitol dehydrogenases are involved in the oxidation of pentitols to their corresponding pentuloses such as xylitol to xylulose, ribitol to ribulose and arabitol to xylulose (9,19). In *Lactobacillus casei*, phosphotransferase was reported to be involved in converting xylitol to xylitol-5-phosphate and ribitol to ribitol-5-phosphate (20).

In eukaryotic microorganisms, the reduction of xylose to xylitol catalyzed by a NADPH-dependent xylose reductase is believed to be the obligatory step for xylose metabolism (21-23). Xylitol thus formed is oxidized to xylulose followed by phosphorylation of xylulose to xylulose-5-phosphate, the key metabolic intermediate. Thus, xylose reductase and xylitol dehydrogenase occupy the strategic positions of xylose metabolism in eukaryotic microorganisms. This is a major important difference between eukaryotic and prokaryotic microorganisms with respect to xylose metabolism.

Our current understanding of the xylose metabolic routes employed by yeasts and mycelial fungi is limited. With the aids of pure xylulose as a substrate and exagenous xylose isomerase, Gong, et al. (24), postulated that yeasts employ a combination of the pentose phosphate and Embeden-Meyerhof pathways to convert xylose to pyruvate and then ultimately to ethanol,glycerol and CO_2.

Some of the enzymes involved in xylose metabolism by yeasts and mycelial fungi that have been studied include xylose reductase, polyol dehydrogenase (21,22,25-35,38), xylose isomerase (36,37), and xylulokinase (22,37,39).

The possible involvement of xylose isomerase in xylose metabolism by yeast has been reported. Tomoyeda and Horitsu (36) purified xylose isomerase from xylose grown *C. utilis*. The enzyme was specific for D-xylose and L-xylose but not for D-arabinose or L-arabinose. It required a bivalent cation. Mn^{++} was most effective followed by Co^{++} and Mg^{++}. Optimal pH and temperature were 6 to 7 and 70°C, respectively. Thus, the properties of this enzyme are similar to those found in bacteria. Höfer, et al. (37), also observed xylose isomerase activity in xylose grown cell-free extracts of an obligatory aerobic yeast, *Rhodotorula gracilis*. They concluded that the direct isomerization of xylose to xylulose catalyzed by xylose isomerase is an obligatory step in this specific yeast for xylose metabolism. In contrast, Chakravorty, et al. (22), failed to detect any xylose isomerase activity in xylose grown cell-free extract of *C. utilis*. In addition, kinase activity in *C. utilis* for xylose was lacking, but not for xylulose kinase. Similar observations were also reported by Chiang and Knight (21), and McCracken and Gong (34). Xylose isomerase activity has not been detected in many yeasts species or in xylose-grown mycelial molds such as *Penicillium, Aspergillus, Fusarium* and *Monilia*. Thus, the definite involvement of xylose isomerase in xylose metabolism by yeasts and mycelial fungi remains to be proven. Further examination of this possibility should be undertaken.

The existence of high activities of pentose phosphate pathway in some yeasts has been demonstrated (39,40), thus enabling them to convert xylulose-5-phosphate through epimerization, transketolation and transaldolation into intermediates of the hexose diphosphate catabolic pathway. The study of the regulation of the pentose phosphate pathway in yeasts during the metabolism of xylose has not been studied. The metabolism of glucose and the regulation of the flow of carbon through various pathways including pentose phosphate pathway has been the subject of intensive study since the early work of Pasteur.

The most intensively studied yeast, *S. cerevisiae,* is known to metabolize glucose primarily through Embden-Meyerhof pathway. However, C^{14}-labeled glucose utilization studies showed that approximately 12% of glucose taken up by aerobically growing yeast is catabolized through the pentose phosphate pathway (41-43). Similar examinations of glucose uptake by resting cells of *S. cerevisiae* also indicate that anywhere from zero to 17% of the glucose is metabolized via the pentose phosphate pathway (44,45). Of the three yeasts, *C. utilis, C. albicans,* and *R. gracilis,* that have been examined for pentose phosphate pathway activity in the catabolism of glucose, *C. utilis* was found to catabolize 35% of glucose via this pathway (39), *C. albicans,* 50% (40), and *R. gracilis,* 100% (39).

It is known that the oxidation of NADPH causes an immediate stimulation of the pentose phosphate pathway in yeasts (46,47). Thus, it is likely that the rate of xylose reduction and activation of the pentose phosphate pathway are controlled by the availability of NADP and NADPH. The generation of NADP through reduction of xylose to xylitol may stimulate the activity of glucose-6-phosphate dehydrogenase. This in turn could stimulate the activation of the pentose phosphate pathway. Jeffries (48) has suggested that the metabolism of xylose in yeasts is probably operated in a coordinated close cycle. In this mode, the regeneration of NADPH required for the reduction of xylose to xylitol is insured. Thus, the metabolism of xylose in yeast through pentose phosphate pathway and subsequent oxidation of hexose phosphate can provide two moles of NADPH per mole of CO_2 released.

Thus, for every mole of xylose metabolized to CO_2, 10 moles of NADPH can be generated. This in turn supplies the reduced NADP to convert 10 moles of xylose to produce 10 moles of xylulose through xylitol as intermediate.

Through this metabolic route, 90% of the theoretical ethanol yield can be expected from xylose as compared to that which can be expected from either glucose or xylulose. The overall theoretical ethanol yields from glucose, xylose, and xylulose are:

$$1 \text{ Glucose} \longrightarrow 2 \text{ Ethanol} + 2\ CO_2$$

$$6 \text{ Xylose} \longrightarrow 9 \text{ Ethanol} + 12\ CO_2$$

$$3 \text{ Xylulose} \longrightarrow 5 \text{ Ethanol} + 5\ CO_2$$

Osmond and ApRees (46) studied the control of the pentose phosphate pathway in *C. utilis* and suggested that the capacity of pentose phosphate pathway in yeast may be controlled by variation in the amounts of glucose-6-phosphate dehydrogenase. This is especially interesting in that during the first step of xylose metabolism NADPH is required and this NADPH can be regenerated from the oxidation of glucose-6-phosphate to glucose-δ-lactone-6-phosphate (Fig. 1).

For the subsequent catabolism of pyruvate via the TCA cycle or the fermentative pathway, two distinct regulatory mechanisms used by yeasts for determining the flow of the carbons into the two pathways have been observed. These regulatory effects are known as Pasteur and Crabtree effects.

Regulation under the Pasteur effect can be described as follows. Under aerobic conditions, yeasts utilize the "energy-rich" TCA cycle, electron transport and the oxidative phosphorylation mechanism and must regulate the quantity of glucose metabolized to limit the flow of energy into the cells. Under anaerobic conditions, electron transport and oxidative phosphorylation are not operating, thus, effectively shutting down the flow of carbons through the TCA cycle. Instead, the carbons flow through the fermentative pathway providing an alternative "energy-poor" route for the oxidation of NADH (for review, see Sols, et al. (49)).

The other regulatory mechanism, Crabtree effect, is observed when high levels of glucose repress respiration under aerobic conditions. In the presence of high concentrations of glucose, the formation of respiratory enzymes is repressed, thus, the carbons flow through the fermentative pathway even in the presence of high oxygen levels. This effect has been extensively studied (50-53) and has been generally accepted as a primary regulatory mechanism in *S. cerevisiae*. Recently, however, Barford, et al. (54), have suggested that the Crabtree effect is simply a long term transient stage of respiratory adaptation.

Under aerobic conditions many yeasts have a high potential to produce polyhydric alcohols as metabolic by-products from glucose and xylose. Glycerol, erythritol, D-arabitol and mannitol are the common polyols produced by yeasts from glucose (55). Xylitol, glycerol and D-arabitol are produced from xylose (56,57). L-Arabitol and ribitol are derived from

L-arabinose and ribose (56), respectively. The type and amounts of polyols produced by yeasts are affected by environmental factors as well as by the type of yeast strains employed (55).

Xylitol and L-arabitol are resulted from the reduction of xylose and L-arabinose catalyzed by aldoreductase. D-Arabitol arises by dephosphorylation of D-ribulose-5-phosphate and subsequent reduction of D-ribulose by a $NADPH_2$-linked D-arabitol dehydrogenase (58,59). Erythritol is probably derived from erythrose-4-phosphate, an intermediate of the pentose phosphate pathway. Mannitol is derived from fructose-6-phosphate through the reduction of fructose-1-phosphate to mannitol-1-phosphate catalyzed by oxidoreductase and subsequent hydrolysis of the resulting mannitol-1-phosphate to mannitol by mannitol-1-phosphatase (60-64). Glycerol is derived from the reduction of dihydroxyacetone phosphate to glycerol-3-phosphate followed by the dephosphorylation of glycerol-3-phosphate.

Other products from xylose metabolism by yeasts include acetic acid (48), citric acid (56), D-glycero-D-idoheptitol (65) and xylonic acid (66).

III. CONVERSION OF D-XYLOSE

Many yeast species utilize xylose as a carbon and energy source. Barnett surveyed a total of 434 strains of yeasts including those commonly used for ethanol fermentation in industries. About half of the yeasts surveyed (214 strains) utilized and metabolized xylose aerobically and non utilized it anaerobically (67). Onishi and Suzuki (56) in their study of 58 yeast strains with respect to aerobic xylose utilization, demonstrated that xylitol is the major product derived from D-xylose. Among the xylose utilizing yeasts, *Candida polymorpha, C. tropicalis, C. guilliermondii, Pichia miso, Hansenulla subpelliculosa, H. anomala, Torulopsis famata,* and *Monilia sitophila* are good xylitol producers from xylose. Other yeast genera such as, *Trichosporon, Cryptococcus* and *Debaryomyces* produce little or no xylitol in spite of their ability to utilize xylose.

Gong, et al. (57), studied xylose utilization by yeasts under both aerobic and oxygen-limited fermentative conditions and confirmed that xylitol is the major metabolic product by yeasts. Smaller quantities of ethanol, glycerol and arabitol are also produced.

Ethanol was not recognized as a metabolic product from xylose by yeasts until recently. Gong, et al. (24), reported that a small quantity of ethanol is produced from xylose by *C. diddensii, C. utilis, S. diastaticus, Saccharomycopsis fibuligera* and *Torula Candida* even though xylitol is the major product formed by those yeasts. This is the first demonstration that yeasts possess the ability to produce ethanol from xylose. Further examination of yeasts by Gong, et al., revealed that a wide range of yeasts including those belonging to the genera, *Saccharomyces* and *Schizosaccharomyces* are able to produce ethanol from xylose (57), even though they are unable to utilize xylose as a carbon and energy source effectively (69).

An unusual yeast, *Pachysolen tannophilus*, originally isolated from wood sulfur liquor by Wickerham (70) was recently reported to produce significant quantities of ethanol from xylose, especially in the presence of oxygen (71). A similar observation was also reported by Slininger, et al. (72). In their brief report, Schneider, et al. (71), described ethanol production by *P. tannophilus* NRRL Y-2460 from xylose when oxygen was present. More systematic study of this yeast strain with respect to xylose conversion to ethanol was reported by Slininger, et al. (72). From their study, they concluded that oxygen was required for cell growth but not for ethanol production. In summary, ethanol production by *P. tannophilus* was known to possess the following characteristics (72):

1. Oxygen is required for cell growth but not for ethanol production from xylose.
2. A lag period of 20 hrs after the onset of cell growth occurs before ethanol production can be detected.
3. Ethanol yield is 0.34 g ethanol per g xylose consumed (60% of the theoretical value).
4. Optimal temperature for cell growth and ethanol production is 32°C.
5. Optimal pH for cell growth is 2.5-4.5 and optimal pH for ethanol production is 2.5.
6. Ethanol production is subject to substrate (55 g/l xylose) and product (30 g/l ethanol) inhibition.

The rate of ethanol formation is slow, three days of incubation were required to reach a final ethanol yield of 0.5-1% from a 5% xylose solution, and a yield of 0.34 g of ethanol per gm of xylose consumed (72). Immobilized *P. tannophilus* was employed in order to overcome this slow rate of ethanol production. The highest concentration of ethanol produced by immobilized yeast was around 0.8% after 3 days of operation (73). The slow rate of ethanol production from xylose and the low yield is, in part, due to the slow initial growth rate, and asci formation during the course of incubation, and the formation of other products such as xylitol, arabitol and acetic acid.

A. *Yeast Mutants*

Yeast mutants with enhanced ethanol production capabilities from xylose have been reported. A UV-induced mutant derived from a high-xylitol producing yeast strain, *Candida sp.* was reported to produce nearly five-times as much ethanol as did the parent strain (74). The parent strain, *Candida sp.* C 2, produces xylitol from xylose as the major product (Fig. 2A). The mutant, *Candida sp.* XF217, produces predominantly ethanol at the expense of xylitol production (Fig. 2D). Several other mutants exhibiting increased ethanol production in conjunction with decreased xylitol production have also been isolated. In one mutant, *Candida sp.* C26, arabitol was produced instead of xylitol (Fig. 2E) (57).

Candida sp. XF217 produces ethanol from xylose aerobically as well as under oxygen-limited fermentative conditions (Fig. 3). The rate of ethanol production under aerobic incubation conditions is greater, indicating that there is an oxygen requirement for rapid rates of xylose uptake and ethanol production. Xylulose is a better substrate for ethanol production by this yeast than xylose under fermentative conditions (34). When xylose isomerase was present with xylose under fermentative conditions, the rates of xylose consumption and ethanol production by *Candida sp.* XF217 were similar to those found under aerobic conditions.

The results of experiments using *Candida sp.* XF217 to ferment xylose to ethanol indicate the following characteristics:

1. Xylose is consumed by XF217 under either aerobic or fermentative incubation conditions resulting in the production of ethanol as the major metabolic product.
2. The presence of oxygen is required for cell growth and for xylose consumption to produce ethanol (75).
3. In addition to the major product, ethanol, xylitol, arabitol and glycerol are also produced.
4. The product, ethanol, could be utilized by the yeast under aerobic incubation conditions after xylose was consumed.
5. When xylose isomerase was present, yeasts utilized xylose under fermentative conditions at a rate similar to that under aerobic conditions.

Enzymatic studies of xylose metabolism by *Candida sp.* C2 and XF217 indicate that the synthesis of xylose reductase, xylitol dehydrogenase and xylulokinase are enhanced by the presence of xylose (34). The specific enzyme activities of xylitol dehydrogenase and xylulokinase increased significantly over those of the parent strain, *Candida sp.* C2. The activity

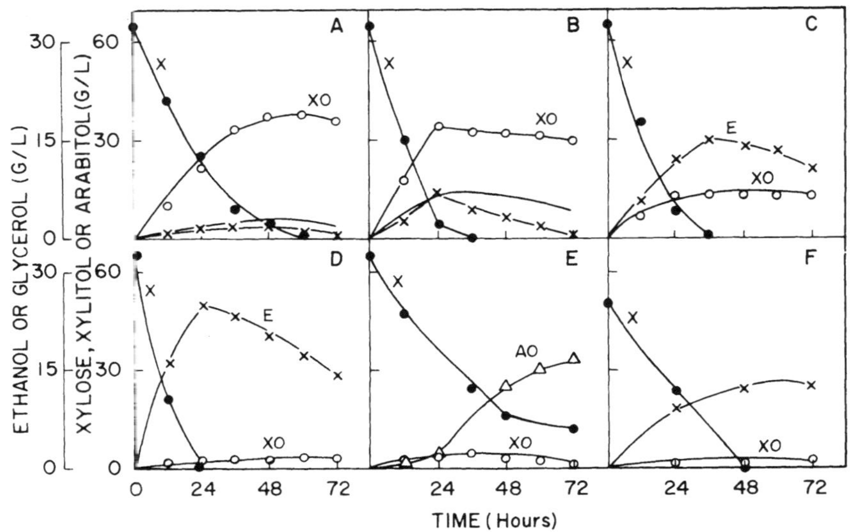

Figure 2. Aerobic conversion of D-xylose by Candida sp. C2 and mutants. A. wild-type, C2; B. C210; C. C28; D. XF217; E. C26; F. 217-Pl. Incubation was carried out at 30°C with an initial pH of 5.6. The initial cell density was 1 x 10^8 cells per ml, and the initial D-xylose concentration was 65 g/l. X, xylose; Xo, xylitol; Ao, arabitol; E. ethanol. For detail see McCracken and Gong (34).

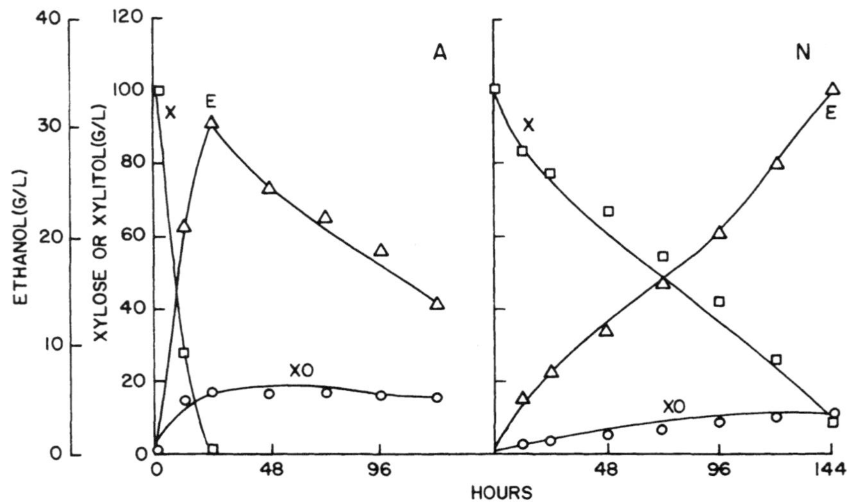

Figure 3. Aerobic and fermentative conversion of D-xylose by Candida sp. XF217. Incubation was carried out at 30°C with an initial D-xylose concentration of 10% (w/v). A, aerobic; N, fermentative; X, xylose; E, ethanol; Xo, xylitol.

of D-xylose reductase remained the same in both strains (Table 1). The increase in enzyme activities of xylitol dehydrogenase and xylulokinase enables XF217 to shift from xylitol production to ethanol production (see Fig. 2A and D). Thus, instead of excretion of xylitol as a final product, more xylitol is converted to xylulose and ultimately to ethanol.

Table 1. D-Xylose Metabolic Enzyme Activities in Candida Sp. C2 and XF217.

Enzyme	Specific Activity* C2	XF217
Xylose Isomerase	0	0
Xylose Reductase	34	24
Xylitol Dehydrogenase	9	17
Xylulose Kinase	4	15

Specific activity: 0.1 unit optical density (A_{340} nm) changed per minute reaction per mg protein at 30°C (34).

The production of arabitol from xylose by the mutant, *Candida sp.* C26, (Fig. 2E), is interesting in that ribulose-5-phosphate is the precursor of arabitol (59), and ribulose-5-phosphate can be derived from xylulose-5-phosphate by phosphoketopentoepimerase (D xylulose-5-phosphate-β-epimerase). *Candida sp.* C26 produces significant quantities of arabitol indicating that this mutant could be high in phosphoketo pentoepimerase, D-ribulose-5-phosphate or arabitol dehydrogenase.

B. Mutants from Yeasts Other than Candida Yeast

Yeast mutants from genera other than *Candida* with enhanced abilities to produce ethanol have also been reported. *S. cerevisiae* SC138 derived from a xylose-utilizing strain of *S. cerevisiae* also exhibited enhanced ethanol production from xylose. Rates of xylose utilization and ethanol production were greater when oxygen was more accessible (57). Detailed information about this yeast mutant is unavilable at this time.

IV. CONVERSION OF D-XYLULOSE

Onishi and Suzuki's (76) examination of xylulose utilization by 128 strains of yeast indicate that xylulose is a good substrate for yeasts. Under aerobic incubation conditions, polyols are the major products. The yeasts studied fall into three major classes which are based on the products formed: those that produce xylitol, those that produce D-arabitol and xylitol, and those that produce D-arabitol.

The production of ethanol from xylulose by yeasts was not recognized until recently. In their brief report, Wang, et al. (77), observed that ethanol was produced from xylulose by *S. pombe, S. cerevisiae, S. amucae,* and *Kluyveromyces lactis.* Prolonged periods of incubation were required to achieve maximal ethanol production. The rate of ethanol production from xylulose by *S. pombe* is much slower than that from glucose. This report is particularly significant in that it represents the first published report of the conversion of a pentose to ethanol by yeasts.

A more detailed report of conversion of xylulose to ethanol was later published by Gong, et al. (24). They observed that many yeasts that convert glucose to ethanol, utilized xylulose readily to produce ethanol and xylitol as metabolic products. The relative amounts of ethanol and xylitol produced varied depending on the type of yeast used. In general, more xylitol was produced by yeasts that utilized xylose aerobically. In addition, small amounts of xylitol were produced from xylulose by strains of *S. cerevisiae* that normally do not utilize xylose effectively. It was also observed that the ethanol yields resulting from yeast fermentations of xylose supplemented with glucose isomerase were the same as those achieved when D-xylulose alone was used as the initial substrate.

Ueng, et al. (78), examined the abilities of forty yeast strains belonging to the genera *Saccharomyces, Candida,* and *Schizosaccharomyces* to utilize xylulose and found that yeast strains that utilize glucose under fermentative conditions to produce ethanol also produced ethanol from xylulose under the same conditions. Fermentation products included ethanol, xylitol and arabitol. Among the yeasts examined, *S. pombe, S. cerevisiae* ATCC 24553, *C. melibiosica* and *C. mogii* were shown to be good ethanol producers from xylulose.

In a survey of 35 randomly-chosen microorganisms, Jeffries (79) found that many yeast species are capable of fermenting xylulose to ethanol. *C. tropicalis* and *S. pombe* utilized xylulose at good rates without an initial lag. Strains of *S. cerevisiae* fermented xylulose at much slower rates.

In a broad survey of the abilities of yeast strains to utilize xylose, xylulose, and glucose aerobically as well as fermentatively, Gong, et al. (57), found that xylulose, in general, is a good fermentation substrate for yeasts. Different strains of yeasts exhibited considerable variation in their rates of xylulose assimilation. The type and relative amounts of products varied depending on the incubation conditions and yeast strains employed. Four commonly observed products were ethanol, xylitol, glycerol and arabitol.

The ability of microorganisms to utilize xylulose is not restricted to yeasts. A diverse group of mycelial fungi including *Aspergillus, Fusarium, Monilia, Mucor* and *Rhizopus* also utilize xylulose to produce ethanol (80). The ability of prokaryotes to utilize xylose has not been demonstrated. Jeffries (79) reported that *Zymomonas mobilis,* a good ethanol producer from glucose and sucrose (81), is not able to utilize xylulose. He attributed this to its lack of transketolase and transaldolase.

A. *Factors Affecting the Conversion of Xylulose by Yeasts*

Wang and Schneider (82) examined the aerobic growth of yeasts on xylulose. These yeasts include six strains capable of utilizing xylose for growth. All six were able to use xylulose for growth. Of the seven strains of yeast that were incapable of growing on xylose, four strains grew on xylulose. These included two strains of *S. cerevisiae,* one strain of *S. carlsbergensis* and one strain of *S. rouxii.* The inability of a respiratory deficient petite mutant to grow on xylulose indicates the possible need for an active respiratory pathway to provide energy for xylulose assimilation.

The rate of xylulose utilization and product formation varies depending upon the strains and species of yeasts employed. *S. cerevisiae,* a typical hexose-fermenting yeast, assimilates xylulose at a much slower rate than glucose (24). When xylose is the sole carbon and energy source, the majority of strains of *S. cerevisiae* exhibit poor growth even after prolonged incubation. When xylulose is the carbon source, *S. cerevisiae* shows relatively good growth. Similar phenomena are also observed in *S. pombe* and in other species of *Saccharomyces* (82).

The rates of xylulose utilization and ethanol production are generally slower than those of glucose in *S. cerevisiae* whereas in *S. pombe* the reverse is true (82). The reason for slower glucose fermentation rates in *S. pombe* is not known. The reduction of xylulose to xylitol by polyol dehydrogenase in *S. pombe* does not occur when xylulose is the

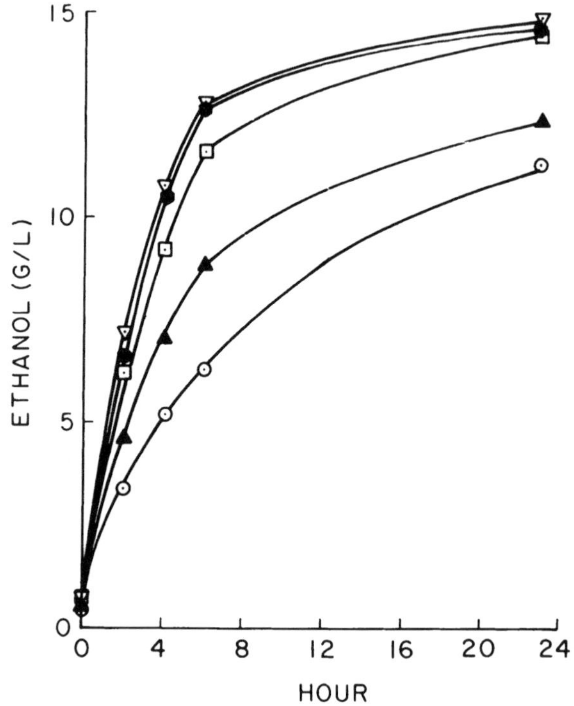

Figure 4. Effect of yeast cell densities on the production of ethanol from xylulose by S. cerevisiae. ⊙ *, 50 g/l;* ▲ *, 100 g/liter;* ⊡ *, 150 g/liter;* ⬟ *, 200 g/liter;* ▽ *, 250 g/liter. Temperature, 37°C; pH, 6. For detail see Chiang, et al. (83).*

substrate. Instead the xylulose is phosphorylated to xylulose-5-phosphate by xylulokinase and enters pathways described in Section II.

Factors affecting ethanol production from xylulose by yeasts have not been studied extensively. In this article, they will be discussed only briefly. Much work in this area is currently underway in our laboratory.

Chiang, et al. (83), conducted a comprehensive study of the factors affecting the conversion of xylulose to ethanol by an active dry yeast cake (*S. cerevisiae*, Universal Food Co.). The factors studied include yeast cell density, temperature, pH, substrate composition, substrate concentrations and the effect of by-products.

In a typical alcohol fermentation by yeasts, the cell density is one of the more important factors determining the rate of substrate utilization and product formation. Higher initial cell densities often result in higher rates of ethanol production. In the case of *S. cerevisiae*, the rates of xylulose utilization and ethanol formation are much slower than glucose. This slow rate of ethanol production can be increased by increasing the initial cell density (Fig. 4). The rate of ethanol production from xylulose was linear over a concentration range from 50 mg to 150 mg of cells per ml. Ethanol production reached its maximum at 150 mg of cells per ml. The specific ethanol production rate was approximately 0.05 g/l hr. This is considerably lower than that in a typical fermentation of glucose.

The optimum pH for xylulose fermentation lies in the region of 4.0-6.0. At pH 7, the initial fermentation rate and the final ethanol yields were 75 and 67% of the values obtained at pH 4, respectively. The reduction in ethanol yields at high pH is probably due to the enhancement of xylitol production at pH values greater than 5.

Temperature has a profound effect on all aspects of growth, metabolism, viability and fermentation capability of yeasts. The optimal temperature for fermentation is usually higher than the optimal growth temperature. As is the case for glucose fermentation, this yeast ferments xylulose faster at temperatures higher than its maximum growth temperature, whereas at temperatures lower than its maximal growth temperature, the capacity to ferment lasts longer. This is because continuous yeast growth is required for fermentation. The effect of temperature on activity as shown in Fig. 5 indicates that the optimal temperature for xylulose fermentation by *S. cerevisiae* is 35°C.

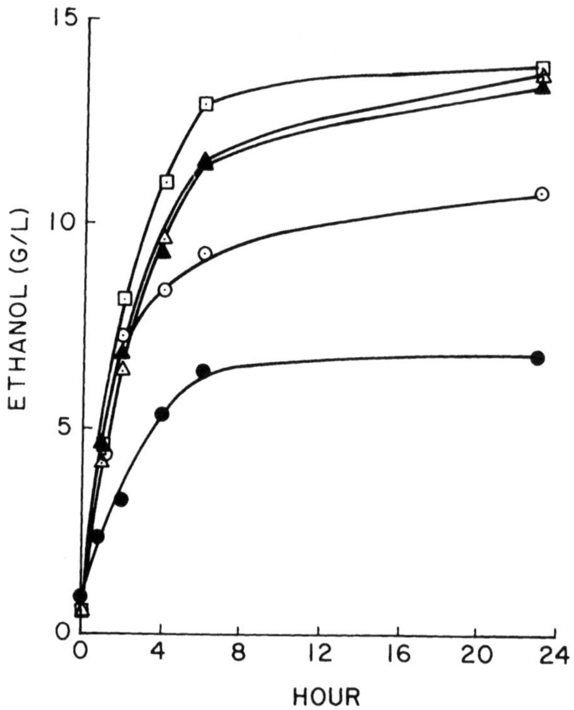

Figure 5. Effect of temperature on the production of ethanol from xylulose by S. cerevisiae. ⊙, 30°C; ⊡ , 35°C; ▲, 37°C; △ , 40°C; ● , 45°C. pH, 6; cell density, 75 g/liter. For detail see Chiang, et al. (83).

The effect of xylulose concentration on ethanol production rate is shown in Fig. 6 indicating that an increase in xylulose concentration results in an increased initial ethanol production rate.

B. Characteristics of Xylulose Conversion by Yeasts

Several characteristics are now established with respect to utilization of xylulose by yeasts:

1. Xylulose is readily fermentable substrate for many yeasts, especially for those yeasts that produce ethanol from glucose.
2. The rate of xylulose assimilation by yeasts is generally slower than that of glucose assimilation.
3. The presence of oxygen is not essential for yeasts to utilize xylulose.
4. The most common products from xylulose are ethanol and xylitol. Other products include arabitol, glycerol, acetic acid, and citric acid.
5. Yeasts that cannot utilize xylose effectively utilize xylulose readily.
6. The theoretical yield of ethanol can be expected to be obtained from xylulose by certain yeasts.

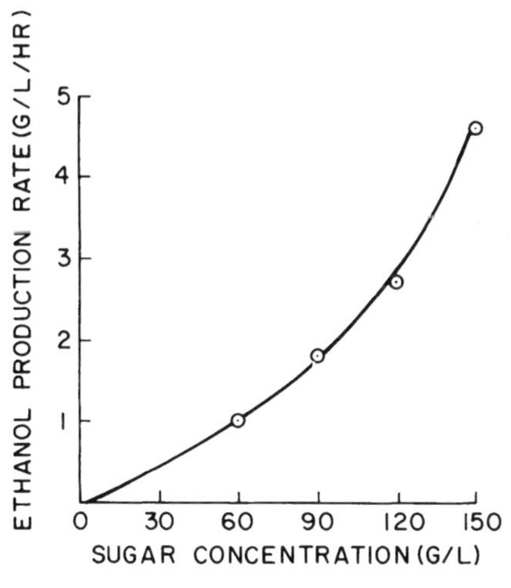

Figure 6. Effect of xylulose concentration on ethanol production rate. For detail see Chiang, et al. (83).

These observations in conjunction with the fact that xylose isomerase isomerizes xylose to xylulose provided the basis for the development of a process to utilize xylose isomerase and yeasts to convert xylose to ethanol.

V. *SCHIZOSACCHAROMYCES POMBE*

Next to *S. cerevisiae, S. pombe* has been subjected to the most intensive study with respect to its genetic characteristics (for recent review, see Egel, et al. (84)), and its cell cycle (for recent article, see Miyata, et al. (85)). Unlike *S. cerevisiae* and most other species of yeast, yeasts of the genera *Schizosaccharomyces* reproduce by fission rather than budding.

Species of fission yeasts include *S. versatilis* (86), *S. pombe, S. japonicus, S. malidevorans,* and *S. octosporus* (87). They are differentiated by the shapes of their spores, the number of spores formed per ascus, by the presence or absence of mycelium, and by their abilities to ferment sugars. Strains belonging to this genus have been isolated from grape wine, fruit wine, and molasses. *S. japonicus* and its variety *S. japonicus* var. *versatilis* are naturally-occurring respiratory deficient yeasts.

The cells of *Schizosaccharomyces* yeasts are cylindrical in shape and are predominantly haploid. However, malt-agar can be used to induce mating and subsequent formation of ascospores. The ascus is formed after somatogamous conjugation. The ascospores are round or oval and may have distinct warts.

S. pombe has strong fermentation capability although the rate of fermentation is about 50% slower than *S. cerevisiae*. *S. pombe* currently has little industrial value except in the wine industry where it is used to reduce acid content during wine making.

VI. PREPARATION OF D-XYLULOSE

High purity xylulose, especially that free from xylose is useful for the study of metabolism by yeasts. Hough and Theobald (88), and Touster (89) provide detailed procedures for preparation of pure xylulose by the epimerization of xylose in hot pyridine (90). Small quantities of pure xylulose can be obtained after elaborate purification schemes.

Pure xylulose can also be obtained from glucose by the biological means. Hann, et al. (91), reported a sequential biological conversion carried out by yeasts and bacteria.

Glucose is first converted to arabitol by yeasts (92) then the arabitol is converted to xylulose by either *Acetobacter suboxydans* (93) or *A. xylinum* (91). Yields of xylulose are 32% of the initial glucose and 94% of the arabitol.

Xylulose can also be obtained from xylose by isomerization catalyzed by xylose isomerase. However, due to the limitation of isomerization equilibrium, only a small portion (up to 25%) of xylose can be isomerized. Further purification to remove xylose is required. Anion exchange resins are usually used to remove xylose from xylulose. To overcome the equilibrium limitation, borate or germanate (94) have been added during the isomerization reaction to form a complex with xylulose thus shifting the reaction equilibrium from 25% conversion of xylose to 73%. The subsequent removal of borate or germanate is difficult to achieve.

Recently, a method to prepare large quantities of pure xylulose from xylose by utilizing a combination of xylose isomerase and microorganisms with differential xylose utilization patterns has been developed (80). First, a pentose syrup containing 25% xylulose and 75% xylose is prepared through enzymatic isomerization. This step is followed by partial removal of xylose through ethanol precipitation to obtain a xylulose enriched pentose syrup (3 to 1 ratio of xylulose to xylose). The complete removal of xylose can then be achieved via differential xylose utilization by a strain of *Fusarium oxysporum* or by other microorganisms that utilize xylose at a faster rate than xylulose. A highly purified xylulose for fermentation studies was obtained (Fig. 7) by utilizing this method. Since no chemical treatment is involved, the application of this method ensures the stereospecificity of xylulose.

VII. ISOMERIZATION-FERMENTATION

The production of ethanol by most yeasts from xylose displays a dependence on the presence of xylose isomerase which converts the xylose to xylulose before fermentation to ethanol can take place. This observation indicates that the production of ethanol from xylose can be accomplished by using yeast and commercially available xylose isomerase in a simultaneous isomerization and fermentation process. Wang, et al. (96), using an immobilized glucose isomerase (Maxazyme GI) and *S. pombe* to convert xylose to ethanol and obtained a 0.2% ethanol after six-days incubation. They attributed the low yield of ethanol to a low conversion rate of xylose to xylulose by isomerase, loss of yeast viability during incubation, deactivation of the isomerase and browning of the media (Millard reaction).

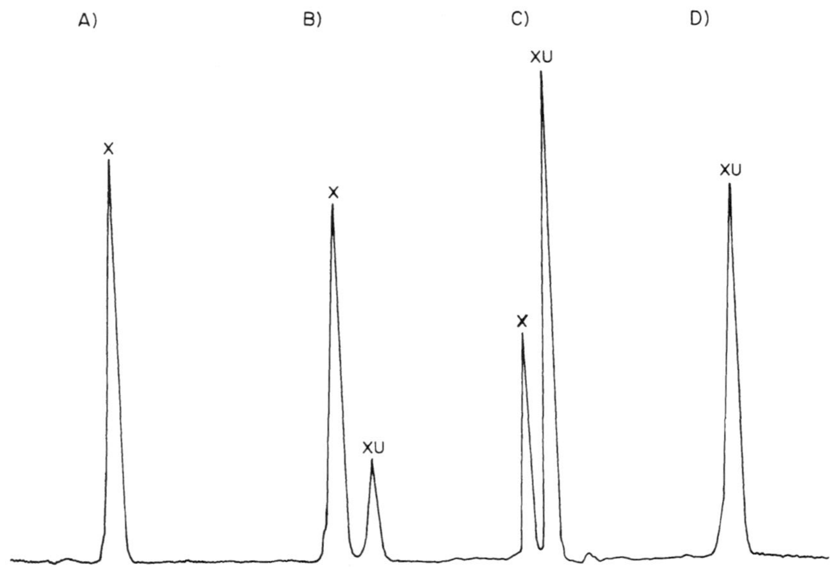

Figure 7. Chromatograms of xylose and xylulose at different stages of preparation. (A) Xylose; (B) Xylose and xylulose after enzymatic isomerization; (C) Xylulose and xylose after ethanol precipitation of xylose; (D) Xylulose, after differential microbial fermentation For detail see Chiang, et al. (80).

Gong, et al. (24), studied the effect of glucose isomerase obtained from various microbial sources on the production of ethanol from xylose by *S. cerevisiae* ATCC 24860 and observed that around 2% (w/v) ethanol can be produced from xylose after 48 hrs of incubation. Results indicated that ethanol can be readily produced from xylose by yeast when isomerase is present (Table 2). The higher rates of ethanol production and higher ethanol yields were probably due to the higher initial yeast density (2 x 10^8 cells/ml) and to the presence of the cofactors, Co^{++} and Mn^{++}, that are required for isomerase activity and stability (97).

A more detailed study of isomerization-fermentation of xylose to ethanol by yeast and isomerase was conducted by Chiang, et al. (98). They compared ethanol production from xylose by a strain of *S. cerevisiae* and *S. pombe* and observed that over 4% (w/v) ethanol was produced by these yeasts within 50 hrs of incubation when isomerase was present. *S. pombe* was a better yeast than *S. cerevisiae* for production of ethanol from xylose. In one instance, over 6% (w/v) ethanol was produced from a simulated hemicellulose hydrolyzate by *S. pombe* in the presence of xylose isomerase (Fig. 8). The sugar composition of this substrate was glucose (4.5%), xylose (7.9%) and L-arabinose (2.5%).

These observations indicated that commercial production of ethanol from xylose containing substrates can be accomplished by using yeast and commercial available xylose isomerase in a simultaneous isomerization-fermentation process. However, the optimal conditions (e.g., pH and temperature) for isomerization and xylulose fermentation are different (24,97), it is necessary to conduct this type of process at a compromise environmental conditions.

Other problems surrounding this isomerization-fermentation process are the unfavorable isomerization equilibrium which leads to xylulose formation of only 20-25% from xylose (99) and the relatively high cost of commercial isomerase. In order to overcome the low conversion of xylose to xylulose, Hsiao, et al. (100), introduced sodium tetraborate into the system to shift the isomerization equilibrium toward xylulose formation. The results of their study indicated a linear relationship between the shift of equilibrium from xylose to xylulose and the amount of sodium tetraborate present. A high concentration of tetraborate (0.3M) is needed to obtain an 80% conversion to xylulose. Increased conversion is due to the formation of a xylulose-borate complex. Subsequent fermentation experiments indicated that *S. pombe* can ferment xylulose to ethanol in the presence of borate. Yet, due to the extra expense of borate and the difficulty of borate recycle, it is doubtful this system could be applied in a commercial process.

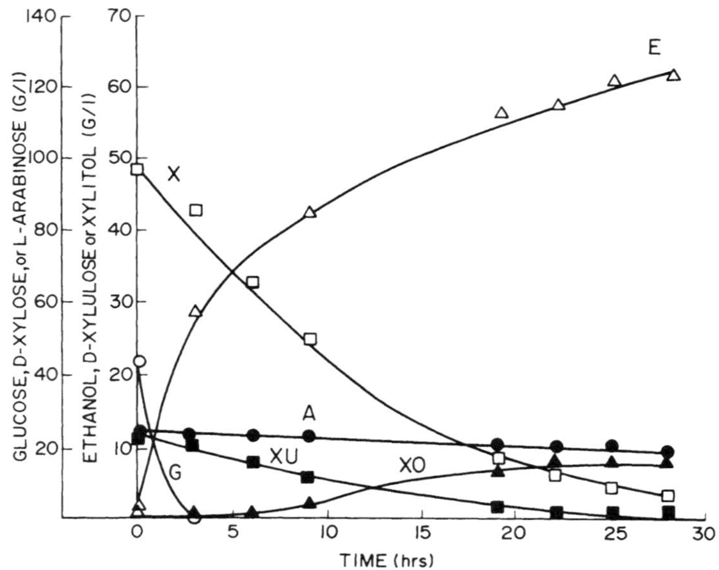

Figure 8. Ethanol production from simulated hemicellulose hydrolyzate by Schizosaccharomyces pombe in the presence of isomerase. The temperature was at 30°C, and the pH was controlled at 6.0. X, xylose; G, glucose; A, arabinose; Xu, xylulose; E, ethanol; Xo, xylitol. For detail see Chiang, et al. (98).

Table 2. *Effect of Glucose Isomerase on D-Xylose Fermentation by S. cerevisiae 24860*

Organism	Isomerase	Source	Percent ethanol $(\frac{w}{v})$
(control)	-	-	-
A. missouriensis	Cell-free	ATCC 14538	1.81
Bacillus sp.	Whole cell	Novo	1.30
Bacillus sp.	Immobilized whole cell	Novo	1.26
Streptomyces sp.	Cell-free	ATCC 21175	1.95
S. olivaceus	Immobilized whole cell	Miles Lab.	2.03

Isomerase, 100 IU; Fermentation time, 24 hr; Temperature, 30°C; pH, 5.8; Cell density, 2×10^8 cells per milliliter; (24).

In order to carry out isomerization-fermentation of xylose by utilizing isomerase and yeasts, several limitations need to be overcome. One of the major obstacles in this process is the incompatibility of the environmental conditions needed to carry out isomerization and those necessary for fermentation. The optimal pH and temperature for isomerization are 7 and 70°C, respectively, however, the optimal pH and temperature for xylulose fermentation by yeasts are much lower. At higher pH values and temperatures the ability of yeast to ferment xylulose is reduced and production of xylitol is enhanced. At lower pH values and temperatures, isomerization activity is decreased. However, since the rate of fermentation is slower than the rate of isomerization in any case a slower rate of isomerization has no major effect on the rate of fermentation. Thus, a compromised pH and temperature ranges have been found. On the basis of results derived from isomerization-fermentation of xylose by *S. cerevisiae*, the optimal pH of 6 and temperature 37°C was established (83).

Separate steps of isomerization and fermentation by isomerase and yeasts can also be employed to produce ethanol from xylose. One method is the "repetitive" enzymatic isomerization and yeast fermentation of xylose. The "repetitive" isomerization and fermentation is conducted in a stepwise fashion. First the xylose is converted to xylulose and then the xylulose is used as a fermentation substrate for conversion to ethanol by yeasts (98). The isomerization reaction was conducted at pH 7 and 60°C to convert 22% of xylose to xylulose. The fermentation was carried out by *S. pombe* at pH 4 and 30°C. Upon exhaustion of xylulose (within 6 hrs), the yeast was removed and the fermented broth was subjected to further isomerization and fermentation. The results indicated that 85% of the original 14% xylose solution was consumed after five alternated runs of isomerization and fermentation. Although, the final ethanol yield was only 55% of the theoretical value based on xylose consumed, this method has been demonstrated to be feasible. The low yield of ethanol is probably due in part to the evaporation of ethanol during the experiment.

A two step isomerization-fermentation using an immobilized column reactor to convert xylose to xylulose followed by batch fermentation of the xylulose to ethanol can also be employed (101). It allows the isomerization reaction and fermentation process to be conducted independently. Isomerization is carried out in a pack-column reactor with immobilized isomerase. The isomerized "pentose syrup" is then fed into a fermentor containing yeast to carry out the fermentation. After fermentation, yeast cells are separated from the broth by a cell-separator. The broth is then pumped back into the isomerase column for further sugar isomerization. This process can be repeated until the xylose is exhausted. Immobilized isomerase derived from *Bacillus coagulans* (Sweetzyme Q, Novo Industries, Inc.) has been used routinely in our laboratory for this purpose.

Other factors determining the feasibility of using yeast and isomerase to convert xylose to ethanol include yeast cell viability, isomerase stability and the cost of the isomerase. The viability of yeast cells during repetitive usage has not been studied systematically. Our laboratory results indicate a loss of fermentation capability by yeast cells after repetitive use. This could be due to the fact that continuous yeast growth is required to maintain the ability of yeast to ferment and tnat the reproduction of the yeast is limited while under oxygen-limited fermentation conditions. The cost of isomerase could conceivably be lowered by recycling. The half-life of immobilized isomerase has been reported to be between 750 hrs to 1200 hrs (102). If the stability of isomerase during the isomerization of xylose to xylulose could be maintained, the cost of this process should be drastically improved. Recycling of isomerase has been shown to be practical especially when the enzymes used as isomerase sources are in the forms of whole cell or immobilized whole-cell.

VIII. XYLOSE ISOMERASE (GLUCOSE ISOMERASE)

Xylose isomerase (EC 5.3.1.5) is produced by many prokaryotic microorganisms. The commercially available products originate from either *Streptomyces* spp. *Actinoplanes missouriensis*, or *Bacillus* spp. (97).

Xylose isomerase catalyzes the reversible isomerization of an aldo-pentose to its corresponding keto-isomer. Some examples of this type of reaction are the isomerization of xylose to xylulose and of arabinose to ribulose (103,104). The enzyme was first reported by Mitsuhashi and Lampen (105) who discovered its presence in *Lactobacillus pentosus*.

The ability of xylose isomerase to catalyze the reversible reaction of glucose to fructose was recognized by Marshall and Kooi (103) in 1957. Since then, the production of high-fructose syrup has been developed into a full-scale commercial process (for review see Abbott (106)).

Xylose isomerase is generally considered a temperature-stable, intracellular enzyme that is synthesized by microorganisms in response to the presence of xylose in the growth environment. Various sources of isomerase can be used to isomerize xylose to xylulose as well as glucose to fructose. For isomerization, the equilibrium concentration of glucose is about the same as that of fructose. At 75°C and pH 7.6, 28% xylose is converted to xylulose and 52.4% glucose is converted to fructose.

Temperature, pH, oxygen sensitivity and the presence of cofactors are the important factors influencing the stability of the enzyme and the reaction rate.

The optimal temperature for isomerase ranges from 45°C for the enzyme obtained from *Lactobacillus brevis* (107) to 90°C for the enzyme derived from *A. missouriensis* (108,109). Generally enzyme activities increase with increasing temperature. Continuous increase in enzyme activity as a function of temperature does not necessarily reflect the thermostability of the enzyme. Generally, the lower the temperature, the more stable the enzyme. Most enzymes have a temperature optima greater than 65°C. Nevertheless, a typical temperature range for commercial high-fructose-syrup production by isomerase is 60-65°C. Temperature exhibits an insignificant effect on the equilibrium concentration of fructose produced from glucose (110,111), but has significant effect on xylulose produced from xylose (100) (Table 3).

Table 3. Equilibrium concentration of Fructose and Xylulose at Various Temperatures

Temperature (°C)	*Concentration at Equilibrium (%)*	
	Xylose → Xylulose (100)	*Glucose → Fructose* (97)
30	-	46.5
40	15.5	-
45	-	48.2
50	18.2	-
60	23.4	49.9
70	28.0	53.1
80	-	54.2

The optimum pH for isomerase is generally greater than 7. However, an enzyme with a lower pH optimum (6.5) is known to be produced by *L. brevis* (107). *S. phaeochromogenes* has an optimum pH range of 9.0 to 9.5 (111). In some cases, optimal pH varies depending on the substrate and on the presence of co-factors. Immobilization of cells can also alter the pH optimum. The optimum pH range for the enzyme derived from *Bacillus coagulans* is 7.0-7.5 when glucose or ribose is used as substrates, and 8.0 to 8.5 when xylose is the substrate (112). The optimum pH range of *S. phaeochromogenes* enzyme be extended to 7.5 if Co^{++} is present (111). The optimum pH range of *A. missouriensis* enzyme becomes broader when it is in immobilized form (102). The optimal pH for cell-free enzyme is 7.5 and for immobilized enzyme ranges from 6.5 to 8. The relatively broad optimum pH ranges for immobilized enzyme is believed to be under the influence of the microenvironments surrounding the immobilized enzymes.

Xylose isomerase, in general, requires a divalent cation for its catalytic activity. Enzyme from *L. brevis* requries Mn^{++} for activity but Co^{++} can partially substitute for Mn^{++}. On the other hand, enzyme from *B. coagulans* requires CO^{++} for activity and Mn^{++} and Mg^{++} can partially replace it (113). The optimum concentrations of divalent cations are usually 0.2-0.7 mM of CO^{++} and Mn^{++} and 1 to 7 mM of Mg^{++}. When cation concentrations higher than the optimal values are used, enzyme activity is inhibited. Cations also have a positive effect on enzyme heat stability (113,114).

IX. UTILIZATION OF MIXED MONOSACCHARIDES

When cellulosic materials are subjected to hydrolysis, a mixture of monosaccharides is produced. The predominant sugars released are glucose, xylose and to a lesser extent arabinose, mannose and galactose. When microorganisms are exposed to this sugar mixture, the phenomena of diauxic growth and differential utilization of sugars are often observed. The differential utilization of sugars by a given microorganism is dictated by the rate of sugar uptake, the subsequent utilization of the sugars and by the degree of catabolite repression affecting specific enzyme biosynthesis.

It has often been observed that microorganisms accumulate a wide variety of monosaccharides against considerable concentration gradients without metabolizing them (115). The red yeast, *Rhodotorula gracilis*, accumulates xylose within the cells (116). Utilization of xylose does not occur until after a considerable period of delay, this is partially due to the biosynthesis of enzymes that are required for metabolism of this sugar and to the preferential utilization of endogenous substrates (37). Similar phenomena have also been observed when other pentoses such as arabinose are used as the substrate (116).

Alcorn and Griffin (117) studied xylose transport in *R. glutinis* and concluded that there are two different carriers involved in xylose transport. These carriers have different kinetic constants. The lower K_m carrier is repressed in rapidly growing cells and becomes derepressed by starvation. This carrier is competitively inhibited by galactose. 2-Deoxy-glucose inhibits both carriers.

Mutual uptake inhibition between sugar pairs, such as L-arabinose and D-glucose, D-xylose and L-xylose, (116) and D-xylose and D-glucose, (118) suggests that these sugars utilize the same carrier. However, Janda, et al. (119), came to a different conclusion. They found that xylose and galactose were transported by the same carrier, and fructose was transported by a different carrier and that glucose interacts with both of the carriers.

Yeasts appeared to employ different transport systems for assimilating xylose and xylulose. This tentative conclusion is drawn from several observations:

1. Many, if not all, yeast species that utilize both xylose and xylulose exhibit different sensitivities to the presence of oxygen. Active xylulose utilization by yeasts occurs under both aerobic and oxygen-limited fermentative condition. Oxygen is required for active xylose utilization (24).
2. Many yeast species, most noticeably strains of *S. cerevisiae* and *S. pombe*, utilize xylulose readily but utilize xylose only sparingly (24).
3. Induction of xylulose utilization is not necessary while xylose utilization undergoes a considerable lag period (120).
4. *C. blankii*, *C. steatolytic* and *P. tannophilus* utilize xylose readily but utilize xylulose only sparingly (57).
5. Rates of xylulose utilization are often faster than xylose when both sugars are present (120).

It is our common observation that yeasts utilize sugars in a sugar mixture in a sequential manner. They prefer glucose over xylulose, followed by xylose, arabinose, xylitol and arabitol when the substrate contains all these carbohydrates. This differential utilization of monosaccharides is due to the different metabolic rates, mutual competition for an uptake carrier and to the induction of uptake enzymes. Different yeasts exhibit different patterns in utilizing sugar mixtures. *S. pombe* takes up glucose and xylulose simultaneously; the presence of glucose only slightly inhibits xylulose utilization. Glucose inhibits xylulose utilization in *R. toruloides* to a greater degree. In *C. utilis* and *R. toruloides*, xylose and xylulose are consumed simultaneously with greater xylulose utilization rates in *C. utilis* than in *R. toruloides* (120).

Flickinger, et al. (121), examined the utilization of a mixture of glucose and xylose in the presence of xylose isomerase by a commercial baker's yeast (*S. cerevisiae*) and observed a rapid sequential fermentaion of glucose and xylulose (from xylose). A 95% ethanol yield based on glucose, xylose and xylulose consumed was obtained. They also observed that the initial rate of xylulose fermentation was directly proportional to the quantity of yeast used as the inoculum. The specific ethanol productivity from xylulose by this yeast is approximately 0.01 g ethanol per g yeast per hour.

X. ETHANOL FROM HEMICELLULOSES

Hemicellulose is defined as the plant cell wall polysaccharide other than cellulose and pectin that is readily extractable by alkaline solutions. It normally constitutes 20 to 30% of the cell wall of annual and perennial plants. It is most abundant in agricultural crop residues, especially in corn cobs, wheat straw and seed hulls. Upon hydrolysis, hemicellulose yields a mixture of carbohydrates of which xylose is the major constituent (Table 4).

Hemicellulose can be hydrolyzed to its sugar constituents by the action of microbial enzymes or by chemical or physical means. The hydrolysis products can be converted to a variety of useful products including ethanol by microorganisms.

Since hemicellulose is more easily hydrolyzed than cellulose, many approaches to selectively hydrolyze this fraction of plant material have been devised. The most common method involves the use of dilute sulfuric or hydrochloric acid (122-124). Reduced size biomass is mixed with acid and then heated to carry out the hydrolysis. Acid hydrolysates obtained this way can then be neutralized and used as fermentation substrates.

Often during acid hydrolysis many potentially toxic compounds are formed. The most frequently encountered are furfural, 5-hydroxymethyl furfural and 2-hydroxymethyl furan, organic acids, terpenes and phenolic compounds may also be present. The types and amounts of these substances in a hydrolysate depend on the hydrolysis conditions and on the type of biomass used as starting material. The toxicity of

Table 4. Neutral Carbohydrates of Agricultural Residue[a] and Woods[b]

Material	Percent (%, $\frac{w}{v}$) Glucose	Xylose	Others[c]
Agricultural Residues			
Corn			
Leaves	58	35	7
Stalks	61	32	7
Husks	54	36	10
Cobs	54	40	6
Fibers	67	29	4
Sugarcane Bagasse[d]	60	32	8
Sun Flower Stalks	70	26	4
Wheat Straw	60	32	8
Hardwood			
Alder (A. incana)	67	27	6
Ash (F. excelsior)	60	32	8
Beech (F. silvatica)	65	28	7
Birch (B. verricosa)	58.5	39	2.5
Elm (U. scabra)	67.5	27	5.5
Linden (T. cordata)	58.5	34.5	7
Maple (A. platanoides)	60.5	32.5	7
Oak (Q. excelsior)	68.5	26	5.5
Poplar (P. tremula)	64.5	30	5.5
Willow (S. alba)	67.5	26	6.5
Softwood			
Larch (L. sibirica)	63	9	28
Spruce (P. excelsa)	65.5	9	25.5
White Pine (P. silvestris)	65	13	21.5

a Compiled from Krull and Inglett (8)
b From Wenzl, H.F.J. (7)
c Mannose, galactose and arabinose
d Data from Laboratory of Renewable Resources Engineering, Purdue University

these compounds also depends on the metabolism of the yeast expected to utilize the hydrolysate as substrate. *S. cerevisiae* is more sensitive to furfural than to 5-hydroxymethyl furfural (125) and formic acid has a more toxic affect on this strain than acetic acid (126). In order for a fermentation to be successfully implemented it is necessary to understand and overcome the inhibitory affects of toxins in the hydrolysate.

The methods employed to alleviate the inhibitory effect of toxic materials include the addition of reducing agents, heating of the substrate, steam distillation, high yeast density for fermentation, activated charcoal treatment and ion-exchange resins pretreatments.

The production of ethanol from woods attracted a great deal of attention during the 1940's. However, most of the ethanol was attained from the hexose portion and the yields were poor (127). The first microorganism found that could convert both the hexose and pentose portions of wood and wheat straw hydrolysates was *Fusarium lini (F. oxysporum F. lini)* (128). This organism has the ability to ferment both sugars to ethanol. However, as is the case for ethanol fermentation by other mycellial fungi, the yield is low. *F. lini* also has a high sensitivity to the inhibitory materials present in hydrolysate (129,130).

Recently much attention has focused on utilizing yeasts to convert sugars of hemicellulose hydrolysates to ethanol.

A. Ethanol From Corn Residues

Corn residue is the most abundant potential hemicellulose source in the United States. It represents about half of the total agricultural residues (131), however, due to wide spread cultivation, efficient and economical collection and transportation is a problem.

Corn residues are characterized by high xylose contents (see Table 4). Sitton, et al. (132), conducted a comprehensive analysis of conversion of corn residues to ethanol and reported that a yield of 63 gallons of 95% ethanol can be produced from a ton of corn stover. Their analysis was based on a 94% xylose yield (from 13.9% hemicellulose content) and a 89% glucose yield (from 36.8% cellulose content). Their calculations were based on ethanol production from both glucose and xylose. *S. cerevisiae* ATCC 24858 would be utilized to carry out glucose conversion and they assumed that xylose could be converted to ethanol by *C. utilus*. Both conversions must be at 100% of the theoretical yields to attain their proposed figures.

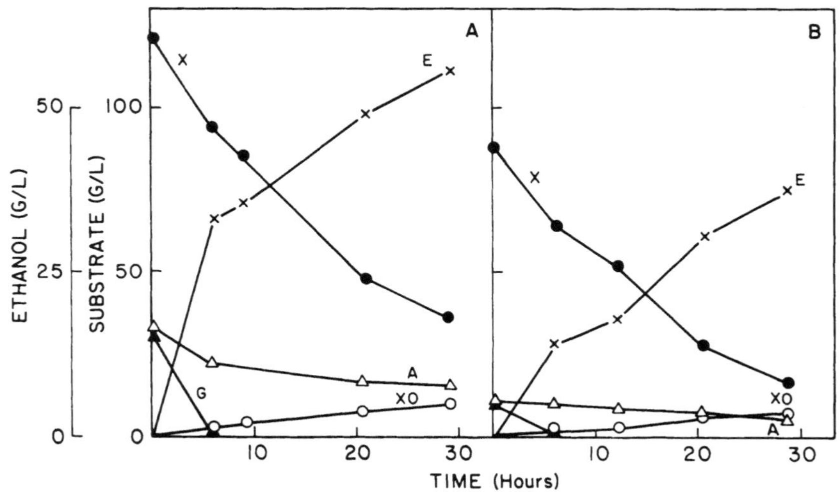

Figure 9. Isomerization-fermentation of sugarcane bagasse and wood hemicellulose hydrolyzates by S. pombe *sp 70. A, sugarcane bagasse hydrolyzate; B, wood hydrolyzate; G, glucose; X, xylose; E, ethanol; A, arabinose; Xo, xylitol. For detail see Gong and Tsao (133).*

B. Ethanol From Sugarcane Bagasse

Sugarcane bagasse hemicellulose can be hydrolyzed by dilute sulfuric acid (5%) at 100°C and atmosphere pressure to yield a solution with a typical sugar composition of glucose, 17%; xylose, 66%; and L-arabinose, 17%.

This hydrolysate has been used as a substrate and converted to ethanol by mycelial fungi (130) and by yeasts (100, 133). It was pretreated by passage through ion-exchange resins, neutralized, and supplemented with small amounts of inorganic salts and organic nutrients before offered as a substrate. It contained 15.1% (w/v) fermentable sugars, glucose and xylose. Fermentation was carried out at 30°C with incubation time of up to 30 hours.

When xylose isomerase was used to convert the xylose to xylulose, *S. pombe* fermented bagasse hydrolysate to a final concentration of 5.5% (w/v) ethanol. Under fermentative incubation conditions, glucose was utilized by this yeast within the first six hours of incubation (Figure 9A) presumably to produce ethanol. Ethanol and small amounts of xylitol were produced by *S. pombe* from xylose. L-Arabinose utilization was slow.

Ion-exchange resin pretreated sugarcane bagasse hemicellulose hydrolyzate has also been used as a substrate and fermented to ethanol by a xylose-fermenting, ethanol-producing yeast mutant, *Candida* sp. XF217 (133).

Under aerobic incubation conditions, a total of 4.8% ethanol was produced from both glucose and xylose portions of sugars after 29 hours of incubation. Xylitol (2.6%) was also produced from xylose (Fig. 10A). When xylose isomerase was present under fermentative conditions the rates of xylose consumption and ethanol production were similar to those under aerobic conditions (Fig. 10B). The amounts of xylitol produced were higher when isomerase was present.

C. Ethanol From Woods

Leonard and Hajny (127) investigated yeast fermentation of wood sugars in 1945 and concluded that pretreatment of wood hydrolyzates to remove inhibitor(s) is required. Methods of pretreatment include: the addition of reducing agents, steam distillation, activated carbon treatment, and ion-exchange treatments. In addition, high yeast cell density and acclimitization of yeast toward hydrolyzates were required to overcome inhibition.

Recently, the production of ethanol from wood-chips hemicellulose hydrolyzates derived from mixed hard woods by yeasts has been reported (133,134). Around 2% (w/v) ethanol was produced from both the glucose (0.4%, w/v) and xylose (1.4%, w/v) of hydrolyzate by *Candida sp.* XF217. The fermentation

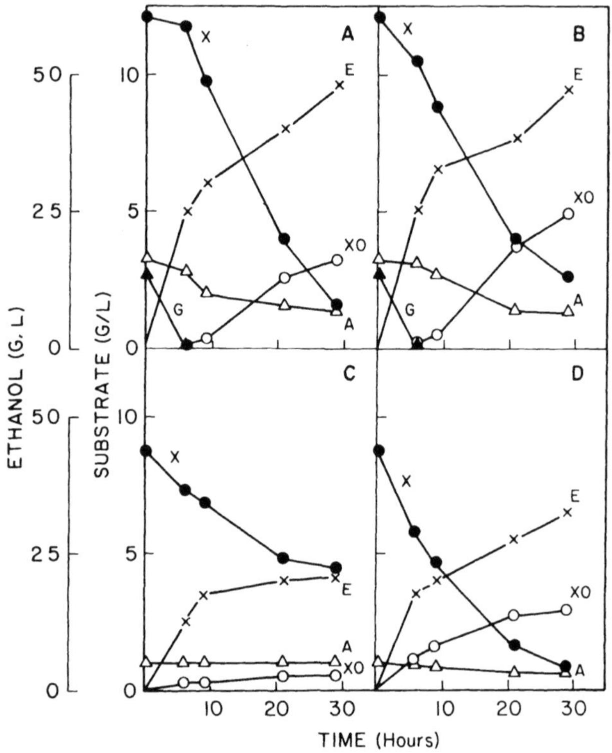

Figure 10. Fermentation of sugarcane bagasse and wood hemicellulose hydrolyzates by Candida sp XF217. A, aerobic fermentation of sugarcane bagasse hydrolyzate; B, isomerization-fermentation of sugarcane bagasse hydrolyzate by yeast and xylose isomerase; C, aerobic fermentation of wood hydrolyzate; D, isomerization-fermentation of wood hydrolyzate by yeast and xylose isomerase. G, glucose; X, xylose; E, ethanol; Xo, xylitol; A, arabinose. For detail see Gong and Tsao (133).

was conducted under aerobic conditions (Fig. 11A). Oxygen-limited fermentative conditions (Fig. 11B) and fermentative conditions in which the media was supplemented with xylose isomerase (Fig. 11C).

Higher ethanol concentrations were also obtained from hydrolyzates comprised of higher initial sugar concentrations of (glucose, 1.2%; xylose, 10%; and L-arabinose 1%). In the presence of xylose isomerase, *S. pombe* produced about 5% (w/v) ethanol (Fig. 9B). While *Candida* sp XF 217 produced lesser amounts of ethanol under either aerobic conditions (Fig. 10C) or under fermentative conditions (Fig. 10D).

XI. CONCLUSION

The pentose sugars, D-xylose and L-arabinose, comprise up to 30% of the neutral carbohydrates contained in the 500 million tons of annual crop residues. If these pentoses could be converted to ethanol, some 4 billion additional gallons of ethanol could be obtained in addition to that derived from the D-glucose portion. Yeasts, in general, are a better choice for fermentation of carbohydrates to ethanol than are other microorganisms due to their high fermentation rate, ethanol tolerance, and high ethanol yield. In the design of a simple industrial process to produce ethanol from biomass-derived carbohydrates, a biological system that could ferment both pentoses and hexoses simultaneously to produce high yields of ethanol with high rates of conversion is essential.

Yeasts, in general, do not convert pentoses to ethanol effectively. The enzyme systems in yeasts normally enable them to assimilate pentoses to produce pentitols. The conversion of pentoses to pentitols by yeasts serves the following purposes: (a) production of reserve carbohydrates to be utilized when needed as carbon and energy sources; (b) storage of reducing power; and (c) osmoregulation.

Recent research in the area of D-xylose utilization by yeasts has demonstrated that many yeasts are able to produce ethanol from D-xylulose (isomerized D-xylose). They also produce ethanol from D-xylose when D-xylose isomerase is present. Since the use of D-xylose isomerase would increase the cost for the conversion of biomass derived sugars to ethanol, reduction in the cost of isomerase and maintenance of isomerase stability during the isomerization-fermentation process is important.

The use of yeast or yeast mutants for the direct ethanol production from D-xylose would not require isomerase, this, could lower the costs considerably. However, further improvement of yeast strains which are able to produce ethanol from

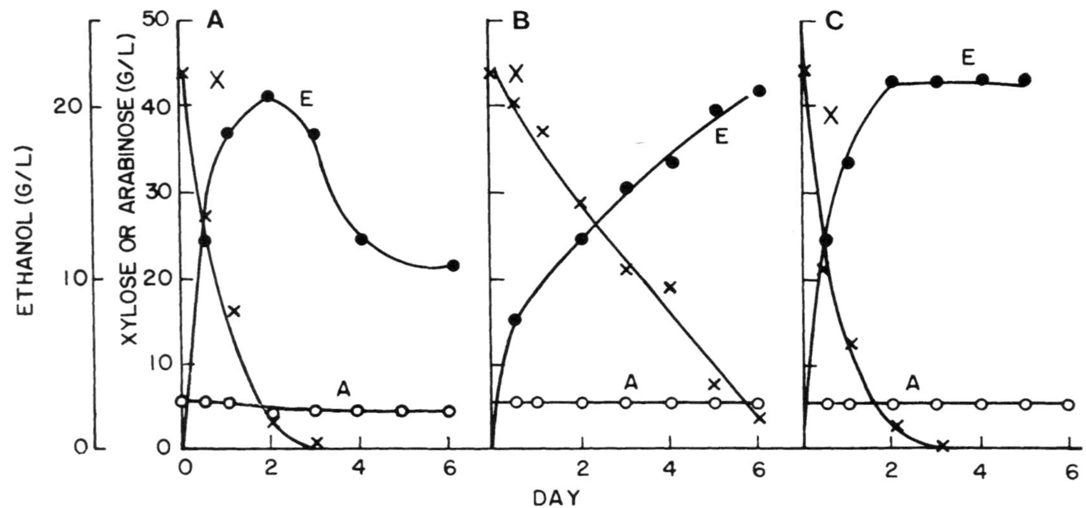

Figure 11. Production of ethanol from wood chips hemicellulose hydrolyzates by Candida sp. XF217. A, aerobic; B, fermentative; C, fermentative with glucose isomerase. (X), xylose; (e) ethanol; (o) L-arabinose. For detail see Gong, et al. (134).

D-xylose is required. These improvements include: (a) capacity to produce ethanol from D-xylose under industrial fermentative conditions; (b) increases in D-xylose fermentation rates and ethanol yields; (c) increased tolerance to high ethanol concentrations, and (d) ability to ferment authentic substrates without elaborate pretreatment of substrates.

The theoretical yield of ethanol from D-xylose is only 90% of that which can be obtained from either glucose or D-xylulose. The lowest yield of ethanol from D-xylose is primarily due to the regeneration of NADPH through the oxidation of hexose phosphate.

In addition to basic studies related to the genetics and the molecular biology of existing yeasts to understand the regulation of pentose metabolism, "new organisms" can be developed through "molecular cloning" so that the limitation of ethanol production from pentoses can be overcome. The cloning of the D-xylose isomerase gene in yeasts to by-pass the oxidation-reduction steps; the cloning of nicotinamide nucleotide transhydrogenase (transhydrogenase EC-1.6.1.1) to carry out the regeneration of NADPH through the interconversion of: "$NADP^+ + NADH \rightleftharpoons NADPH + NAD^+$"(135,136); and the cloning of D-xylulokinase to shift the equilibrium of conversion of D-xylose to D-xylulose-5-phosphate are few of the possible approaches.

Ethanol can be produced from both D-glucose and D-xylose portions of cellulosic materials by yeasts. Thus, large scale production of "pentose alcohol" in addition to ethanol derived from hexoses contained in agricultural residues, woods, celluloses and industrial wastes will be realized in the near future.

REFERENCES

1. Aspinall, G.O. *Adv. Carbohydr. Chem. 14,* 429 (1959).
2. Siegel, S.M. In "International Series of Monographs on Pure and Applied Biology," (Wareing, P.F. and Galston, A.W., eds.), Vol. 2, "The Plant Cell Wall."
3. Timell, T.E. *Adv. Carbohydr. Chem. 19,* 247 (1964).
4. Timell, T.E. *Adv. Carbohydr. Chem. 20,* 409 (1965).
5. Wilkie, K.C.B. *Adv. Carbohydro. Chem. 36,* 215 (1979).
6. Chang, M.M., Chou, T.Y.C., and Tsao, G.T. *Adv. Biochem. Engineering 20,* 15 (1981).
7. Wenzl, H.F.J. "The Acid Hydrolysis of Wood," in "The Chemical Technology of Wood," New York, Acad. Press, p. 157 (1970).
8. Krull, L.H. and Inglett, G.E. *J. Agric. Food Chem. 28,* 917 (1980).

9. Mortlock, R.P. and Wood, W.A. *J. Bacteriol. 88,* 838 (1964).
10. Mortlock, R.P. and Wood, W.A. *J. Bacteriol. 88,* 845 (1964).
11. David, J. and Wiesmeyer, H. *Biochim. Biophys. Acta 201,* 497 (1970).
12. Shamanna, D.K. and Sanderson, K.E. *J. Bacteriol. 139,* 64 (1979).
13. Shamanna, D.E. and Sanderson, K.E. *J. Bacteriol. 139,* 71 (1979).
14. Rosenberg, S.L. *Enzyme Microb. Technol. 2,* 185 (1980).
15. Zeikus, J.G. *Ann. Rev. Microbiol. 34,* 423 (1980).
16. David, J. and Wiesmeyer, H. *Biochim. Biophys. Acta 208,* 45 (1970).
17. Yoshitake, J., Ishizaki, H., Shimamura, M., and Imai, T. *Agr. Biol. Chem. 37,* 2261 (1973).
18. Yoshitake, J., Shimamura, M., and Imai, T. *Agr. Biol. Chem. 37,* 2251 (1973).
19. Mortlock, R.P. "Catabolism of Unnatural Carbohydrates by Microorganisms," in Adv. Microbiol. Physiology, (Rose, A.H. and Tempest, D.W., eds.) Vol. 13, p. 1-53 (1976).
20. London, J. and Chace, N.M. *Natl. Acad. Sci. USA 74,* 4096 (1977).
21. Chiang, C. and Knight, S.G. *Nature (London) 188,* 79 (1960).
22. Chakravorty, M., Viega, L.A., Bacila, Horecker, B.L. *J. Biol. Chem. 237,* 1014 (1962).
23. Barnett, J.A. In "The Fungi," (Ainsworth, G.C. and Sussman, A.S., eds.), Vol. III, New York, Acad. Press, p. 557 (1968).
24. Gong, C.S., Chen, L.F., Flickinger, M.C., Chiang, L.C., and Tsao, G.T. *Appl. Environ. Microbiol. 41,* 430 (1981).
25. Arcus, A.C. and Edson, N.L. *Biochem. J. 64,* 385 (1956).
26. Chiang, C. and Knight, S.C. *Biochem. Biophys. Res. Commu. 3,* 554 (1960).
27. Veiga, L.A., Bacila, M., and Horecker, B.L. *Biochem. Biophys. Res. Commu. 2,* 440 (1960).
28. Moret, V. and Sperti, S. *Arch. Biochem. Biophys. 46,* 124 (1962).
29. Veiga, L.A. *J. Gen. Appl. Microbiol. 14,* 65 (1968).
30. Veiga, L.A. *J. Gen. Appl. Microbiol. 14,* 79 (1968).
31. Horitsu, H. and Tomoeda, M. *Agr. Biol. Chem. 30,* 962 (1968).
32. Horitsu, H., Tomoeda, M., and Kumagai, K. *Agr. Biol. Chem. 32,* 514 (1968).
33. Suzuki, T. and Onishi, H. *Appl. Microbiol. 25,* 850 (1973).

34. McCracken, L.D. and Gong, C.S. *Adv. Biochem. Eng. 22*, (In press).
35. Clancy, F.G. and Coffey, M.D. *J. Gen. Microbiol. 120*, 85 (1980).
36. Tomoyeda, M. and Horitsu, H. *Agr. Biol. Chem. 28*, 139 (1964).
37. Höfer, M., Betz, A., and Kotyk, A. *Biochim. Biophys. Acta 252*, 1 (1971).
38. Peixoto, B.G. and Veige, L.A. *Arq. Biol. Technol. 17*, 87 (1974).
39. Höfer, M.K., Brand, K., Decker, K., and Becker, J.U. *Biochem. J. 123*, 855 (1971).
40. Chattaway, F.W., Bishop, R., Holmes, M.R., Odds, F.C., and Barlow, A.J.E. *J. Gen. Microbiol. 75*, 97 (1973).
41. Wang, C.H., Stern, I.J., Gilmour, C.M., Klungsoyr, S., Reed, D.J., Bailley, J.J., Christensen, B.E., and Cheldelin, V.H. *J. Bacteriol. 76*, 207 (1958).
42. Chen, S.L. *Biochem. Biophys. Acta 32*, 470 (1959).
43. Wang, C.H., Gregg, G.T., Forbusch, I.A., Christensen, B.E., and Cheldelin, V.H. *J. Am. Chem. Soc. 78*, 1869 (1965).
44. Blumenthal, H.J., Lewis, K.F., and Weinhouse, S. *J. Am. Chem. Soc. 76*, 6093 (1954).
45. Eaton, N.R. and Klein, H.P. *Biochem. J. 67*, 373 (1957).
46. Osmond, C.B. and ApRees, T. *Biochim. Biophys. Acta 184*, 35 (1969).
47. Holzer, H. and Witt, I. *Biochim. Biophys. Acta 38*, 163 (1960).
48. Jeffries, T.W. "Conversion of Xylose to Ethanol by Yeasts," Am. Soc. Microbiol. Annual Meeting (1982).
49. Sols, A., Gancedo, C., and DelaFuente, G. "Energy-Yielding Metabolism in Yeasts," in, "The Yeasts," (Rose, A.H. and Harrison, J.S., eds.) Vol. 2, Acad. Press, New York, p. 271 (1971).
50. Beck, C. and Von Meyenberg, H.K. *J. Bacteriol. 96*, 479 (1968).
51. Polakis, E.S. and Bartley, W. *Biochem. J. 97*, 284 (1965).
52. Ferdouse, M., Richard, P.H., Moss, F.J., and Blanch, H.W. *Biotechnol. Bioeng. 14*, 1007 (1972).
53. Fiechter, A., Fuhrmann, G.F., and Kappeli, O. *Adv. Microbial. Physiol. 22*, 123 (1981).
54. Barford, J.P., Jeffrey, P.M., and Hall, R.J. "The Crabtree Effect in *Saccharomyces cerevisiae* - Primary Control Mechanism or Transient," Presented at VI International Fermentation Symposium, London, Canada (1980).
55. Onishi, H. *Adv. Food Res. 12*, 53 (1963).
56. Onishi, H. and Suzuki, T. *Agr. Biol. Chem. 30*, 1139 (1966).

57. Gong, C.S., Claypool, T.A., McCracken, L.D., Maun, C.M., Ueng, P.P., and Tsao, G.T. *Biotechnol. Bioeng. Suppl. II*, (In press).
58. Spencer, J.F.T., Neish, A.C., Blackwood, A.C., and Sallans, H.R. *Can. J. Biochem. Physiol. 34*, 495 (1956).
59. Ingram, J.M. and Wood, W.A. *J. Bacteriol. 89*, 1186 (1965).
60. Lee, W.H. *Appl. Microbiol. 15*, 1206 (1967).
61. Lee, W.H. *Can. J. Biochem. 16*, 363 (1970).
62. Wang, S.C. and Tourneau, D.L. *Arch. Mikrobiol. 81*, 91 (1972).
63. Ikawa, T., Watanabe, T., and Nisizawa, K. *Plant Cell Physiol. 13*, 1017 (1972).
64. Boonsaeng, V., Sullivan, P.A., and Shepherd, M.G. *Can. J. Microbiol. 22*, 808 (1976).
65. Onishi, H. and Perry, M.B. *Can. J. Microbiol. 11*, 929 (1965).
66. Kiessling, H., Lindberg, B., and McKay, J. *Acta Chemica. Scand. 16*, 1858 (1962).
67. Barnett, J.A. *Adv. Carbohydr. Chem. Biochem. 32*, 125 (1976).
68. Jeffries, T.W. *Biotechnol. Letters 3*, 213 (1981).
69. Barnett, J.A., Payne, R.W., and Yarrow, D. "A Guide to Identifying and Classifying Yeast," Cambridge University Press, London, New York, Melbourne (1979).
70. Wickerham, L.J. In "The Yeasts," (Lodder, J., ed.), North Holland Publ. Co., Amsterdam, p. 448 (1971).
71. Schneider, H., Wang, P.Y., Chan, Y.K., and Maleszka, R. *Biotechnol. Letters 3*, 89 (1981).
72. Slininger, P.J., Bothast, R.J., Van Cauwenberge, J.E., and Kurtzman, C.P. *Biotechnol. Bioeng. 24*, 371 (1982).
73. Maleszke, R., Veliky, I.A., and Schneider, H. *Biotechnol. Letters 3*, 415 (1981).
74. Gong, C.S., McCracken, L.D., and Tsao, G.T. *Biotechnol. Letters 3*, 245 (1981).
75. Baillargeon, M.W., Jansen, N.B., Gong, C.S., and Tsao, G.T. *Biotechnol. Letters* (In press).
76. Onishi, H. and Suzuki, T. *Appl. Microbiol. 18*, 1031 (1969).
77. Wang, P.Y., Shopsis, C., and Schneider, H. *Biochem. Biophys. Res. Commu. 94*, 248 (1980).
78. Ueng, P.P., Hunger, C.A., Gong, C.S., and Tsao, G.T. *Biotechnol. Letters 3*, 315 (1981).
79. Jeffries, T.W. *Biotechnol. Bioeng. Symp. 11*, 315 (1981).
80. Chiang, L.C., Hsiao, H.Y., Ueng, P.P., and Tsao, G.T. *Appl. Environ. Microbiol. 42*, 66 (1981).
81. Swings, J. and DeLey, J. *Bact. Rev. 41*, 1 (1977).

82. Wang, P.Y. and Schneider, H. *Can. J. Microbiol. 26,* 1165 (1980).
83. Chiang, L.C., Gong, C.S., Chen, L.F., and Tsao, G.T. *Appl. Environ. Microbiol. 42,* 284 (1981).
84. Egel, R., Kohli, J., Thuriaux, P., and Wolf, K. *Ann. Rev. Genet. 14,* 77 (1980).
85. Miyata, H., Miyata, M., and Ito, M. *Cell Struc. Funct. 6,* 629 (1981).
86. Wickerham, L.J. and Duprat, E. *J. Bact. 50,* 597 (1945).
87. Sloff, W.C. In "The Yeasts--A Taxonomic Study," (Lodder, J., ed.), North-Holland Publishing Co., Amsterdam, London (1970).
88. Hough, L. and Theobald, R.S. In "Methods in Carbohydrates Chemistry: Analysis and Preparation of Sugars," (Whistler, R.L. and Wolfrom, M.L., eds.), Vol. 1, Acad. Press, p. 94 (1962).
89. Touster, O. In "Methods in Carbohydrates Chemistry: Analysis and Preparation in Sugars," (Whistler, R.L. and Wolfrom, M.L., eds.), Vol. 1, Acad. Press, p. 98 (1962).
90. Schmidt, O.T. and Treiber, R. *Ber. 66,* 1765 (1933).
91. Hann, R.M., Tilden, E.B., and Hudson, C.S. *J. Am. Chem. Soc. 60,* 1201 (1938).
92. Prince, R. and Rechstein, T. *Helv. Chim. Acta 20,* 101 (1937).
93. Moses, V. and Ferrier, R.J. *Biochem. J. 83,* 8 (1962).
94. Barker, S.A. and Somer, P.J. In "Topics in Enzyme and Fermentation Biotechnology," (Wiseman, A., ed.), Vol. 3, John Wiley and Sons, p. 12 (1979).
95. Campello, J.P. and Veiga, L.A. *Ann. Acad. Cienc. 45,* 309 (1973).
96. Wang, P.Y., Johnson, B.F., and Schneider, H. *Biotechnol. Letters 2,* 273 (1980).
97. Antrim, R.L., Colilla, W., and Schnyder, B.J. In "Applied Biochemistry and Bioengineering: Enzyme Technology," (Wingrad, L.B., Katchalski-Katzier, E., and Goldstein, L., eds.), Vol. 2, Acad. Press, New York, p. 97 (1979).
98. Chiang, L.C., Hsiao, H.Y., Ueng, P.P., Chen, L.F., and Tsao, G.T. *Biotechnol. Bioeng. Symp. 11,* 263 (1981).
99. Hochster, R.M. and Watson, R.W. *Arch. Biochem. Biophys. 48,* 120 (1954).
100. Hsiao, H.Y., Chiang, L.C., Chen, L.F., and Tsao, G.T. *Enzyme Microb. Technol. 4,* 25 (1982).
101. Gong, C.S., Chen, L.F., Flickinger, M.C., and Tsao, G.T. *Adv. Biochem. Engineering 20,* 93 (1981).
102. Chen, L.F., Gong, C.S., and Tsao, G.T. *Starch 33,* 58 (1981).

103. Marshall, R.O. and Kooi, E.R. *Science 125*, 648 (1957).
104. Sanchez, S. and Smiley, K.L. *Appl. Microbiol. 29*, 745 (1975).
105. Mitsuhasi, S. and Lampen, J.O. *J. Am. Chem. Soc. 204*, 1011 (1953).
106. Abbott, B.J. *Ann. Report Ferment. Processes 2*, 91 (1978).
107. Yamanaka, K. *Biochim. Biophys. Acta 151*, 670 (1968).
108. Scallet, B.L., Shieh, K., Ehrenthal, I., and Slapshak, L. *Starch 26*, 405 (1974).
109. Gong, C.S., Chen, L.F., and Tsao, G.T. *Biotechnol. Bioeng. 22*, 833 (1980).
110. Lloyd, N.E. and Khaleeluddin, K. *Cereal Chem. 53*, 270 (1976).
111. Tsumura, N. and Sato, T. *Agric. Biol. Chem. 24*, 1123 (1965).
112. Danno, G. *Agric. Biol. Chem. 34*, 1795 (1970).
113. Danno, G., Yoshimura, S., and Natake, M. *Agric. Biol. Chem. 31*, 284 (1967).
114. Aschengren, N.H. *Process Biochem. 10*, 17 (1975).
115. Cirillo, V.P. *Ann. Rev. Microbiol. 15*, 197 (1961).
116. Kotyk, A. and Höfer, M. *Biochim. Biophys. Acta 102*, 410 (1965).
117. Alcorn, M.E. and Griffin, C.C. *Biochim. Biophys. Acta 510*, 361 (1978).
118. Höfer, M. and Dahle, P. *Eur. J. Biochem. 29*, 326 (1972).
119. Janda, S., Kotyk, A., and Tauchova, R. *Arch. Microbiol. 111*, 151 (1976).
120. Hsiao, H.Y., Chiang, L.C., Ueng, P.P., and Tsao, G.T. *Appl. Environ. Microbiol. 43*, 840 (1982).
121. Flickinger, M.C., Chiang, L.C., Bremmon, C., and Tetzlaff, C. "Process Development Studies on the Conversion of Mixtures of Glucose and Xylose to Ethanol by Yeasts," Paper No. 12, Am. Chem. Scc. Annual Meeting, Las Vegas, NE (1980).
122. Dunning, J.W. and Lathrop, E.C. *Ind. Eng. Chem. 37*, 24 (1945).
123. Lee, Y.Y., Lin, L.M., Johnson, T., and Chamber, R.P. *Biotechnol. Bioeng. Symp. 8*, 75 (1978).
124. Ladisch, M.R. *Proc. Biochem. 14*, 21 (1979).
125. Azhar, A.F., Bery, M.K., Colcord, A.R., Roberts, R.S., and Corbitt, G.V. *Biotechnol. Bioeng. Symp. 11*, 293 (1981).
126. Katagir, W. *Biochem. J. 20*, 427 (1944).
127. Leonard, R.H. and Hajny, G.J. *Ind. Eng. Chem. 37*, 390 (1945).
128. Nord, F.F. and Mull, R.P. *Adv. Enzymol. 5*, 165 (1966).

129. White, M.G. and Willaman, J.J. *Biochem. J. 22,* 583 (1928).
130. Ueng, P.P. and Gong, C.S. *Enzyme Microb. Technol. 4,* 169 (1982).
131. Detroy, R.W. and Hesseltine, C.W. *Proc. Biochem. 13,* 2 (1978).
132. Sitton, O.C., Foutch, G.L., Book, N.L., and Gaddy, J.L. *Proc. Biochem. 14,* 7 (1979).
133. Gong, C.S. and Tsao, G.T. In "International Symposium on Ethanol from Biomass," The Royal Society of Canada, Winnepeg, Canada (1982).
134. Gong, C.S., Ladisch, M.R., and Tsao, G.T. *Biotechnol. Lett. 3,* 657 (1981).
135. Chung, A.E. *J. Bacteriol. 102,* 438 (1970).
136. Rydstroem, J., Hoek, J.B., and Ernster, L. *The Enzyme 13,* 51 (1976).

CHAPTER 11

EFFECT OF HIGH-ENERGY RADIATION ON LIGNOCELLULOSE CONVERSION

Youn W. Han
Alex Ciegler

United States Department of Agriculture
New Orleans, Louisiana

The practice of high energy irradiation of biomass to increase its utility is reviewed. The effects of high energy radiation on lignocellulosics with respect to changes in chemical and physical properties, sugar yield, digestibility to acid and enzymatic hydrolysis, and possible utilization of irradiated biomass were investigated. The primary effect of irradiation on cellulose is chain cleavage with ensuing decomposition of the formed carbohydrates resulting in formation of acid and reducing groups. The solubility of irradiated cellulose is increased in water and in alkaline solution. Solubilized products include a homologous series of cellodextrins and other low molecular weight compounds but very little glucose. The radiation effect on cellulose is generally noticeable at or above 1 Mrad and is proportional to the radiation dose. The possible use of nuclear wastes in the form of Cs-137 in conversion of lignocellulose is also discussed.

I. INTRODUCTION

The inevitable depletion of the world's petroleum resources has focused attention on alternate energy sources. Conflicting political trends tend to obscure this inevitability but little logic is required to realize that petroleum is not continuously regenerated *de novo*. Considerable attention has been focused on conversion of lignocellulosic biomass to fermentable sugars for production of food and liquid fuels. These alternate fuels have been, for the most part, limited to relatively few substances including ethanol,

ISBN 0-12-040306-4

methanol, butanol and acetone although the approaches to bio-production has ranged from the application of standard fermentation technology to sophisticated genetic engineering. The broader concept of chemical feedstock production from biomass has received scant attention.

In 1962 Pritchard et al. (1) and later Pigden et al. (2) indicated that gamma radiation effectively disrupts lignocellulosic polymers in forages thereby making them more susceptible to microbial attack and concomitantly increasing the available energy content of these feeds. This suggested a process of utilizing waste radioactive material (primarily cesium-137) from nuclear power plants as a source to hydrolyze or degrade recalcitrant lignocellulosic molecules into fermentable sugars and/or chemical feed stocks.

Initial research along these lines has been reported by Han et al. (3,4,5,40,61,62,64) and by Timpa (6). Their process has involved chemical pretreatment of lignocellulosics, such as sugarcane bagasse, followed by high-energy gamma irradiation using either a cobalt or cesium source; dosages up to 50 Mrad were required to effect solubilization of the substrate. Extensive analyses of the product has not been reported although preliminary results indicated that lactic acid, additional unidentified organic acids and ketones, low molecular weight oligosaccharides and phenolics were produced. Other investigators (7,8,9,10,11) have reported alterations in the physical properties of cellulose following irradiation; these include loss of mechanical strength and decreases in crystallinity. Irradiation of cellulose and hemicellulose results in random cleavage of biopolymers giving a variety of products including xylose, arabinose, glucuronic acid, formic acid, malondial dehyde and low molecular weight products such as water, CO, CO_2, and H_2 (12,13).

This information coupled with the problem of disposing of large quantities of gamma-emitting cesium-137, a waste product of the nuclear industry, suggests a potential application of waste radioactive materials for radiolysis of various sugars and complex sugar polymers to produce chemical feed stocks and fermentation substrates.

II. BIOMASS RESOURCES

Biomass resources are the depots for all the energy stored up by photosynthesis, through which about 1% of solar energy incident on earth is fixed by the green plants. Biomass resources are constantly generated in the form of trees, bush plants, farm crops, aquatic plants and other vegetation. The world supply of biomass is estimated to be 1.8×10^{12} tons and annual net growth or productivity about 4 x 10^{10} tons (14). Renewable biomass, thereby, amounts to about 10 tons annually for each individual on earth.

These resources are useful in many ways besides serving as fiber and fuel. In the past, wastes created in the production of food and fiber have been largely ignored as a potential source of food and energy. In the future, biomass could be essential and economically attractive materials for conversion to these uses. If cellulose could be broken down into glucose economically a large amount of fermentable sugar would be available which then could be used for production of a wide variety of products that have been available until now only from agriculture or from petroleum.

Although technically feasible, cellulose saccharification is not yet commercially exploited. A large portion of native cellulose is in a highly crystalline form and is also protected by a sheath of lignin. This makes cellulose resistant to attack by hydrolytic agents. Enzymatic and acid hydrolysis technology exist but the lignocellulosic materials must be pretreated to disrupt the cellulose structure before attempting hydrolysis. The need for pretreatment, coupled with the complexity of the saccharification process, makes sugar production costs from cellulose prohibitively high. Many physical, chemical and biological treatments have been tried (15,16,17,18,19,20,21,22,23) but because of economic constraints, none of these methods have been successfully used by commercial ventures. Therefore, most of the earths' biomass resources still virtually remain unexploited.

III. HIGH ENERGY RADIATION SOURCES

High energy radiation comprises beams of accelerated particles and high energy electromagnetic radiation. Radiation effects are produced when energy is absorbed from the beams by the constituents of irradiated materials. High energy radiation may be supplied by particle accelerators which produces cathode rays (accelerated electrons) or by nuclear reactors which produces neutrons. Interaction of the neutrons with reactor materials yield beams of photons or gamma rays, electrons, and a variety of charge particles with higher masses. Material under study may be irradiated simultaneously by neutrons and their interaction products. Elements may be made radioactive by the absorption of the neutrons and in turn serve as a source of high energy photons (radioisotopes).

High energy radiation causes chemical changes in a system. One of the characteristic resultant effects is ionization, whereby the incident high energy particle removes an electron from the parent atom or molecule of the matter on which it is incident and leaves charged species, namely ions or free radicals. The ions and free radicals are not stable but cause further reactions which may result in either breakdown of the molecule or in the formation of new molecules by combination between themselves or with molecules of other substances.

A widely used gamma source is Co-60 which is formed by the capture of a neutron by the Co-59 nucleus. In this isotope half of the atoms decay within 5.3 years by the emission of gamma and beta rays to non-radioactive nickel. Another source of gamma rays is Cs-137, produced as a by-product from nuclear fission. Large quantities of Cs-137 are found in fission product wastes stored since the initiation of plutonium-239 production during World War II; much larger quantities are potentially available from the operation of nuclear power reactors.

The task of disposing of the radioactive wastes produced by nuclear power plant is a technical and social problem which hinders the use of nuclear fission for generation of electric power. At present, nuclear waste materials are buried deeply underground as a means of disposal. Separation and utilization of Cs-137 would alleviate the waste disposal problem by removing a long lived isotope from the mixture of radioisotopes resulting from fission. For this reason, DOE at Hanford, Washington began isolating and packing Cs-137 in 1974 and total of 38 Mci of Cs-137 have been processed and encapsulated in stainless steel; an additional 121 Mci will be encapsulated by 1990 (24).

Cs-137 has many characteristics advantageous for use as a gamma source. It has a long half life of 30 years compared to that of 5.3 years for Co-60. It decays to stable barium with the emission of beta particles and 0.66 Mev gamma rays. The level of energy is about one-half of that emitted from Co-60, thus requiring less shielding for comparable dose rates. Although the energy of the gamma rays emitted by either of these isotopes is high it is not sufficient to cause nuclear changes in the irradiated materials. Threshold energy levels required to activate the materials are about 5-10 Mev. Therefore, radiation hazards from treated materials are not a concern (no radioactivity is induced through treatment with these isotopes). The inexpensiveness and the ready availability of large quantities of Cs-137 from nuclear power plant wastes are additional advantages of its use as a gamma source. A continuous Cs-137 (1 Mci) irradiator has been built and operating at Sandia National Laboratory (25). This facility has been used for irradiating dried sewage sludge to reduce the pathogen load. Irradiated sludge has been used for fertilizer and as an animal feed supplement. The irradiator was also used to treat fruits and vegetables for microbial disinfection.

IV. EFFECT OF HIGH ENERGY RADIATION ON LIGNOCELLULOSE

A. *Depolymerization*

When high doses of radiation were applied to lignocellulose the main effects noted were random scission of molecular chains, thus reducing the degree of polymerization (DP) of the biopolymers (8,26,27,28,29,30,31,32,33,34,57,58). The degree of reduction in DP was a function of the radiation dose applied (Fig. 1). In general, high energy radiation has little effect on the chain length of cellulose at or below 1 Mrad, but produces a large decrease in DP at or above 10 Mrad. The depolymerization effect was enhanced by increasing radiation doses up to 50 Mrad, at which point the cellulose became water soluble. With dosages between 1 to 100 Mrad, the DP of cellulose could be approximated by the formula:

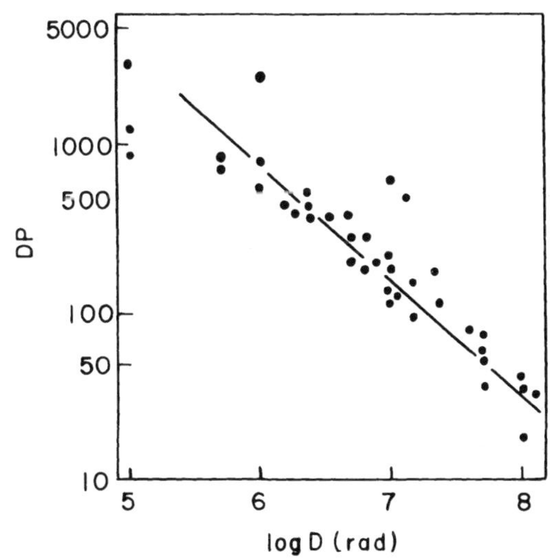

Figure 1. Degree of polymerization of cellulose as a function of radiation dose (10).

$$\log DP = -\log R + \log K$$

where R is the radiation dose and K is a constant (27).

Radiation depolymerizes purified cellulose more readily than native plant biomass in which cellulose is intimately bonded with lignin. Smith and Mixer (57) suggested the lignin may protect the cellulose when wood is exposed to radiation. They studied the effects of gamma radiation on the degree of polymerization in redwood and reported that the number of cellulose chain breaks per molecule was always higher when the holocellulose (cellulose plus hemicellulose) was separated from lignin and other extractives.

Although the mechanism of the degradation is not fully elucidated, it is believed that radiation-induced chain scission of cellulose and oxidation processes occur. It is not known whether depolymerization of cellulose results from cleavage of the β-1,4 glucosidic bonds. However, it is postulated that initially the radiation energy is deposited on the molecule in a random fashion but due to very quick processes of inter- and intra-molecular transfer of energy, soon becomes localized. This causes activation of long-lived excited sites, formation of free radicals, degradation of the molecule and other chemical changes. The penetrating power of gamma rays makes this oxidation reaction unique when compared to other chemical oxidations. The ionization induced in cellulose by high energy radiation is not hindered by the crystalline-amorphous nature of celluloses' structure (7,8).

Radiation of cellulose also causes the formation of reducing and acid groups (9,12). At higher radiation doses the cellulose molecule breaks into smaller units such as pentoses, hexoses and cellodextrins. It also increases in water and alkali solubility, decreases in tensile strength and in the capacity of moisture regain of cellulose fibers. Even though their previous work indicated that only small differences in properties were produced by irradiation in different atmosphere, Blouin and Arthur (26) later reported that the atmosphere under which the irradiation is carried out is very important in respect to the exact chemical nature of the groups produced. The differences noted were not particularly significant, because the methods used were too gross in nature, and the conditions of irradiation and analysis were not precise enough to detect these differences. A considerable quantity of oxygen is consumed during irradiation in the presence of air; oxygen consumption is directly proportional to the irradiation dose. The absolute pressure change measured at a constant volume and constant temperature during irradiation of cellulose showed that gas was evolved during irradiation under nitrogen atmosphere and that pressure increased. In air there is an initial slight decline and then a subsequent build up of pressure whereas, there was a decrease in pressure in an oxygen atmosphere that was eventually reversed with increasing dosage (Fig. 2). The increase of pressure during irradiation under a nitrogen atmosphere indicates that gas is constantly formed from the breakdown of cellulose, and the initial decrease in gas pressure during irradiation under oxygen indicates that oxygen is reacting with ionized cellulose at a faster rate than gas formed from cellulose breakdown. The initial decrease and subsequent increase of pressure in an air atmosphere indicates a reaction of substrate with oxygen and evolution of gas.

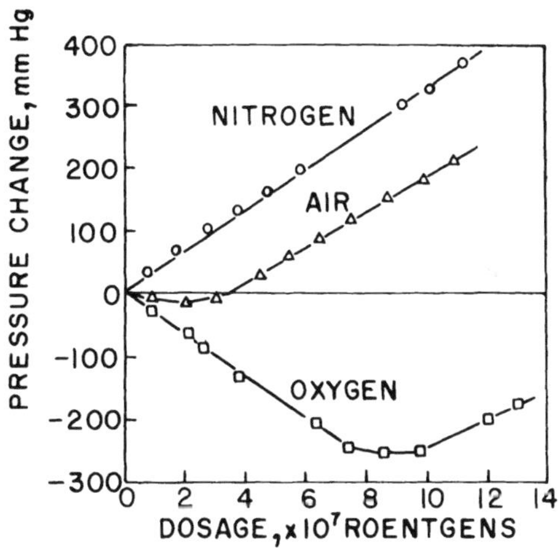

Figure 2. Pressure changes during irradiation of cellulose in various gaseous atmosphere (7).

B. Free Radicals

Gamma radiation generates free radicals throughout the cellulose matrix. Arthur and co-workers have made detailed electron spin resonance (ESR) studies or irradiated cellulose (7,9,35). They found that gamma-ray irradiated cellulose I gave a three-line ESR spectrum probably due to a free radical site at the C_5 position. In the case of cellulose II, a five-line spectrum was observed which was ascribed to the formation of two types of radicals, one at the C_5 and the other at the C_6 carbon atom. The formation of free radicals at C_6 in cellulose II has been attributed to the increase in delocalizing energy resulting from rearrangement in hydrogen bonding at C_6 on changing the lattice type from I to II. Free radicals trapped in crystalline regions are found inaccessible to water (36). Hinojosa et al. (35), however, reported that most of the free radicals formed in irradiated cellulose were trapped in the accessible regions. About 80% of the radicals formed in irradiated cellulose I and 90% in irradiated cellulose II were scavenged by the absorbed water when allowed to regain moisture. Further, they observed that the decay of free radicals in the irradiated cellulose in pure methanol was less than in water. The life time of free radicals produced in a gamma irradiated cellulose lasted several days of post irradiation storage (37). Even when the sample was in contact with air the decay was quite slow. Various types of high energy radiation appeared to produce the same general effects on the cellulose. The total dosage, and not the dose rate, seemed to be the critical factor.

C. Solubility

Irradiation causes solubilization of considerable portions of lignocellulose. Majumdar and Rapson (38), while describing the radiation effect of gamma rays on defatted jute, found that alkali solubility was very high at higher doses and the loss of holocellulose was also appreciable. Further, they found that the presence of lignin did not offer protection, but on the contrary caused degradation. Saeman et al. (8) and Charlesby (27) showed that irradiation of woodpulp and cotton linters with cathode rays resulted in chain cleavage which was evident from the decrease in viscosity and the increase in water solubility. They also reported that the course of degradation during irradiation was similar to that in cellulose hydrolysis. The amount of soluble materials in irradiated cellulose increased in direct proportion to the applied dose (Fig. 3).

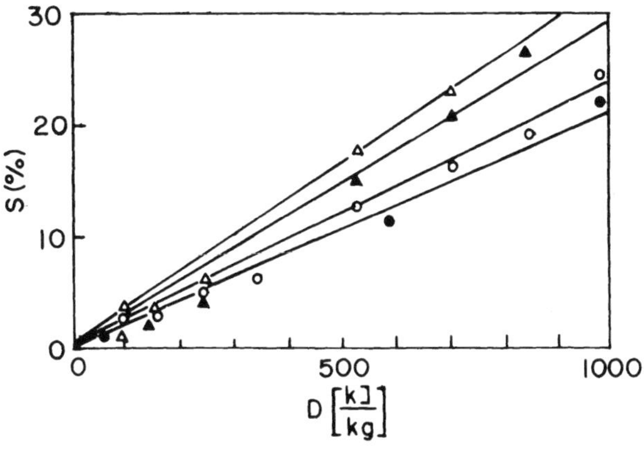

Figure 3. Effect of radiation dose (D) and influence of medium on the solubility of irradiated cellulose (S). ● *, dried sample;* ▲*, acidic medium; o, neutral water;* △ *, alkaline medium (39).*

Han et al. (62) observed a profound solubilization effect when the substrate was suspended in alkaline solution during irradiation. The concentration of soluble substances formed during irradiation of cellulose in).1N NaOH, 0.1N H_2SO_4 and water was higher (about 50%, 45%, and 5%, respectively) than the levels observed after irradiation of dried samples (39). Pretreatment with swelling agents increased the solubility of lignocellulose by gamma irradiation (3). The sample treated with 17% NaOH, upon irradiation with 100 Mrad, completely lost the physical integrity of the fiber and formed a dark brownish paste or liquid, depending on the amount of liquid used. Other swelling agents also increased solubilization, especially at lower dosages, but the effect was greatest with NaOH. Newspaper, cotton linters, cotton cloth, saw dust, and alpha cellulose powder could be completely solubilized by NaOH treatment and irradiation at 50 Mrad or greater.

Table I. Chemical Composition of Irradiated Sugarcane Bagasse (% Dry Matter).

Composition	Radiation (Mrad)					
	0	1	10	50	100	300
Cell-soluble matter	4.2	12.4	12.8	24.2	48.9	93.7
Hemicellulose	28.5	21.2	22.6	10.0	6.1	0
Cellulose	50.4	46.2	46.0	47.5	31.9	6.1
Lignin	14.9	18.1	15.9	16.0	10.7	2.1
Ash	2.0	2.0	2.3	2.1	2.1	1.8

From Han et al. (4).

Table II. Degree of Solubilization and Sugar Production from Sugarcane Bagasse Treated with NaOH (17%) and Irradiated with Cesium-137.

Treatment	Solubilization (%)	Sugar (% Dry Matter)		
		Total Sugar	Reducing Sugar	Glucose
0 rad	44.0	5.5	0	0
24.5 Mrad	61.0	22.5	2.0	0.11
54.2 Mrad	91.0	30.0	3.8	0.24
84.7 Mrad	100.0	35.0	5.0	0.31

From Han et al. (3).

The degree of solubilization by gamma irradiation of sugarcane bagasse components varied with dose (Table I). At 50 Mrad the hemicellulose content decreased and the cell soluble matter increased. It is apparent that the increase in cell soluble matter mainly results from solubilization of hemicellulose. Cellulose, lignin and ash were little affected at this dose level. At 100 Mrad, cellulose and lignin also decomposed and with 300 Mrad most of the sugarcane bagasse became water soluble. Although the ionizing action of gamma radiation took place in both crystalline and amorphous regions of the cellulose, the more highly ordered regions retained their integrity in a manner similar to that seen in heterogeneous degradation by acid, enzyme or oxidizing agents (7).

Table II shows the solubility and sugar production from sugarcane bagasse that was pretreated with 17% NaOH and irradiated with Cs-137. The bagasse that received 24.5 Mrad of irradiation was 61% water soluble and contained 22.5% total sugar, whereas the sample irradiated with 54.2 Mrad was 91% soluble and contained 30% total sugar. The bagasse irradiated with 84.7 Mrad was completely solubilized and contained 35% total sugar. The dosage level required for Cs-137 and Co-60 to solubilize sugarcane bagasse was about the same. Although the energy of gamma rays from Co-60 and Cs-137 differ by a factor of two, their ability to degrade lignocellulose was about the same. The radiation effects are thus similar among the different sources of radiation as long as the total absorbed dosage is the same.

D. *Crystallinity and Physical Properties*

Crystallinity of cellulose is believed to be one of the major deterrents in effective conversion of lignocellulose to fermentable sugars. Many attempts have been made to alter the crystallinity of lignocellulose, thereby increasing the digestibility (4,15,16,17,19,20,21,22,23). However, because of technical and economic constraints few have been commercially exploited. Radiation causes an apparent decrease in crystallinity and increases the digestibility of lignocellulose materials (10,40,44,45,46). For instance, the crystallinity index of 66.6% for sugarcane bagasse was reduced to 44.4% by 100 Mrad gamma irradiation (40). The irradiation effect on crystallinity became apparent at 50 Mrad or above.

The effect of high energy irradiation on cellulose crystallinity is the subject of controversy. Some reported that gamma irradiation had no effect (7,8,34,41,42,43) while others reported that there were significant changes in crystallinity of cellulose by high energy radiation (10,40, 44,45,46). Arthur et al. (7) argued that the x-ray diffraction pattern and infrared absorption curves of irradiated samples retained all of the characteristics given by purified cotton without any lessening of intensity. It has been shown that a loss in crystallinity of cellulose due to vibratory ball milling was clearly indicated by a broadening and smoothing out of the characteristic peaks. The gamma irradiated samples did not show this behavior, suggesting that irradiation did not produce a decrease in crystallinity of cellulose. Saeman et al. (8) reported that the rate of depolymerization by radiation of cotton linters and woodpulp appeared to be the same. This was not the case for acid hydrolysis of these materials where the reaction rate for the crystalline portion of the woodpulp was approximately twice that for the linters. This observation suggested that the radiation-induced depolymerization was independent of the extent of organization of the crystalline regions. Relative crystallineity indices of wood and cellulose as a function of radiation dosage has been investigated (Fig. 4). The crystallinity index is unchanged by irradiation up to 50 Mrad, but

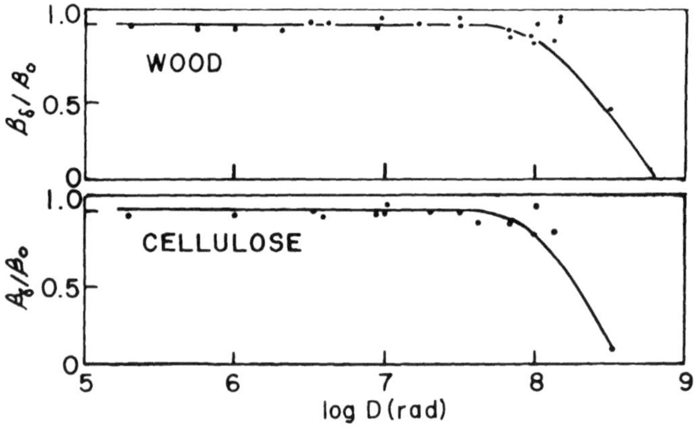

Figure 4. Relative crystallinity index ($\beta r/\beta_o$) of wood and cellulose as a function of irradiation dose (D) (10).

starts to decrease rapidly at about 80 Mrad. A similar tendency was observed in the infrared spectra in which a broadening of the so-called crystalline-sensitive bands occurred at 97 Mrad irradiation (46). This controversy probably results from the crudeness of the method used for determining crystallinity and the difference in radiation doses applied. It is certain that at higher dosages, for example, over 300 Mrad, crystallinity would change because cellulose irradiated at that dosage level would become water soluble.

Irradiation caused considerable changes in fiber strength of sugarcane bagasse (40). As the irradiation dose increased the fiber lost its strength and became mushy. When a probe was forced to penetrate the irradiated samples, the degree of compression, recovery, and energy required to compress were closely correlated to the level of dosage. There is some evidence that low dosages of radiation (up to 10^4 rad) results in increased yields of pulp in subsequent pulping operations (58,59,60). This implies some stabilization of the wood structure against chemical degradation in the pulping process. Crosslinking between the various components of the wood (cellulose, hemicellulose and lignins) seem one possible way in which irradiation could produce stabilization. The resultant Kraft pulp had an increased lignin content compared to the untreated controls. These results suggest that cellulose chain fractures may be offset by the information of new crosslinks with the lignin, which is in intimate contact with the cellulose in the wood.

E. Digestibility

Irradiation of lignocellulose altered structure sufficiently to make some insoluble carbohydrate available to rumen bacteria (1,2,11,48). The digestibility in rumen fermentation of wood increased with increase in dosage level within the range of 6.5 Mrad to 100 Mrad, but irradiation had no apparent effect at a dosage below 6.5 Mrad. With radiation doses greater than 100 Mrad, the carbohydrate fraction was converted to a form less readily usable by rumen bacteria. Samples of barley straw, pea straw, sugarcane bagasse, sunflower hulls and pine sawdust irradiated up to 10 Mrad showed little change whereas doses of 100 Mrad or more caused substantial losses of the fiber component and resulted in increased rumen digestibility (11). Irradiation was apparently more effective for sunflower hulls and sawdust than barley and pea straw and sugarcane bagasse. Thus, they suggested that if gamma irradiation had any role in the treatment of by-products, it would be more useful for highly refractory materials, such as sunflower hulls and sawdust which

Table III. Glucose Production Cellulosic Materials by Irradiation and Acid Hydrolysis

Substrate	Irradiation (Mrad)	Acid Hydrolysis	Glucose (%)	Ref
Filter paper	10	7.5% HCl, 90C, 20 hr	27 (19)[1]	(51)
Sugarcane bagasse	10	4.9% H_2SO_4, 121C, 30 min	31 (25)	(40)
Newspaper	10	0.5% H_2SO_4, 232C, 10 sec	50 (35)	(50)
Cellulose pulp	15	1% H_2SO_4, 190C, 30 min	32 (26)	(52)
Sawdust	50	0.8% H_2SO_4, 180C, 30 min	48 (25)	(49)

[1] Figures in the parenthesis is the amount of glucose produced from untreated control.

normally do not respond appreciably to other physical and chemical treatments. Pigden et al. (2) irradiated forages up to 280 Mrad and found that radiation slightly decreased rumen digestibility of immature forage but markedly increased the digestibility of intermediate and mature forages. When sugarcane bagasse was irradiated, *in vitro* rumen digestibility increased from 38.3% for untreated samples to 50.1% for 100 Mrad-irradiated samples. The increase in *in vitro* rumen digestibility was correlated with decrease in crystallinity (4).

F. Glucose Production

In addition to depolymerization, irradiation produces changes in the susceptibility of the cellulose to subsequent acid and enzymatic hydrolysis (1,2,3,11,41,49,50,51,52). When samples of irradiated materials were hydrolyzed the degree of hydrolyzability increased as the radiation dose increased. Table III summarizes the results of glucose production from various substrates that were gamma-irradiated and acid-hydrolyzed. Although the maximum glucose yield was different due to differences in raw material and differences in the conditions for irradiation and acid hydrolysis, radiation pretreatment invariably increased the glucose yield. In general, the radiation effect was apparent at or above 5 Mrad.

When sugarcane bagasse was irradiated and subsequently hydrolyzed by 1N H_2SO_4 the sugar yield was proportional to the radiation dosage (3). The sample irradiated with 100 Mrad produced almost 3 times more sugar than the untreated control; about 30% total sugar (dry matter basis) was obtained by acid hydrolysis. On further irradiation up to 300 Mrad, significant increase in sugar liberation was observed but the yield of sugar based on bagasse weight was less, apparently reaching the maximum level of sugar yield. Brenner et al. (50) reported that a combination hydro-pulping/irradiation pretreatment followed by a high-temperature short-time acid hydrolysis process was capable of producing a glucose yield up to 50% based on the available cellulose charged. They further claimed that their procedure had reached a point where a technically and economically viable process is within reach.

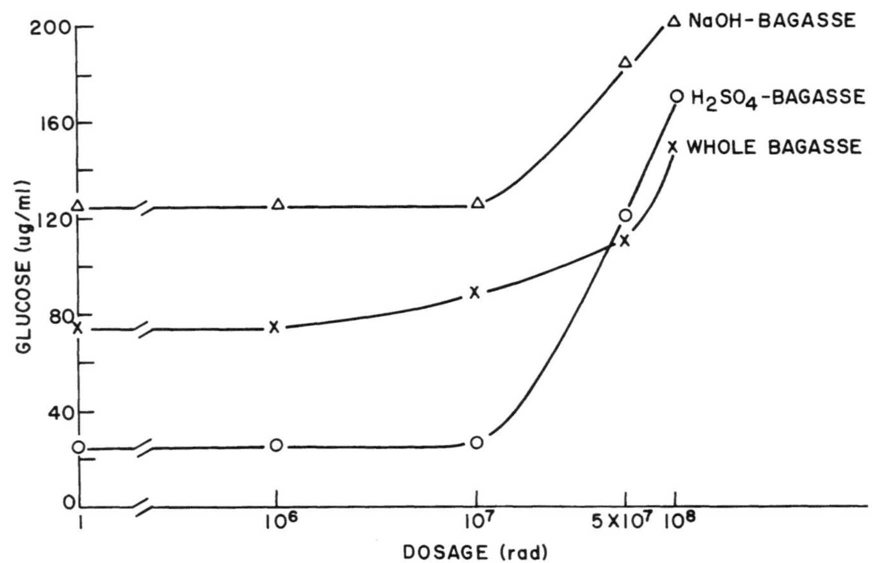

Figure 5. Glucose yield from acid hydrolysis of irradiated sugarcane bagasse and hemicellulose-free bagasses (3).

Since lignocellulosic biomass consists of heterogenous components it could be advantageous to fractionate the components prior to acid hydrolysis. The hemicellulose component could be readily extracted by acid or alkali. As much as 33% of hemicellulose per dry weight of sugarcane bagasse was obtained by treating it with 10% NaOH for 1 hr at 121C (3). Figure 5 shows the pattern of glucose production by gamma irradiation followed by acid hydrolysis from various fractionated samples of sugarcane bagasse. NaOH-treated (hemicellulose-free) bagasse produced more glucose than untreated sugarcane bagasse, whereas the H_2SO_4-treated (hemicellulose-free) bagasse produced less glucose than untreated bagasse at a low radiation dosage but the treated material produced more glucose than the whole bagasse at higher dosages. The effect of radiation was noticeable at or above 50 Mrad, at which hydrolyzability of cellulose drastically increased. Sugar production from sugarcane bagasse was greatly enhanced by NaOH treatment. About 5-7 times more glucose and reducing sugar were produced from NaOH-extracted bagasse than from untreated bagasse. This feature is especially noteworthy because by NaOH treatment most of the hemicellulose could be separated and utilized and the residual material produces even more glucose than the untreated bagasse.

The effect of gamma irradiation of cellulose on enzymatic hydrolysis is similar to that of acid hydrolysis. Susceptibility of cellulose to enzymatic hydrolysis was significantly increased by pretreatment with gamma radiation (3,41,49,56). Sugar yield by gamma irradiation followed by enzymatic hydrolysis was higher with increasing radiation dose and the effect became apparent when the dosage level was about 10 Mrad or greater. The amount of reducing sugar produced from the sample irradiated with 100 Mrad was more than 3 times that of the unirradiated control (3).

Kumakura and Kaetsu (56) reported that glucose yields from the enzymatic hydrolysis of raw cellulosic materials such as straw chaff and sawdust were greatly increased by preliminary treatment with irradiation. The irradiated materials contained considerable amounts of large molecular weight carbohydrates and subsequent enzymatic hydrolysis produced about 50% glucose in the total reducing sugar. Irradiation doses of more than 100 Mrad were required to produce a noticeable effect on sugar production. The required dose for effective acceleration of enzymatic hydrolysis could be much reduced by the addition of chlorine during irradiation. Although the total quantity of glucose in the hydrolysate was increased with increasing dosage, the glucose fraction in the total sugar decreased with higher dose levels, indicating decomposition of the sugar by irradiation.

Beardmore et al. (41) found that the amount of reducing sugar produced during irradiation was proportional to the dosage applied and that the relative hydrolysis rate significantly increased. At 100 Mrad, the relative hydrolysis rate was more than doubled compared to the untreated substrate. The surface area of the irradiated cellulose drastically increased.

G. *Degradation Products*

During irradiation of cellulose, molecular chains are cleaved with a concomitant decrease in the degree of polymerization. Since the energy from gamma photons randomly cleaves the cellulose molecule a variety of compounds including pentoses, saccharide acids, deoxysaccharides, malondialdehyde, dihydroxy acetone and low molecular weight products such as H_2O, CO, CO_2, H_2 HCHO, HCOOH, etc. are formed (7,12,13,39,49,53,54). Malondialdehyde is most commonly found especially in a neutral or alkaline environment (12,13). Carbon monoxides is probably formed from decomposition of aldehydes and ketones and CO_2 from the decarboxylation of carboxy groups. A considerable quantity of oxygen is consumed during irradiation in the presence of oxygen to form carboxyl and carbonyl groups. The oxygen consumption is directly proportional to the radiation dose. Any oxidized cellulose that contains carbonyl groups at any position other than the end group is very labile to alkali; under very mild conditions (e.g., 0.05N NaOH at room temperature for 1 hr) almost complete scission may be anticipated of the cellulose molecule at any unit that contains the carbonyl group (12). Thus, the yield of individual products considerably depends on the type of irradiated material and the amount of irradiation but essentially does not depend on the type of radiation source.

The water-soluble degradation products from 100 Mrad-irradiated cellulose has an average molecular weight of 347 as determined by freezing point depression (7). When the water-soluble products were dried and extracted with 90% ethanol, about 35% of the solids dissolved in aqueous ethanol. Acid groups were predominantly present in the alcohol-soluble fraction, while the reducing groups were evenly distributed between the two fractions. The alcohol-soluble fraction contained glucose, cellobiose, homologous series of cellodextrins, arabinose, 2 ketogluconic acid, glyoxal and other oligosaccharides. 2-Ketogluconic acid was one of the major acid monosaccharide units produced. The other major acid components were partially identified as the lactone of

2-ketogluconic acid and 2-ketocellobionic acid. The ultraviolet spectral data indicated that 2-ketogluconic acid was present both in the low molecular weight fraction and the higher molecular weight fraction in approximately the same ratio. The infrared spectra also showed a similar pattern for the distribution of acid groups.

When sugarcane bagasse in alakline solution was irradiated a substantial amount of total carbohydrate was produced but very little glucose (3). The low glucose level was attributed to the instability of glucose to alkali and high energy radiation. Glucose was degraded in alkali solution at an exponential rate with increase in contact time. About 60% of the initial glucose was destroyed during the first 24 hr in the presence of alkali and almost complete destruction occurred within a 6 day period. Although the total absorbed dosage is important, the rate of radiation also plays an important role in glucose production from biomass by NaOH treatment followed by gamma irradiation. The degradation pattern of cellobiose was similar to that of glucose. The NaOH concentration within the range of 5%-17% had about the same effect on sugar decomposition. Sugar degradation by gamma radiation was minimal up to 1 Mrad, but, almost complete destruction of glucose and cellobiose was noted at or above 50 Mrad (Table IV).

Table IV. Sugar Decomposition by Gamma Radiation.

Dosage (Mrad)	*Decomposition (% of original)*	
	Glucose	*Cellobiose*
0	*0*	*0*
1	*6.1*	*2.0*
10	*30.5*	*16.0*
50	*100.0*	*99.0*
100	*100.0*	*99.0*

From Han et al. (5).

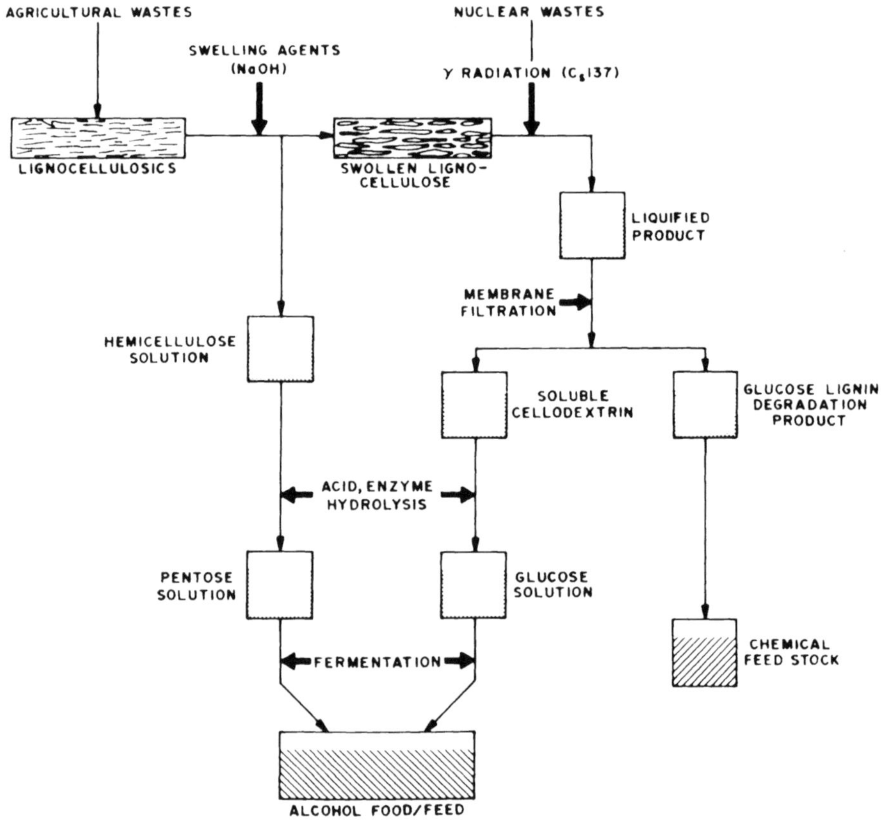

Figure 6. Proposed scheme for using agricultural wastes and nuclear wastes to produce useful products (4).

With respect to the stability of sugar to high energy radiation, Watanabe and Sato (54) reported that sugar content in undiluted molasses was stable to gamma radiation up to 3 Mrad: only 2% of the sucrose was converted to glucose and fructose at this level of radiation. Temperature had no apparent effect but low pH enhanced the sugar decomposition. Molten et al. (55) identified the products derived from alkali degradation of cellulose as aliphatic alicyclic hydrocarbons, aldehydes, ketones, alcohols and furans. They also postulated that alkali treatment not only forms these compounds but also results in resynthesis of molecules larger than 6 carbon atoms. Studying radiation-alkali degradation products, Han et al. (3) reported that the degradation product contained reducing di- and trisaccharide-like compounds as revealed by paper chromatography. An alcohol extract of the hydrolysate analyzed by thin-layer chromatography showed a series of U.V. light-absorbing zones. A visible U.V. scan of an alcohol extract of the product showed a very intense absorption maximum at 270 nm, strongly indicating the presence of aromatic materials. No glucose was revealed on any chromatograms.

V. USE OF GAMMA RADIATION IN BIOMASS CONVERSION

Even though cellulose radiolysis by high energy radiation has been attempted in the past, the level of dosage required to solubilize cellulose was too high (300-500 Mrad range) for practical use. By combination of chemical pretreatment and irradiation, the dosage requirement to solubilize or modify the structure can be reduced considerably. From the economic and technical standpoint, the use of Cs-137 is advantageous over the use of Co-60 or high energy cathode rays.

In Fig. 6, a scheme for utilization of biomass by gamma irradiation in producing renewable energy and other chemical feedstocks is proposed. Once lignocellulosic biomass is solubilized, fractionation of the components and their use becomes simpler compared to the use of solid substrates. The hemicellulose portion of biomass can be separated during the alkali pretreatment step by simply washing out the alkali. The solubilized hemicellulose can be converted to monomeric sugar by dilute acid or enzymatic treatments. The predominant sugar in plant hemicellulose is xylose (63), which can be fermented to alcohol or used as a chemical feedstock. The hemicellulose-free biomass can then be irradiated to solubilize the lignocellulose polymers. The solubilized cellulose may be directly fermented using a mixed culture of further hydrolyzed by acid or enzyme to produce fermentable sugars. The hydrolysis of solubilized cellulose should be easier than that of native cellulose. A mild (less than 1 Mrad) irradiation of biomass to produce substrates for use as ruminant feed or for biogas generation is another promising approach.

Development of a process such as those described above could open the way for the potential use of a vase amount of plant biomass and a large quantity of nuclear wastes. Disposal and utilization of biomass and nuclear wastes are currently of great concern in many countries.

REFERENCES

1. Pritchard, G.I., Pigden, W.J., and Minson, D.J. *Can. J. Animal Sci. 42*: 215 (1962).
2. Pigden, W.J., Pritchard, G.I., and Heaney, D.P. *Proc. Int. Grassland Congr. Paper No. 11* (1966).
3. Han, Y.W., Timpa, J., Ciegler, A., Courtney, J., Curry, W.J., and Lambremont, E.N. *Biotech. Bioeng.* 23: 2525 (1981).
4. Han, Y.W. and Ciegler, A. *Process Biochem. 17(1),* 32 (1982).
5. Han, Y.W. and Ciegler, A. Proc. 4th Sympo. Biotechnol. in Energy production and conservation (in press) (1982).
6. Timpa, J. Abstract ACS National Meeting Atlanta, GA (1981).
7. Arthur, J.C., Blouin, F.A., and Demint, R.J. *American Dyestuff Reporter 49:* 21 (1960).
8. Saeman, J.F., Millett, M.A., and Lawton, E.J. *Ind. Eng. Chem. 44:* 2848 (1952).
9. Blouin, F. and Arthur, J.C. *J. Chem. Eng. Data 5,* 470 (1960).
10. Aoki, T., Norimoto, M., and Yamada, T. *Wood Research 62,* 19 (1977).
11. Ibrahim, M.N.M. and Pearce, G.R. *Agricultural Wastes Vol. 2,* 253 (1980).
12. Bludovsky, B. and Duchacek, V. *Radiochem. Radional. Letters 38:* (1), 21 (1979).
13. Scherz, H. *Carbohydrate Res. 14:* 417 (1970).
14. Stephens, G.R. and Heichel, G.H. *Biotechnol. and Bioeng. Symp. No. 5,* 27 (1975).
15. Detroy, R.W., Lindenfelser, L.A., St. Julian, G. Jr., and Orton, W.L. *Biotechnol. Bioeng. Symp. No. 10,* 135 (1980).
16. Dunlap, C.E. and Chiang, L.C. in Utilization and Recycle of Agricultural Wastes, M.L. Schler, ed., 19-65 (1981).

17. Fan, L.T., Lee, Y.H., and Beardmore, D.H. *Biotechnol. Bioeng. 22*, 177 (1980).
18. Han, Y.W., Chen, W.P., and Miles, T.R. *Biotechnol. Bioeng. 20*, 567 (1978).
19. Johnson, D.C., Nicholson, M.D., and Haigh, F.C. *Appl. Polymer Sympo. No. 28*, 931 (1976).
20. Kleinert, R.N. *Tappi 59*, 122 (1976).
21. Millett, M.A., Baker, A.J., and Satter, L.D. *Biotechnol. Bioeng. Sympo. No. 6*, 125 (1976).
22. Tassainari, T. and Macy, C. *Biotechnol. Bioeng. 19*, 1321 (1977).
23. Tsao, G.T., Ladisch, M., Ladisch, C., Hsu, T.A., Dale, B., and Chou, T. in Ann. Rep. on Ferm. Process., D. Perlman, ed., Academic Press 2, 1 (1978).
24. Lamb, Z. in Use of cesium-137 to process sludge for further reduction of pathogens SAND80-2744. Sandia National Laboratories, Albuquerque, NM (1980).
25. Morris, M.E., Pierce, J.D., and Whitfield, W.J. U.S. Patent 4,151,419 (1977).
26. Blouin, F.A. and Arthur, J.C., Jr. *Text. Res. J. 28*, 198 (1958).
27. Charlsby, A.J. *Polymer Sci. 15:* 263 (1955).
28. Horio, M., Imamura, R., and Mizukami, H. *Bull. Inst. Chem. Res. Kyoto Univ. 41, No. 1*, 17 (1963).
29. Huang, R.Y.M., Immergut, B., Immergut, E.H., and Rapson, W.H. *J. Polymer Sci. A-1*, 1257 (1963).
30. Ifju, G. *Forest Prod. J. 14*, 366 (1964).
31. Mizukami, H. Effect of gamma irradiation upon cellulose, Thesis, Kyoto Univ. (1961).
32. Neal, J.L. and Kreassig, H.A. *Tappi 46*, 70 (1963).
33. Sugimatsu, A., Sanda, S., and Harada, Y. *J. Chem. Soc. Japan 62*, 90 (1959).
34. Takamuku, S., Miyamoto, Y., and Hachihama, Y. *Wood Res. 63:* 1043 (1961).
35. Hinojosa, O., Nakmura, Y., and Arthur, J.C., Jr. *J. Poly. Sci. Part C 37*, 27 (1972).
36. Dilli, S., Ernst, I.T., and Garnett, J.L. *Aus. J. Chem. 20*, 911 (1967).
37. Florin, R.Z. and Wall, C.A. *J. Poly. Sci. A. I*, 1163 (1963).
38. Majumdar, S.K. and Rapson, W.H. *Textile Res. J. 33*, 1 (1963).
39. Duchacek, V. and Bludovsky, R. *Radiochem. Radioanal. Letters 38(1)*, 31 (1979).
40. Han, Y.W., Catalano, E.D., and Ciegler, A. *J. Agr. Food Chem.* in press) (1982).
41. Beardmore, D.M., Fan, L.T., and Lee. Y.H. *Biotechnol. Letters 2(40)*, 435 (1980).

42. Hirai, N., Date, M., and Fakuda, E. *Mokuzai Gakkaishi 14,* 252 (1968).
43. Murayama, T. *Mokuzai Kohgyo 18:* 69 (1963).
44. Seifert, K. *Holz als Roh-und Werk Stuff 22,* 267 (1964).
45. Ishiguro, T., Hirai, N., Takemura, T., and Asano, M. *Mokuzai Kohgyo 28*: 449 (1973).
46. Goto, T., Harada, H., and Saiki, H. *Bull. Kyoto Univ. Forest 46,* 153 (1974).
47. Taniguchi, T., Harada, H., and Nakato, K. *Mokuzai Gakkaishi 11,* 142 (1965).
48. Lawton, E.J., Bellamy, W.D., Hungate, R.Z., Bryant, M.P., and Hall, E. *Science 113:* 380 (1951).
49. Kamakura, M. and Kaetsu, I. *Innl. J. Appl. Radiation and Isotopes 30:* 139 (1979).
50. Brenner, W., Rugg, B., Arnon, J., and Rogers, C. *Radiat. Phys. Chem. 14,* 299 (1979).
51. Kunz, N.D., Gaimer, J.L., and Kelly, J.L. *Nuclear Technol. 16,* 556 (1972).
52. David, C., Daro, A., and Monnoye, D. *European Polymer J. 16,* 1159 (1980).
53. Sundaram, V., Bhatt, I.G., and Chidambaresmaran, P.K. *Ind. Polym. Radiat. Proc. Symp. 194* (1979).
54. Watanbe, M. and Sato, T. *J. Ferment. Technol. 58:* 363-366 (1980).
55. Molton, P.M., Miller, R.K., Donovan, J.M., and Demmitt, T.T. *Carbohydrate Res. 75:* 199 (1979).
56. Kumakura, M. and Kaetsu, I. *Biotech. Bioeng. 20,* 1309 (1978).
57. Smith, D.M. and Mixer, R.Y. *Radiat. Res. 11,* 776 (1959).
58. McLaren, K.G. *Intl. J. Appl. Radiat. Isotopes 29,* 631 (1978).
59. Nicholls, G.A., Linderall, L., and Mitchel, G.W. *Tappi 57,* (1974).
60. Guillermina, B. De V., Guadalupe, A.S. Ma., Mauricio, A., Jose, S., Loyola, M., and Hoffman, A.S. *J. Radiat. Curing 18* (1976).
61. Han, Y.W., Lillehoj, E., Timpa, J., and Ciegler, A. *Biotechnol. Letters 2,* 397-402 (1980).
62. Han, Y.W., Lillehoj, E., and Ciegler, A. U.S. Patent No. 4,304,649 (1981).
63. Han, Y.W., Lee, J.S., and Anderson, A.W. *J. Agr. Food Chem. 23:* 928 (1975).
64. Han, Y.W., Timpa, J., Bryan, W., Lillehoj, E., Lembremont, A., and Ciegler, A. *Proc. Bio-Energy '80,* 569 (1980).

CHAPTER 12

INTERACTION BETWEEN SOLID SUBSTRATE AND CELLULASE ENZYMES IN CELLULOSE HYDROLYSIS

Ivan G. Gilbert
George T. Tsao

Purdue University
West Lafayette, Indiana

I. INTRODUCTION

Cellulose, a high polymer of glucose, is the most abundant organic material on the earth. Because it is renewable and abundant, much research has been undertaken to transform it to food, fuel, and chemicals. One avenue of research has dealt with the use of cellulolytic enzymes (cellulases) to breakdown the cellulose to glucose. Many reviews of cellulase research have appeared in the literature. Among the most comprehensive are reviews by Reese and Mandels (1), Nisizawa (2), Emert, et al. (3), Wood (4), Ghose (5), and Lee and Fan (6). A review of cellulases and their biosynthesis by Gong has appeared in an earlier volume of this series (7).

The fact that cellulase enzymes are strongly adsorbed by cellulose was observed over 20 years ago (8), but until recently studies looking directly at this phenomenon have been few. Now that the literature related to cellulase adsorption by cellulose is beginning to accumulate, a review of the findings pertinent to this subject is in order. Since the adsorption characteristics of cellulase onto cellulose is intimately related to the structure and properties of both the enzymes and the solid substrate, these topics will be reviewed before enzyme adsorption.

ISBN 0-12-040306-4

II. THE SOLID SUBSTRATE

1. Cellulose Structure and Properties

Cellulose is a high molecular weight linear polymer of cellobiose units connected by β-1,4-glucosidic linkages. Each cellobiose unit consists of two anhydroglucose molecules as illustrated in Fig. 1. The degree of polymerization (D.P.) of D-glucose units in some native celluloses has been found to be as high as fifteen thousand. Most cellulose (depending on its source and prior treatment) has an average D.P. between fifty and a few thousand. Cellodextrins with a degree of polymerization from 1 to 6 are soluble in water. However, the solubility decreases with increasing D.P., cellohexose (D.P. = 6) being only slightly soluble.

As the length of the cellulose chains increase, the chains in an aqueous environment will associate. The hydrogen bonds and Van der Waals forces which are quite weak individually become a strong associating force between cellulose molecules as the D.P. increases. These forces can cause neighboring chains to coalesce into crystalline regions which are so tightly packed that even small water molecules can not enter.

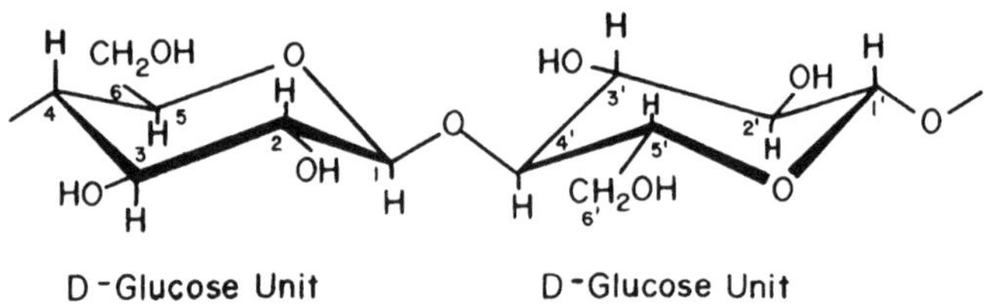

Figure 1. The molecular structure of the cellobiose unit.

2. *Cellulose as it Exists in Nature*

Common sources of cellulose fibers are cotton and wood. Cotton is about 90% cellulose, while wood is typically about 45% cellulose. Most of the cellulose fibers found in wood are mixed with hemicellulose and lignin. Hemicellulose is a heteropolymer which is mainly made up of xylose, galactose, mannose, glucose, and arabinose. Lignin is a complex, three dimensional, phenolic polymer which associates itself with the carbohydrate to form a tough rigid structure.

The cellulose-containing cell walls of both cotton and wood have an external layer called the primary wall. This layer consists of structural units oriented around the fiber at approximately right angles to the fiber direction. In wood this primary layer consists of approximately 20 to 25% cellulose. The secondary wall which is just inside the primary wall, as shown in Fig. 2, is made up of three layers designated, S1, S2, and S3. Each layer is made during a particular growth stage of the cell and each contains structural units with a different orientation resulting in a laminated appearance. Of these layers, most of the cellulose is located in the thick S2 portion. Inside the secondary wall, there is a central region called the lumen. This is the

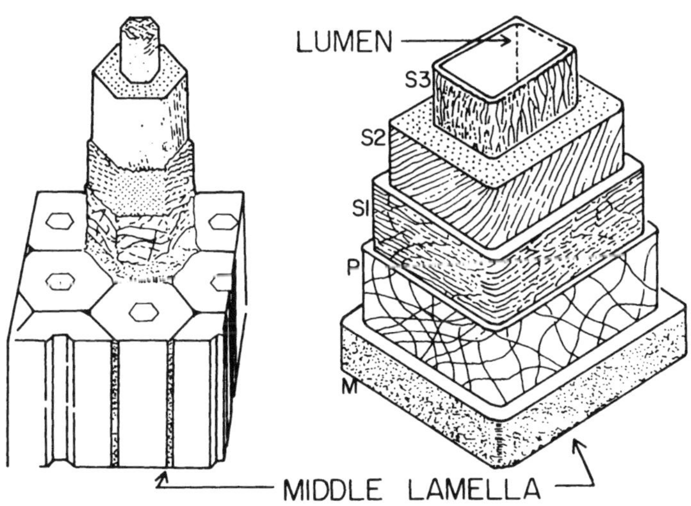

Figure 2. Architecture of the cell wall.

portion of the cell which when living contains the cellular constituents, but when dead is hollow and is used by the plant to transport water. The primary layer of the cell wall in wood is surrounded by the middle lamella. The middle lamella consists of 50% lignin and 20% hemicellulose which acts as a glue to hold the wood cells together. Analogously, a cotton fiber possesses an external coating called the cuticle around its primary cell wall. The layer along with the primary wall contain most of the 10% non-cellulosic extractables found in cotton.

Native cellulosic fibers are made up of distinct structural units called microfibrils, as shown in Fig 3, which are about 120 angstroms in cross-section and around 300 angstroms long. Each microfibril contains several elementary fibrils of about 30 angstroms cross-section. Each elementary fibril consists of several dozen linear cellulose molecules. It has been established by X-ray diffraction studies that the linear cellulose molecules of wood and cotton run antiparallel to adjacent cellulose molecules, and are bound to them by hydrogen bonds, specifically, two hydrogen bonds per glucose molecule. However, the precise manner in which native cellulose molecules are aggregated to form the elementary fibril structure is not well understood. Three of the concepts which have been proposed are presented in Fig. 4.

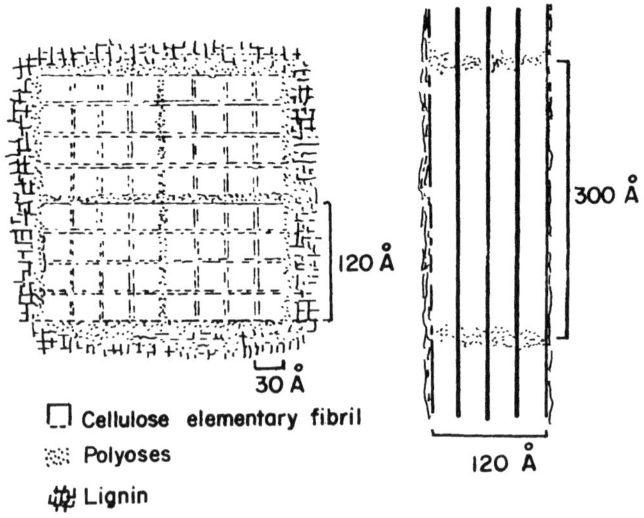

Figure 3. Arrangement of the cellulose microfibrils shown in cross section on the left and longitudinal section on the right (9).

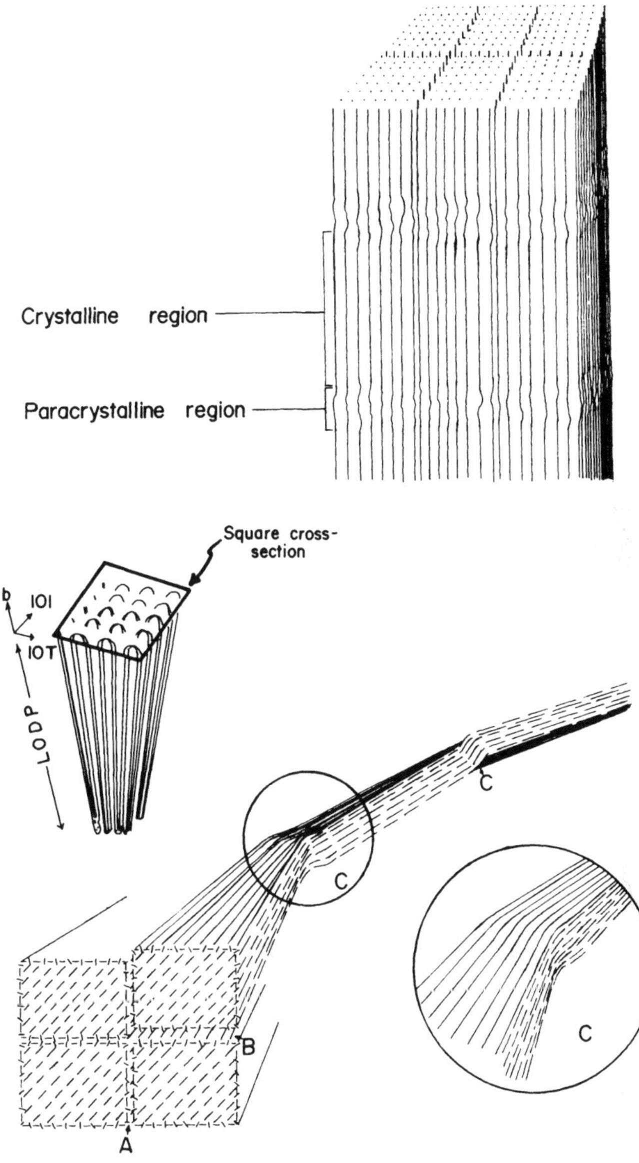

Figure 4. Three postulated representations of elementary fibrils in native cellulose. (a) Diagram of six elementary fibrils according to Hess, Mahl, and Gutter (10). (b) Elementary fibrils according to Rowland and Roberts (11). (c) View of an elementary cellulose crystallite according to the folding chain model (12,13).

Figure 4a depicts a possible picture of several elementary fibrils. Here the extended cellulose crystallites are segmented into crystalline and paracrystalline or amorphous regions. This is known as the fringe-micellar model. A similar model proposed by Rowland and Roberts (11) is shown in Fig. 4b. This model was proposed after a study was made of the different types of cellulose surfaces in cotton. It has also been proposed that the elementary fibril is made up of folding chains as in Fig. 4c. The amorphous fraction for this model has been postulated to be the regions where the chains are folded. In this model, each cellulose chain is not extended through a series of several crystalline and amorphous regions; instead, it is predominantly folded within a single cellulose crystallite (12,13).

3. *Unit Cell Structure of Cellulose*

The unit cell structure of crystalline cellulose has been elucidated by X-ray diffraction studies. Cellulose I (native crystalline cellulose) has a monoclinic unit cell structure which contains four anhydroglucose units per cell with dimensions in Fig. 5. The structure of crystalline cellulose II is similar, except the unit cell dimensions are

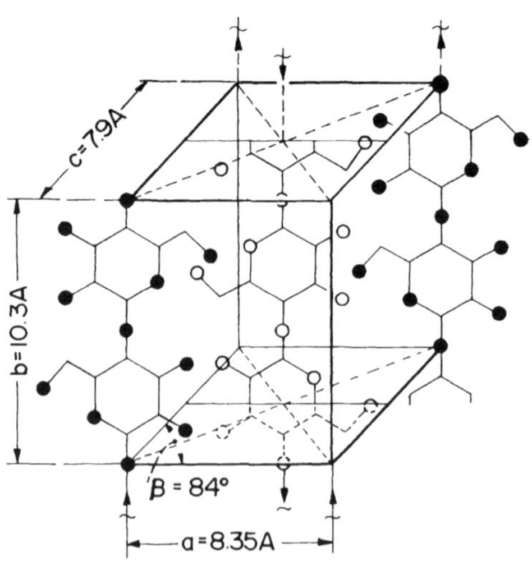

Figure 5. Meyer and Misch model of the crystal unit cell of cellulose (14).

different. Cellulose which contains crystalline regions with the cellulose II unit cell structure is called regenerated or mercerized cellulose depending upon whether it was formed by reprecipitating the cellulose from a solution of dissolved cellulose or by mercerizing it, e.g., swelling the cellulose in a solution of NaOH. A cellulose solution can be made by dissolving cellulose I or II in a cellulose solvent such as cadoxen, cupraammonium hydroxide, zinc chloride, or calcium thiocyanate solutions. A thorough review of cellulose solvent systems has recently been written by Hudson and Cuculo (15). The cellulose II structure is formed upon reprecipitation of the cellulose from a cellulose solution. Cellulose II can also be made by converting the initial cellulose to a soluble derivative such as carboxymethyl cellulose (CMC) or cellulose xanthate (viscose) and then regenerating cellulose by removing the derivative. The reason cellulose I crystalline regions are not formed upon reprecipitation is because the cellulose II conformation allows for increased hydrogen bonding between adjacent cellulose chains and is, therefore, thermodynamically favored. Of course, if the solubilization or mercerization treatment is incomplete, then there are likely to be regions of cellulose I structure as well as regions of cellulose II. Celluloses III, IV, and V, which are formed by other treatments, have been found using X-ray diffraction to have different unit cell geometries. These conformations, however, are encountered less often so no further mention will be made of them here.

4. *The Hydrolysis Reaction*

The actual hydrolysis involves the addition of a water molecule to the β-1,4-linkage connecting two adjacent glucose molecules, breaking the cellulose chain. This reaction proceeds in the presence of two types of catalyst: cellulases and mineral acids.

A difference between the mode of enzymatic and acid hydrolysis can be observed by measuring the change in the degree of polymerization of the cellulose substrate during the course of hydrolysis. Treatment of a cellulose pulp (initial average degree of polymerization of 900) with 4N hydrochloric acid causes a decrease in its average D.P. to 200, while only a 15% decrease in the substrate weight is observed. Treatment of this same pulp with a 1.8% cellulase solution only lowers the average D.P. to 800, while 40% of the cellulosic substrate is converted to soluble sugar (2). In general, with acid hydrolysis there is considerable depolymerization without appreciable weight loss. Whereas, enzymatic solubilization of native cellulose proceeds with only a slight decrease in the degree of polymerization. These facts can be easily explained in terms of the relative sizes of the two catalysts. Enzymatic hydrolysis of cellulose is restricted to the bulk surfaces which are available to these large molecules. Acids, on the other hand, are able to penetrate and diffuse into the amorphous regions of the polymer and break it down from within.

When the relative activities of acid and enzyme catalysts were compared using three cellulosic substrates at 50°C, it was calculated that about 100,000 times as much acid is required compared to enzyme to obtain a similar degree of hydrolysis (16). Because of the larger molecular weight difference between a mineral acid (Mw = 36 for HCl) and a cellulase enzyme (Mw = 60,000), the difference in molecular activity is even larger. It takes approximately 10^8 HCl molecules to do the work of one enzyme molecule under some circumstances (16).

For any enzymatic reaction to occur, direct physical contact must be made between the enzyme and its substrate. This is also true, of course, for the enzymatic hydrolysis of cellulose. Therefore the rate of reaction should be a function of the cellulose surface area accessible to the cellulase.

In a porous material such as cellulose, the surface available to a penetrating molecule depends on the relative size of the diffusing molecule to the pore size and the number of these pores. Cowling (17) has summarized the two main types of capillary pores which exist in wood and cotton as follows: 1) *Gross Capillaries*, such as the cell lumina, pit aperatures, and pit-membrane are pores that are visible in the light microscope and, thus, are in the range between 2000 Angstroms and 10 or more microns in diameter; 2) *Cell-wall capillaries*, such as the spaces between microfibrils and the cellulose molecules in the amorphous regions. Most of the cell-wall capillaries are closed when the walls are free of water, but open when moisture is absorbed. When wood and cotton are fully saturated with water the cell-wall capillaries attain their maximum dimensions. Some expand to about 200 angstroms in diameter, but most are substantially smaller. Grinding to expose all of the gross capillaries increases the surface area. Continued grinding, however, will only increase the accessible area slightly. Beyond this point a chemical treatment which can swell the pores of the cell wall is more likely to be effective in increasing the accessible surface area. The total surface area exposed in the gross capillaries is approximately one square meter per gram of wood or cotton. However, this is much smaller than the potentially available 300 square meters per gram which would be accessible to small molecules with the water-swollen cell-wall capillaries.

Several methods have been developed which can examine the structure of porous materials. Methods like electron microscopy, gas adsorption, and mercury porosimetry are powerful techniques, however, all require that the measurements be carried out in the absence of water and when water is removed from cellulose its pore structure collapses. Therefore, a good pore size measurement needs to be made on water-swollen cellulose samples directly because the water-swollen capillaries are the pores that the cellulases encounter.

A method for evaluating the accessibility of the capillary voids in water-swollen cellulose to enzyme molecules is the solute exclusion technique. This method uses different size polymer molecules as "molecular probes" to measure pore size according to the ability of these probes to penetrate into the cellulose. The polymer molecules commonly used are different molecular weight fractions of dextran and polyethylene glycol. By using these molecular probes, several studies (18,19,20,21) have reported the size characteristics of pores within different water-swollen cellulose samples. Stone, et al. (18), and Van Dyke (20) have used this approach to investigate the cellulose pore size in relation to its enzymatic accessibility and digestibility.

Stone, et al., made their measurements on phosphoric acid-swollen cotton. A linear relationship was found between the initial rate of enzymatic digestion and the surface area initially accessible to dextran molecule 40 angstroms in diameter. A molecule of this size is similar to the diameters which have been calculated for the cellulase enzymes.

Van Dyke has reported that untreated Solka Floc cellulose has pores which allow only 15% of its total internal volume to be accessible to a 35 x 210 angstrom enzyme. With this accessibility, his calculations estimate that about 35 square meters of surface would be available to this enzyme per gram of cellulose. This calculation assumes that the pores have a slit-like shape such as could be imagined for the spaces between sheets of microfibrils. Stone used this assumption in his surface area calculations as well.

Other factors are believed to affect enzymatic hydrolysis. Cowling (17) reviewed eight factors which he feels influence the susceptibility of cellulose to enzymatic degradation. These factors are: 1) the moisture content of the fiber, 2) the size and diffusibility of the enzyme molecules in relation to the size and surface properties of the gross capillaries, and the spaces between microfibrils and the cellulose molecules in the amorphous regions, 3) the degree of crystallinity of the cellulose, 4) its unit-cell dimensions, 5) the conformation and steric rigidity of the anhydroglucose units, 6) the degree of polymerization of the cellulose, 7) the nature of the substances with which the cellulose is associated, and 8) the nature, concentration, and distribution of substituent groups.

III. THE ENZYMES

1. *Review of Cellulase Research*

Cellulase refers to a group of enzymes which catalyze the hydrolysis of the β-1,4-linkages of the cellulose molecule. Experiments have shown that complete enzymatic cellulose hydrolysis of native cellulose to glucose actually occurs as a result of the presence of different enzyme components. A model involving only two components C_1 and C_x was proposed in 1950 by Reese, Siu, and Levinson (22) to describe the mechanism of the attack of cellulose by the cellulase system. The C_1 component, according to their theory, was responsible for modifying the native cellulose of high crystallinity in some way to make it more susceptible to the C_x component. The C_x component was believed to be responsible for the actual depolymerization reaction. This model was based on the observation that some cellulase systems, those containing both the C_1 and C_x components, could degrade native cellulose, soluble cellulose derivatives, and acid or base swollen celluloses, while other cellulases that contained only the C_x component, could only hydrolyze the soluble and swollen celluloses. In 1952, Reese and Levinson found it necessary to add a third component to this theory, a β-glucosidase which was found to catalyze the hydrolysis of cellobiose to glucose (23).

This theory, however, could not be put to a test at that time because no cell-free culture filtrates of cellulolytic organisms had been found which were capable of hydrolyzing crystalline cellulose. That is, no organism had been discovered which released the necessary C_1 enzyme into its environment. Then, in the early 1960's, Mandels and Reese (24) discovered that certain strains of the fungi *Trichoderma viride* and *Trichoderma koningii* produced culture filtrates which were highly active toward native cellulose. Following this discovery, these investigators and others (25,26) began to fractionate the proteins in culture filtrates from *T. viride* and *T. koingii* and other organisms which were subsequently found to produce an active extracellular enzyme complex. Gel filtration columns, ion-exchange columns, affinity columns, and isoelectric focusing (27) were fractionation techniques used in an attempt to isolate and purify the postulated C_1 component.

Considerable evidence to support the existence of a C_1 component was found. However, its mode of action was unclear and, therefore, many speculations were proposed to explain how native cellulose was rapidly hydrolyzed by the C_1, C_x, and cellobiase mixture. During the early 60's, it was suggested that the C_1 component could be an enzyme which ruptured the hydrogen bonds holding the cellulose chains together (28). Similarly, it was speculated that the C_1 component was just a biolotical factor that caused a relaxation of the hydrogen bonds (perhaps by binding more tightly to cellulose than cellulose binds to itself) enabling rapid hydrolysis by the C_x component. Another mechanism, proposed by Leatherwood (29), attributes the observed synergistic action* between the C_1 and C_x components to be a result of an affinity factor C_1 and a hydrolytic factor C_x. These factors, when combined, form a complete cellulase complex which is active towards insoluble crystalline cellulose. His experiments with cellulase from *Ruminococcus albus* prompted the following proposal: "The separate hydrolytic factor can hydrolyze soluble derivatives. However, to effectively hydrolyze insoluble cellulose, the hydrolytic factor must be held in position on the insoluble cellulose by the affinity factor."

Selby, on the other hand, experimented with his fractionated components from *T. viride* to see what effect the addition of the C_1 component had upon the adsorption of C_x onto Avicel. No effect was observed (30).

In the 1969 ACS symposium on cellulases, K.E. Eriksson (31) discussed some results that he and co-workers obtained during fractionation studies of the extracellular enzymes from *Chrysosporium lignorum*. His conclusion was that the C_1 component was actually an exoenzyme which attacked cellulose in an endwise manner to split off cellobiose units.

**The phenomenon that native cellulose is extensively hydrolyzed when both the C_1 and C_x components are present, but very little hydrolysis can be achieved by either component alone.*

The synergistic effect between a random-acting* cellulase C_X and this endwise-acting cellulase C_1 was postulated to occur in the following manner: Random-acting C_X enzymes first hydrolyze β-1,4-glucosidic linkages of the cellulose. Then, the endwise-acting C_1 enzymes associate themselves with the free chain-ends which were generated by the C_X component and proceed to cut off cellobiose units. This mechanism, contrary to Reese, et al., assigns the component which attacks amorphous cellulose (i.e., C_X) to the role of initiating the attack of native cellulose. Then the crystalline active component C_1 follows to release cellobiose as the major hydrolysis product. This mechanism has since gained general support. Wood and McCrae (4,32) have concluded similarly and gone on to propose, that for the successful hydrolysis of crystalline surfaces, adsorption of the C_1 enzyme must proceed the approach of C_X (33). Thus, when a C_X enzyme generates a chain end, the C_1 component can quickly associate with the chain and thereby prevent the reformation of the β-1,4-linkage just cleaved. Reformation of the β-1,4-linkage could occur in crystalline regions where the glucose units are firmly held in position by the hydrogen bonds of neighboring chains.

Other studies (34,35,36,37) have confirmed that the so called C_1 factor behaves like a β-1,4-glucan cellobiohydrolase (i.e., an enzyme which produces only cellobiose as the hydrolysis product) and that its mode of action is characteristic of an exoenzyme. As a result, most cellulase workers believe that Eriksson's mechanism is correct. However, King and Vessal (38) brought up the following important concept which should be considered when trying to understand the mode of enzyme hydrolysis of crystalline cellulose:

> "A close study of the oligosaccharide chains shows that only alternate glucosyl bonds are exposed to attack by enzyme, the other glucosyls being heavily masked and facing the opposite direction, into the body of the micells. Attack by enzymes onto the mass of such molecules that a crystal lattics represents would therefore be restricted to exposed glucosyl bonds, freeing fragments of D.P. = 2, 4, 6, 8, etc. These fragments would still be held in the crystal lattice by hydrogen bonds, two such bonds per glucosyl. Kinetic energy sufficient

**A random-acting cellulase will decrease the degree of polymerization of the cellulose during hydrolysis at a significantly greater rate than an endwise-acting enzyme.*

to break four hydrogen bonds would be needed to free a cellobiosyl into solution. Freeing a tetrasaccharide would involve rupturing eight hydrogen bonds.

Considering in addition the increasing mass as the fragment becomes longer, it is evident that a completely non-discriminating random glucanase acting on a crystalline substrate would yield more than twice as much cellobiose as cellotetraose, and the other oligosaccharides would be freed in even lesser amounts. Such a system could give all the appearances of an endwise acting glucanase producing essentially only cellobiose as a reaction product. For this reason the exclusive production of cellobiose from attack on such substrates cannot legitimately be interpreted as evidence of endwise attack. Inherently, then, the nature of the substrate invalidates an experimental approach to cellulase studies which is perfectly valid with the amylases and many other enzymes attacking water soluble polysaccharides."

An example of the type of experimental evidence for assigning the C_x cellulase to be an endoenzyme (random-cleaving) and the C_1 cellulase to be an exoenzyme (endwise-cleaving) is found in an article by Petterson, Axio-Fredricksson, and Berghem (37). They prepared an amorphous cellulose by dissolving the cellulose in cadoxen and precipitating it from solution with isopropanol. The change in the degree of polymerization of the reprecipitated cellulose, as determined by viscosity measurement, was plotted against the reaction time for both of their fractionated C_1 and C_x components. This graph is reproduced in Fig. 6. The amounts of the two enzymes were matched to give approximately the same amount of reducing sugar. Thus, it can be seen that both the C_1 and C_x enzymes are individually capable of hydrolyzing reprecipitated cellulose and that the C_x component causes a rapid decrease in the degree of polymerization while the C_1 component decreases the D.P. in a slow linear fashion. Neither fraction alone was able to significantly attack crystalline cellulose, whereas together they had no problem in transforming much of this same crystalline cellulose to soluble sugar.

It is easy to become confused by the multiplicity of names which have been introduced by different cellulase investigators for these three components. Table 1 summarizes many of the names of these components.

Table 1. A list of the names commonly used for the three cellulase components

General Category	Other names
β- Glucosidase (EC 3.2.1.21)	Cellobiase
β- Glucan cellobiohydrolase	C_1 Exoglucanase Exocellulase Avicelase Cellobiohydrolase
β- Glucan glucanohydrolase	C_x Endoglucanase Endo-β-glucanase CMCase

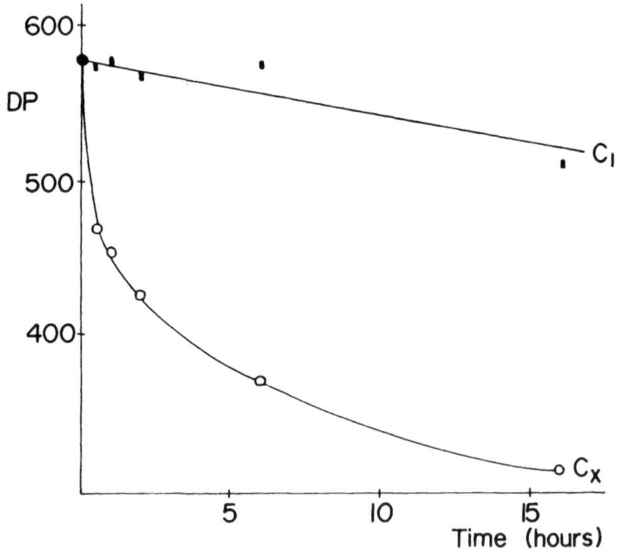

Figure 6. The change in degree of polymerization with time using C_1 and C_x cellulases (37)

2. *Structural Properties of Cellulases*

A detailed characterization of the molecular size and shape of a cellulase enzyme has been done by Whitaker, Calvin, and Cook (39) on a cellulase produced by the mold *Myrothecium verrucaria*. From sedimentation and diffusion experiments they determined the molecular weight of their enzyme to be approximately 63,000. Possible axial ratios of this enzyme were obtained from the measured frictional coefficient and the volume intrinsic viscosity. Then, assuming that this cellulase protein was 30% hydrated, which is commonly assumed for globular proteins, a probable axial ratio was determined to be six. An enzyme with a molecular weight of 63,000 and having an axial ratio of six, would have the approximate dimensions of 200 angstroms by 30 angstroms. If these results are accurate, the elipsoid cellulase would cover nearly 20 cellobiose units of an extended cellulose chain.

Although the shape of the cellulases from the *Trichoderma* genus have not been characterized, several reports have given the molecular weights of the enzyme fractions which have been isolated from the culture filtrates of this fungus. Table 2 lists some of the more recent molecular weight evaluations along with the corresponding carbohydrate content (when determined). The table also indicates whether or not the amino acid composition of the particular cellulase fraction was reported.

The largest component of the *Trichoderma viride* cellulase complex is the cellobiase, although it is not readily apparent from Table 2. The next largest component is the enzyme β-1,4-glucan cellobiohydrolase (C_1). The smallest cellulolytic enzymes of the complex are the β-1,4-glucan glucanohydrolases (C_x). A few investigators report the existence of a low-molecular-weight C_x component in addition to the more commonly identified high-molecular-weight variety. Multiple forms of each component are typically isolated from the culture filtrates of *Trichoderma viride*. There have been up to 3 cellobiase fractions isolated in some studies. Others report finding 4 cellobiohydrolases in a single culture. Up to 5 different endoglucanase proteins with different physical properties have been observed at times. Often, the components of the same type will have slightly different activities and modes of action towards a given cellulosic substrate. Gong, et al. (40), and Nakayama, et al. (41), show evidence which can explain why so many fractions of a particular component are often found in a culture filtrate. By analyzing the peaks with cellulase activity eluting from ion-exchange columns as a function of the age of the culture it was found that as the culture became more mature, more cellulase peaks appeared. Thus, the numerous cellulase components which have

Table 2. Physical Properties of *T. viride* cellulase.

Component	Estimated Molecular Weight*	pI**	Carbohydrate Content	Amino Acid Composition Given	Ref.
β-Glucosidase	76,000 c			No	40
	47,000	5.74	0%	Yes	50
β-1,4-Glucan cellobiohydrolase	71,000 c			No	40
	46,000 b		3.3%	No	49
	41,800 b	3.79	9.2%	Yes	51
	48,900		11.3%	Yes	52
	53,000		A 1.4%		
	53,000 a		B 5.8%		
	53,000		C 10.4%	Yes	42
	53,000		D 6.7%		
β-1,4-Glucan glucanohydrolase	18,000			No	40
(low molecular weight)	12,500 a	4.60	21%	No	48
(high molecular weight)	52,000			No	40
	58,900		17%	Yes	
	62,400 ì		16%	Yes	54
	60,200 b		10%	Yes	
	50,000	3.39	12%	No	48
	37,000		4.5%	Yes	
	52,000 d		15%	Yes	53
	49,000		15.2%	Yes	

*Method used for estimating the molecular weight: a) Gel filtration, b) Sedimentation Equil. c) Gel electrophoresis, d) Meniscus depletion.

**pI, the isoelectric pH, is the pH at which a protein molecule has no net electrical charge.

often been isolated could have resulted from protease activity on just a few precursor enzymes consisting of at least one cellobiase, one cellobiohydrolase, and one endoglucanase. Protease activity has been found in cellulase preparations (41). Others (42,43) have suggested that the multiplicity of components is due to differences in the sugar moieties on the enzymes.

Cowling (17) has summarized the molecular weights of cellulases from many different organisms which have been reported in the literature prior to 1975. He notes that if the enzymes are spherical, then enzymes with molecular weights ranging from 12,500 to 76,000 would correspondingly have diameters from 35 angstroms to 76 angstroms. If the enzymes are ellipsoid with an axial ratio of 6 as shown by Whitaker, et al., then they would vary from about 19 by 115 angstroms to 42 by 250 angstroms in width and length, respectively.

The amino acid composition of the cellulase components from *Trichoderma viride* have been determined (see Table 2 for references). In general, all of the cellulase components are rich in the acidic amino acids, aspartic and glutamic acid, and the polar amino acids, serine, threonine, and glycine, while only small amounts of basic amino acids and sulfur containing amino acids are found.

The carbohydrate content of these glycoproteins have been reported (references in Table 2). The wide variance of values probably reflects differences in techniques used as well as differences which might actually exist between the enzymes. The reports of the composition of the carbohydrate material are in good agreement. Most reports conclude that the major constituent is mannose which is combined with smaller amounts of glucose, galactose, and glucosamine. Arabinose and xylose are also detected on occasion.

3. Fragmentation of Cellulose Fibers by Cellulase

Although early studies (26,44) attributed the disintegration of cellulose fibers into small insoluble fragments to the action of the C_1 component, more recent observations using highly purified enzyme fractions assign this fragmentation to the C_x enzyme (45,46). Wood and McCrae (47) have shown that C_x components isolated from *T. koningii* were ten times more effective than the C_1 component in reducing the tensile strength of cotton yarn. Berghem (48) has purified two C_x enzymes from *T. viride* with markedly different molecular weights. One had a molecular weight of 12,500 while

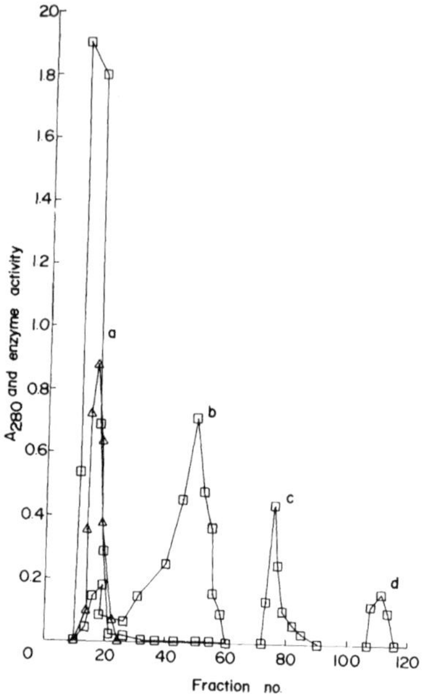

Figure 7. Affinity separation of cellulase using a Cellex cellulose column. (a) Peak containing cellobiase and Cm-cellulase, (b) peak containing CM-cellulase, (c) peak containing component C_2, and (d) peak containing cellobiohydrolase (56).

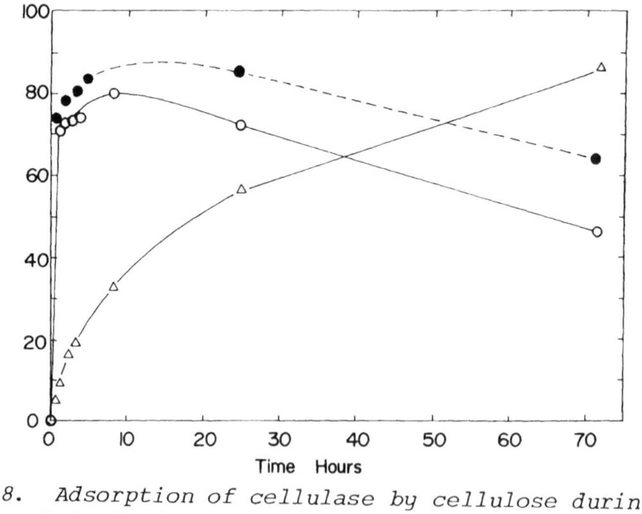

Figure 8. Adsorption of cellulase by cellulose during a 70 hour hydrolysis. (o——o) is % protein adsorbed. (o--o) is % filter paper activity adsorbed. (Δ——Δ) is the level of glucose produced in mg/ml (57).

the other was 50,000. Both were found to actively release free fibers from filter paper, but the low-molecular-weight C_x was estimated to be twice as effective as the high-molecular-weight component.

IV. THE INTERACTION BETWEEN THE SOLID SUBSTRATE AND THE ENZYME: ADSORPTION

In 1965, Li, Flora, and King (25) wrote that they passed a crude *T. viride* cellulase solution (10% w/v in 0.05 M sodium citrate buffer at pH 4.8) through a column of Avicel equal in volume to the enzyme sample at 3°C. Irrigation of the same buffer through the column was continued until ultraviolet absorbing material was no longer detected in the elute. Under these conditions all of the hydrocellulase (C_1) and approximately 5% of the CMCase activity were retained on the Avicel, while 95% of the CMCase activity and all of the β-glucosidase activity were recovered in the buffer effluent. They were then able to elute the adsorbed cellulase with distilled water.

Similar results were obtained by Ogawa and Toyama (55) when they passed a solution of *T. viride* cellulase through a column of Gauze cellulose. The nonadsorbed fraction contained two components haveing CMCase activity. The adsorbed protein consisted of three enzymes which could actively hydrolyze Avicel, filter paper, and CMC.

During a cellulase purification procedure, Pattersson, et al. (37), were able to remove the last traces of C_x enzyme from a C_1 preparation by passing this solution through a short column containing a cellulose gel prepared by cross-linking viscose followed by regeneration. They reported that the gel strongly adsorbed the C_x enzymes while the C_1 enzymes were only slightly adsorbed.

A procedure for affinity chromatographic fractionation of cellulases from *Trichoderma koningii* has been given by Halliwell and Griffin (56). Using the cellulose column system and conditions described in their article, they were able to separate cellobiase, carboxymethyl cellulase, cellobiohydrolase, and component C_2 (a cellulase which is highly effective in promoting the formation of short fibers). These components are released as highly purified fractions by decreasing the ionic strength of the eluent. Fig. 7 is a reproduction showing the results of their affinity separation procedure.

In 1971, Mandels, Kostick, and Parizek (57) reported some observations about cellulase adsorption. At the concentrations which they used, most of the enzyme was removed from solution in 30 minutes at 25°C, but a slower adsorption continued for several hours. At 50°C, the maximum adsorption was achieved at about 8 hours. By 24 hours the enzyme was returning to the solution due to extensive degradation of the cellulose (Fig. 8). They were unable to desorb and recover the adsorbed cellulase from the finely milled Solka Floc cellulose they were using with distilled water as Li, Flora, and King had done. The maximum amount of protein adsorbed using concentrated cellulase solutions was 0.064 mg protein/mg cellulose. The effect of particle size on adsorption was studied and a decrease in the particle size from 50μ to 6.7μ caused an increase in enzyme adsorption.

Leatherwood (29) reported the adsorptive behavior that he observed of two major cellulolytic components from *Ruminococcus albus* obtained by gel filtration on Sephadex G-100. The smaller component had an approximate molecular weight of 41,000, according to the gel filtration. The larger component was excluded from the G-100 Sephadex. Treatment of the larger component with β-mercaptoethanol caused it to disassociate into an active hydrolytic enzyme of the same size as the smaller component. The large component had a strong affinity for cellulose, whereas the small component did not adsorb. Leatherwood observed that the cellulase in these cultures adsorbed strongly to cellulose prior to any purification step. Also, the portion of hydrolytic factor that is adsorbed to cellulose decreased from over 90% in a young culture to less than 50% in the late stationary phase. This was explained in terms of a natural degradation of the cellulase resulting in a free, non-adsorbable hydrolytic factor.

An adsorption study of unfractionated *Trichoderma viride* QM 9414 cellulase on three different celluloses has been communicated by Peitersen, Medeiros, and Mandels (58). Milligrams of protein adsorbed per milligram of cellulose varied from 0.019 to 0.088 depending on the cellulose and the conditions for adsorption. Adsorption isotherms were reported for both protein adsorption and the adsorption of cellulase activity which resembled a Langmuir isotherm. The effect of pH and temperature on adsorption were evaluated. Their results indicated that changes in pH from 3.8 to 5.3 did not affect the adsorption of protein or enzyme activity. However, an increase in temperature from 20 to 50°C caused a decrease in adsorbed protein and in adsorbed enzyme activity. The specific adsorption, activity adsorbed/mg protein, also decreased.

Ghose and Bisaria (59,60) also found that adsorption of cellulase components decreased with increasing temperature (from 5 to 50°C), however, they observed that the specific adsorption of the enzyme components was highest at 50°C. They also observed that the adsorption of enzyme components decreased with an increase or decrease of pH from 4.8. This temperature and pH are the optimum conditions for the enzymatic cellulose hydrolysis reaction. The effects of pH and temperature on desorption were also reported. Their study shows that enzymatically active proteins are preferentially adsorbed to the substrate, compared to inactive protein. An interesting result of Ghose and Bisaria's investigation of the adsorption behavior of cellulase and xylanase onto bagasse, the cellulosic material of sugar cane, was their measurement of the rates and amount of adsorption of the activity of individual enzyme components when they were adsorbed alone and in combination. The four enzyme components studied were an endoglucanase and an exoglucanase from *Trichoderma koningii* IMI73022, a β-glucosidase from *F. solani* IMI95994, and a xylanase from *Trichoderma reesei* QM9414. The xylanase component was included to hydrolyze the hemicellulose in the bagasse. The rates of adsorption are shown in Figs. 9a and b. Maximum adsorption values for this system were achieved in about 10 minutes at 5°C. There did not appear to be any competition between the three adsorbing components* for the surface area (assuming adsorption was not controlled by mass transfer) because the rates of adsorption were almost identical for both individual and combined adsorption. Also the percent adsorption of the three components when they were combined was not much less than when they were alone, although there do appear to be some surface sites at which all three components can adsorb.

Wilke and Yang (61) have exposed 5% suspensions of newsprint to *T. viride* cellulase. They likewise observed a very strong and rapid initial adsorption of the enzyme and that it was later gradually released. Their study revealed that C_x and C_1 activities are adsorbed to about the same extent on this substrate. An attempt was made to release the adsorbed enzyme by resuspension of the hydrolyzed cellulosic residue in fresh buffer, but none was released.

Castanon and Wilke (62) have further investigated the adsorption of *T. viride* onto newspaper. The C_1, C_x, and filter paper activities and total protein adsorbed were measured at intervals during the incubation and the results plotted verses time as shown in Fig. 10. Again notice the rapid rate at which the initial adsorption occurred. This was followed by a continued slower adsorption. As evident from the figure, not all the cellulase activities were adsorbed at the same rate. This indicated that selective adsorption was occurring.

**The β-glucosidase did not adsorb.*

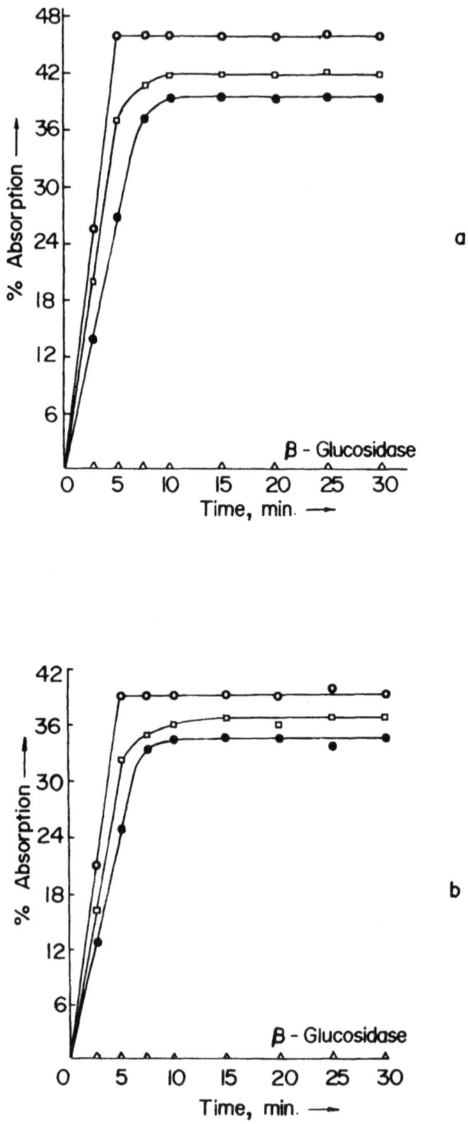

Figure 9. Adsorption of cellulase components on bagasse at 5°C. (a) Adsorption of individual components when present alone, (b) adsorption of the components when present together. (o——o) is the % xylanase activity adsorbed. (□——□) is the % endoglucanase adsorbed. (o——o) is the % exoglucanase adsorbed. (Δ——Δ) is the % β-Glucosidase adsorbed (60).

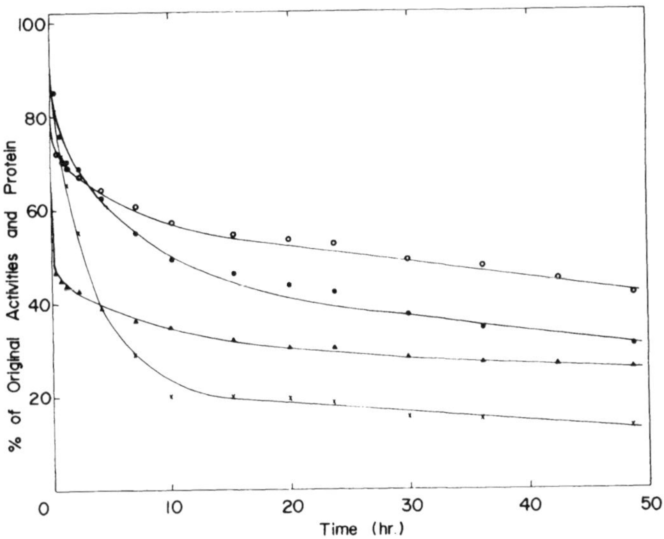

Figure 10. Adsorption of cellulase onto cellulose with time. (o—o) is the % C_x activity adsorbed. (●——●) is the % C_1 activity adsorbed. (Δ——Δ) is the % filter paper activity adsorbed. (x——x) is the % total protein adsorbed (62).

During the first two hours of incubation, enzymes with C_x activity were adsorbed preferentially over cellulases with C_1 activity was greater. Castanon and Wilke attributed this selective adsorption to the fact that initially there was more amorphous cellulose while later crystalline cellulose prevailed. Fig. 10 also shows that no cellulase was released back into solution during the 50 hour incubation. Even more extensive digestion of the cellulose in the newspaper failed to release any enzyme. This evidence indicated that once the cellulases were adsorbed, they were permanently immobilized on the substrate.

The individual protein fractions of their cellulase preparation were analyzed by separating them according to their molecular weight using SDS-polyacrylamide gel electrophoresis. This was done to monitor the adsorption of the individual protein components during the hydrolysis. The results from this analysis showed that different protein fractions were adsorbed at different rates and to different amounts and that no release of any of these fractions occurred as the hydrolysis of the newspaper proceeded. However, when adsorption measurements are made using pure cellulose, a release of

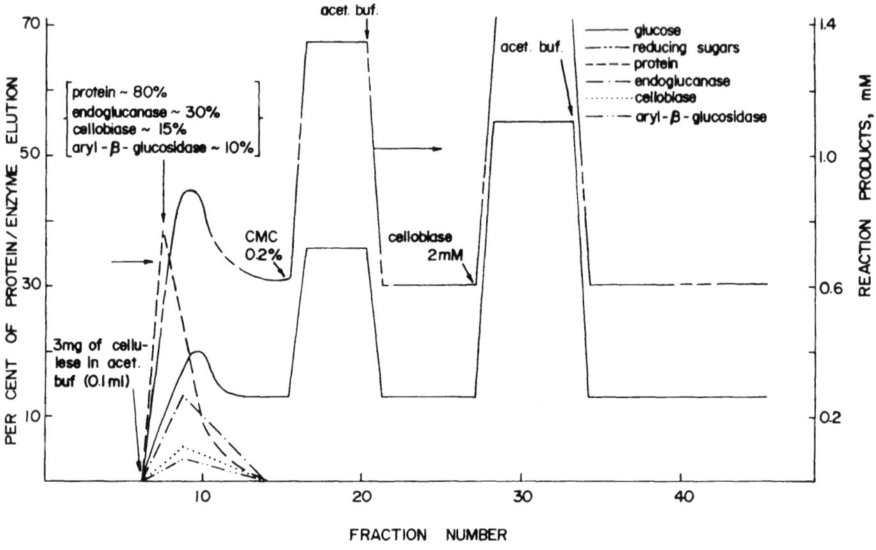

Figure 11. Elution of cellulase and reaction products from a microcrystalline cellulose column loaded with a small volume of cellulase. The conditions were: pH 4.5, 25°C (63).

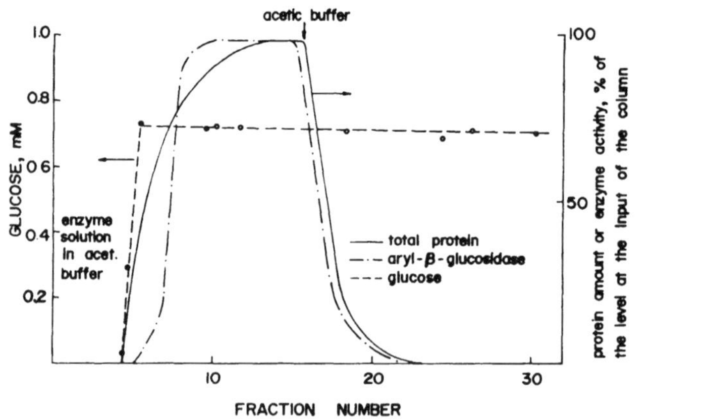

Figure 12. Elution of cellulase and reaction products from a column of microcrystalline cellulose initially saturated with enzyme before washing with buffer. The conditions were: pH 4.5, 15°C (63).

cellulase was observed. The reasoning suggested for this was that pure cellulose can be hydrolyzed much more extensively, i.e., leaving behind much less non-reactive residue, than the lignin containing newsprint.

Experiments were conducted to recover the unadsorbed cellulase left in solution after hydrolysis by adsorbing it onto fresh newspaper. It was found that recovery depended upon the length of hydrolysis. Approximately 40% of the cellulase present in a batch of enzyme could be recovered from a 10 hour hydrolysis. A solution from a 48 hour hydrolysis could only provide about 20% of its original enzyme to the fresh newspaper.

A recent review by Klyosov and Rabinowitch (63) gives a summary of the work in the Soviet Union being done on the enzyme conversion of cellulose to glucose. Their intensive review contains a section describing experiments of adsorption of cellulolytic enzymes onto cellulose. Some experimentally determined isotherms for the cellulase components are given which were adeauately described by the Langmuir equation. The kinetics of adsorption were investigated in a batch reactor showing again that the enzymes are rapidly adsorbed (in approximately 10 to 15 minutes). They observed that removal of unbound enzymes from solution did not lead to any significant release of enzyme back into solution. Their experiments indicated that there were both specific* and non-specific adsorption occurring. They felt that this non-specific adsorption onto the solid interface dominated over specific adsorption and that both types of adsorbed enzyme remain in the adsorption layer and retain their capacity to form specific complexes with the substrate. They called this phenomenon "delocalized adsorption of enzymes; which means that the enzymes may migrate on the surface of the solid particles and initiate hydrolytic reactions at different points on the substrate."

The Soviets also did an interesting experiment with a column reactor of microcrystalline cellulose powder. The result of this experiment is reproduced in Fig. 11. The column was loaded with a small volume of a highly concentrated cellulase solution and followed by elution with buffer. They found that the product concentration eluting from the column was maximal in the fractions which contained the unadsorbed cellulases passing through the column. This concentration

* Enzymes adsorbed specifically are those which are directly responsible for the formation of an enzyme-substrate complex.

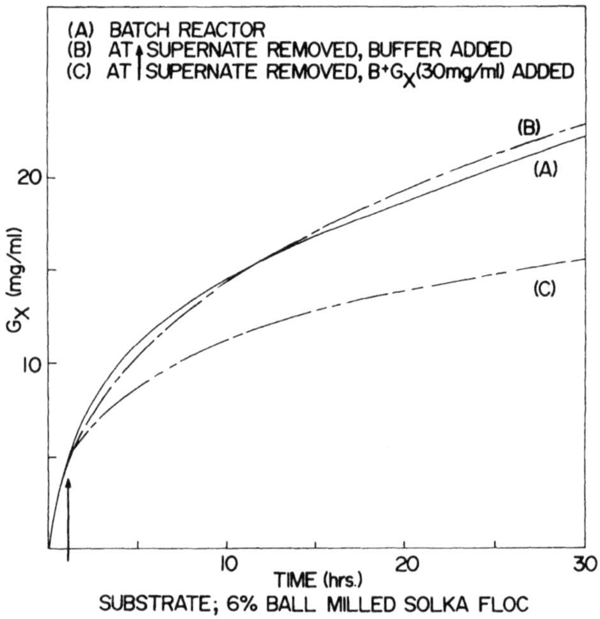

Figure 13. Product concentration curves from the digestion of 6% ball milled Solka Floc in a batch reactor (64).

peak came shortly after the peak for inert protein in the column elute. This result indicated that there is a type of cellulase adsorption which is reversible. (It is possible that the cellulose is initially more reactive.) However, the irreversibility of enzyme adsorption was illustrated, too, because in the course of further elution, little or no desorption of the bound cellulase occurred and the concentration of products leaving the column decreased to a constant level. Washing the column with CMC and cellobiose solutions did not displace the cellulase from the Avicel; instead, the adsorbed cellulase catalyzed the hydrolysis of the CMC and cellobiose*. Fig. 12 shows the result of saturating the cellulose column with cellulase before washing with buffer. The bound enzyme continues to actively produce glucose as the glucose concentration exiting the column is not decreased by the buffer wash.

From this result and the results of the batch experiments, it was postulated that some enzyme is initially adsorbed reversibly and that it gradually becomes bound irreversibly. This gradual immobilization was attributed to the progressive modification of the cellulose surface by the cellulase action. Furthermore, they feel that the nature of the effect of the irreversibly bound cellulase protein on the solid crystalline cellulose might be similar to the effect of a surfactant at an interface. Therefore, the resulting decrease in strength of the solid cellulose particles by could be caused by a phenomenon known as adsorption-induced strength lowering. This type of action appears to be related to the early speculation that the C_1 component might be a protein which binds more tightly to cellulose than cellulose binds to itself.

Kim (64) has given evidence with a batch reactor that the adsorbed cellulase enzymes are primarily responsible for the hydrolysis reaction. Curve A of Fig. 13 shows a typical digestion curve obtained by adsorbing the cellulase from the QM 9414 mutant of *T. viride* onto a 6% suspension of ball milled Solka Floc at 50°C, pH 4.8. Curve B was obtained by removing a portion of the hydrolysis mixture after one hour of reaction, centrifuging, resuspending the cellulose solids in fresh buffer, and incubating this new mixture. Notice that the production of reducing sugar for curve B is slightly higher than that of curve A. This increase for curve B was attributed to a decrease in product inhibition as

**This supports the postulate that not all of the enzymes are specifically adsorbed. Notice that cellobiase activity has been adsorbed.*

approximately 4 mg/ml reducing sugar was removed in the supernate. Curve C illustrates more clearly the effect of product inhibition on sugar production. For this case, the solids (from centrifugation) were resuspended in fresh buffer containing glucose at a concentration of 30 mg/ml.

Cellulase Desorption

A closely related subject to cellulase adsorption, and one that has received even less attention, is the desorption and recovery of active cellulase from spent cellulose residue. As cellulase production represents the most expensive aspect of the enzymatic conversion of cellulose to alcohol, it would be a big breakthrough for the economics of this process if substantial quantities of the catalyst could be reused. To do this, conditions need to be found which desorb cellulase without inactivating it.

Several reports have indicated the successful application of distilled water, i.e., a reduction in the solution ionic strength, to remove cellulase from its adsorbed state. Yet others have not. The conclusions of Klyosov and Rabinowitch would suggest that the extent of cellulose digestion would be an important factor influencing desorption by a distilled water wash. Thus, if contact is brief and is done at low temperatures (0 to 5°C), then water or even buffer washings would desorb the cellulase.

A recent article by Reese (65) reports the results of desorption experiments using cellulase from *Trichoderma reesei*. Over 35 different compounds were screened to evaluate their effectiveness in eluting cellulase from Avicel. Alkaline solutions were found to be very effective agents for desorbing the cellulase protein from Avicel, even at low concentrations (0.02 N), however, much inactivation of the cellulase also occurred. With care, some success was still obtained with an alkaline solution of $Ca(OH)_2$. The best eluting agent tested was 6M urea. The advantages of this agent were: 1) 90% of the adsorbed enzyme was eluted containing much of its original activity and, 2) the enzyme was readily recovered from the 6M urea by precipitation with cold acetone. An even better agent for the elution of cellulase from Avicel was 4M guanidine, but it inactivated the enzyme.

Reese found that the alcohols: methanol, ethanol, and propanol had increasing cellulase eluting power as the alcohol became less polar. Because this ordering of alcohols, methanol < ethanol < propanol, also has an increasing tendency to unfold, i.e., denature, protein molecules. Reese speculates that a conformation change may be required to desorb cellulase from cellulose. Thus, perhaps, only denaturants can be good eluents.

V. THE INTERACTION BETWEEN THE SOLID SUBSTRATE AND THE ENZYME: HYDROLYSIS REACTION KINETICS

Evidence presented earlier in this review suggests that hydrolysis reaction kinetics should be based on the amount of cellulase adsorbed. However, very little data exist which relate the rate of product formation to the amount of adsorbed cellulase protein. Most sutides relate the hydrolysis kinetics to the weight of cellulose and the level of cellulase activity initially added to a reaction vessel. Thus, little has been learned about the specific hydrolysis rates resulting from the cellulase directly in contact with cellulose. Even if a certain percentage of enzyme adsorbed is not active or its active site is not directed toward the substrate, a direct relationship is still likely to exist.

Actually, a complete description of the kinetics of this reaction is much more involved than just relating the rate of product formation as a simple function of adsorbed enzyme. Ideally, kinetic modeling should be structured to describe the rate limiting step(s) of the physical events occurring. A list of physical events occurring individually for both C_x and C_1 enzymes might include:

- Mass transfer of the enzymes from the bulk solution to the surfaces on and within the porous solid substrate.
- Enzyme adsorption on the cellulose surface. Contributing factors may include: 1) cellulose crystallinity, 2) availability of chain ends, and 3) previous enzyme conditioning.
- The actual kinetics of the hydrolysis reaction. Contributing factors may include: 1) cellulose crystallinity, 2) positioning of cellobiose units, e.g., extent of hydrogen bonding to neighboring chains, and 3) length of cellulose chains.
- Mobility of cellulase in the adsorbed layer. Contributing factors may include: 1) number of enzymes in adsorbed layer (possibility of collisions), 2) temperature, and 3) physical agitation of the reaction mixture.

The kinetics resulting from these physical events should be tempered with

Product inhibition of the adsorbed cellulase. Contributing factors may include: 1) a dependence on the rate of desorption and diffusion of product away from the reaction site, and 2) presence of cellobiase to reduce cellobiose inhibition of adsorbed cellobiohydrolase.

Table 3. A Summary of Assumptions Which Have Been Made to Construct Kinetic Models for Enzymatic Reactions on Solid Substrates.

Model Assumptions	Amemura & Terui (74)	McLaren & Packer (75)	Ghose & Das (76)	Ross & Updegraff (77)	Kim (68)	Suga van Dedem & Moo-Young (78)	Suga van Dedem & Moo-Young (79)
1) Diffusion of Enzyme into the Substrate	No	No	No	Yes	No	Yes	No
2) Enzyme Adsorption (Isotherm)	No	Yes	No	No	Yes	No	No
3) Enzyme Multiplicity (C_1 and C_x Components)	No	No	No	No	Yes	No	Yes
4) Substrate Multiplicity a) Crystalline and Amorphous b) Degree of Polymerization	Yes	No	No	No	Yes (a)	No	Yes (b)
5) Decrease in Adsorption Sites	No	No	No	No	Yes	No	No
6) Product Inhibition a) Competitive b) Noncompetitive	Yes (a)	No	Yes (a)	No	Yes	No	No
7) Enzyme Deactivation	No	No	No	No	Yes	No	No
Use of Michaelis-Menten Kinetics	Yes	No	Yes	Yes	No	Yes	Yes
Numerical Solution Required	No	No	No	No	Yes	Yes	Yes

Table 3. A Summary of Assumptions Which Have Been Made to Construct Kinetic Models for Enzymatic Reactions on Solid Substrates (continued)

Model Assumptions	Huang (80)	Howell & Stuck (81)	Humphrey (82)	Howell & Mangat (83)	Okazaki & Moo-Young (84)	Lee et al. (85)	Peitersen & Ross (86)	Ryu et al. (87)
1) Diffusion of Enzyme into the Substrate	No	No	No	No	No	No	No	No
2) Enzyme Adsorption (Isotherm)	Yes	No	Yes	No	No	No	No	Yes
3) Enzyme Multiplicity (C_1 and C_x Components)	No	No	Yes	No	Yes	Yes	No	No
4) Substrate Multiplicity	No	No	No	No	Yes	Yes	Yes	Yes
a) Crystalline and Amorphous							(a)	(a)
b) Degree of Polymerization					(b)	(b)		
5) Decrease in Adsorption Sites	No	No	Yes	No	No	No	No	No
6) Product Inhibition	Yes	Yes	Yes	Yes	Yes	Yes	Yes	Yes
a) Competitive	(a)			(a)				
b) Noncompetitive		(b)						
7) Enzyme Deactivation	No	No	No	Yes	No	No	No	No
Use of Michaelis-Menten Kinetics	Yes	Yes	No	Yes	Yes	Yes	Yes	Yes
Numerican Solution Required	No	No	Yes	No	Yes	Yes	Yes	Yes

Gradual deactivation and/or desorption of enzyme as hydrolysis proceeds.

Of course, each of these steps could potentially be pH and temperature sensitive, some combinations of enzyme and substrate might have different reaction controlling steps than others, and the rate limiting step could change as the hydrolysis proceeds. The presence of inert, non-digestible material would also have its contributing effects.

The first column of Table 3 lists seven assumptions having physical meaning which have been used to formulate kinetic expressions for the description of this heterogeneous reaction. Each subsequent column lists whether or not a particular assumption was made in the model proposed by the authors shown at the top. The equations for most of these models have been reviewed by Lee, et al. (66).

The kinetic models proposed by McClaren and Packer, Kim, Huang, Humphrey, et al., Lee, and Ryu, et al. (see table for references) incorporate a Langmuir isotherm to relate the adsorbed cellulase concentration to that in solution. Possibly the most extensive model proposed was that given by Kim. Of course, a numerical solution was necessary to solve the large set of rate equations he was postulating.

Many of the models of Table 3 have relied extensively on Mechaelis-Menten kinetics for their derivations. Some assumptions associated with these kinetics are:

1. the amount of substrate available for enzyme attack can be expressed in terms of its mass concentration.
2. the substrate concentration greatly exceeds the enzyme concentration, and
3. when integrated to express the product formed as a function of time, it is assumed that $(S) = (S_o) - (P)$, where (S), (S_o), and (P) are the concentrations of substrate, initial substrate, and product, respectively.

However, these assumptions do not apply to enzymatic reactions on a solid substrate.

Most models proposed are tested by their ability to predict the time course of product formation during enzymatic digestion of cellulose. To gain a better understanding of the hydrolysis reaction, models need to be constructed and tested which describe certain segments of the reaction, i.e., hydrolysis during initial, intermediate, and latter stages of incubation. Good convenient methods of experimental analysis are needed to test these models. Then, because of the complexities of thie reaction, numerical solutions may be required to couple the individual equations (which contain the known rate constants, capacities, equilibrium constants, etc., for each state) to correctly describe the interaction between the cellulose enzymes and the solid cellulose substrate.

REFERENCES

1. Reese, E.T. and Mandels, M. In High Polymers (Cellulose and Cellulose Derivatives)," (Segal, L., ed.), Wiley Interscience, New York, p. 1079 (1971).
2. Nisizaws, K. *J. Ferment. Technol. 51*(4), 267 (1973).
3. Emerg, G.H., Gum, E.K., Jr., Lang, J.A., Liu, T.H., and Brown, R.D., Jr. *Adv. Chem. Ser. 136,* 79 (1974).
4. Wood, T.M. *Biotechnol. Bioeng. Symp. 5,* 111 (1975).
5. Ghose, T.K. In "Adv. in Biochem. Eng." (Fiechter, A., ed.), Springer-Verlag, Berlin, p. 39 (1977).
6. Lee, Y.H. and Fan, L.T. In "Adv. in Biochem. Eng.," (Fiechter, A., ed.), Springer-Verlag, Berlin, p. 101 (1980).
7. Gong, C.S. and Tsao, G.T. In "Ann. Rev. of Ferment. Proc.," (Perlman, D., ed.), Academic Press, New York, p. 111 (1979).
8. Halliwell, G. *Biochem. J. 79,* 185 (1961).
9. Fengel, D. *J. Polym. Sci.: Part C 36,* 383 (1971).
10. Hess, K., Mahl, H., and Gutter, E. *Kolloid Z. 155,* 1 (1957).
11. Rowland, S.P. and Roberts, E.J. *J. Polym. Sci. Part A-1 10,* 2447 (1972).
12. Chang, M. *J. Polym. Sci. Part C 36,* 343 (1971).
13. Chang, M. *J. Polym. Sci. 12,* 1349 (1974).
14. Clark, G.L. In "Applied X-Rays 4th Ed." McGraw Hill, New York, p. 782 (1955).
15. Hudson, S.M. and Cuculo, J.A. *J. Macromol. Sci. - Rev. Macromol. Chem. C18*(1), 1 (1980).
16. Ghose, T.K. and Kostick, J.A. *Adv. Chem. Ser. 95,* 415 (1969).
17. Cowling, E.B. *Biotechnol. Bioeng. Symp. 5,* 163 (1975).
18. Stone, J.E., Scallan, A.M., Donefer, E., and Ahlgren, E. *Adv. Chem. Ser. 95,* 219 (1969).
19. Nelson, R. and Oliver, D.W. *J. Polym. Sci. Part C36,* 305 (1971).
20. Van Dyke, B.H., Jr. Ph.D. Dissertation, MIT, Cambridge, MA (1972).
21. Kuga, S. *J. Chromatog. 195,* 221 (1980).
22. Reese, E.T., Siu, R.G.H., and Levinson, H.S. *J. Bacteriol. 59,* 485 (1950).
23. Reese, E.T. and Levinson, H.S. *Physiol. Plantarum 5,* 345 (1952).
24. Mandels, M. and Reese, E.T. *Develop. Ind. Microbiol. 5,* 5 (1964).
25. Li, L.H., Flora, R.M., and King, K.W. *Arch. Biochem. Biophys. 111,* 439 (1965).

26. Selby, K. and Maitland, C.C. *Biochem. J. 104,* 716 (1967).
27. Eriksson, K.E. *Adv. Chem. Ser. 95,* 83 (1969).
28. Siu, R.G.H. In "Advances in Enzymatic Hydrolysis of Cellulose and Related Materials," (Reese, E.T., ed.), Pergamon Press, London, p. 257 (1963).
29. Leatherwood, J.M. *Adv. Chem. Ser. 95,* 53 (1969).
30. Selby, K. *Adv. Chem. Ser. 95,* 34 (1969).
31. Eriksson, K.E. *Adv. Chem. Ser. 95,* 59 (1969).
32. Wood, T.M. and McCrae, S.I. *Biochem. J. 128,* 1183 (1972).
33. Wood, T.M. and McCrae, S.I. *Adv. Chem. Ser. 181,* 181 (1979).
34. Halliwell, G. and Griffen, M. *Biochem. J. 135,* 587 (1973).
35. Nisizawa, K., Tomita, Y., Kanda, T., Suzuki, H., and Wakabayashi, K. *Proc. IV Int. Ferment. Symp.* 719 (1972).
36. Wood, T.M. *Proc. IV Int. Ferment. Symp.,* 711 (1972).
37. Pettersson, L.G., Axio-Fredriksson, U.B., and Berghem, L.E.R. *Proc. IV Int. Ferment. Symp.* 727 (1972).
38. King, K.W. and Vessel, M.I. *Adv. Chem. Ser. 95,* 7 (1969).
39. Whitaker, D.R., Colvin, J.R., and Cook, W.H. *Arch. Biochem. Biophys. 49,* 259 (1954).
40. Gong, C.S., Ladisch, M.R., and Tsao, G.T. *Adv. Chem. Ser. 181,* 261 (1979).
41. Nakayama, M., Tomita, Y., Suzuli, H., and Nisizawa, K. *J. Biochem. 79,* 955 (1976).
42. Gum, E.K., Jr. and Brown, R.D., Jr. *Biochem. Biophys. Acta 492,* 225 (1977).
43. Eriksson, K.E. and Petersson, B. *Arch. Biochem. Biophys. 124,* 142 (1968).
44. Liu, T.H. and King, K.W. *Arch. Biochem. Biophys. 120,* 462 (1967).
45. Halliwell, G. and Riaz, M. *Biochem. J. 116,* 35 (1970).
46. Pettersson, L.G. In "Proc. Symp. Enz. Hydrol. Cellulose," (Linko, M., ed.), Aulanko, Finaldn, p. 255 (1975).
47. Wood, T.M. and McCrae, S.I. In "Proc. Symp. Enz. Hydrol. Cellulose," (Linko, H., ed.), Aulanko, Finald, p. 231 (1975).
48. Berghem, L.E.R., Pettersson, L.G., and Axio-Fredricksson, U.B. *Eur. J. Biochem. 61,* 621 (1976).
49. Berghem, L.E.R., Pettersson, L.G., and Axio-Fredricksson, U.B. *Eur. J. Biochem. 37,* 21 (1973).
50. Berghem, L.E.R. and Pettersson, L.G. *Eur. J. Biochem. 46,* 295 (1974).
51. Berghem, L.E.R., Pettersson, L.G., and Axio-Fredricksson, U.B. *Eur. J. Biochem. 53,* 55 (1975).
52. Gum, E.K., Jr. and Brown, R.D., Jr. *Biochim. Biophys. Acta 446,* 371 (1976).

53. Shoemaker, S.P. and Brown, R.D., Jr. *Biochim. Biophys. Acta 523,* 147 (1978).
54. Okada, G., Nisizawa, K., and Suzuki, H. *J. Biochem. 63,* 591 (1968).
55. Ogawa, K. and Toyama, N. *J. Ferment. Technol. 50,* 236 (1972).
56. Halliwell, G. and Griffin, M. *Biochem. J. 169,* 713 (1978).
57. Mandels, M., Kostick, J., and Parizek, R. *J. Polym. Sci. Part C36,* 445 (1971).
58. Peiterson, N., Medeiros, J., and Mandels, M. *Biotechnol. Bioeng. 19,* 1091 (1977).
59. Bisaria, V.S. and Ghose, T.K. In "Proc. Bioconversion Symp.," IIT Delhi, India, p. (1977).
60. Ghose, T.K. and Bisaris, V.S. *Biotechnol. Bioeng. 21,* 131 (1979).
61. Wilke, C.R. and Yang, R.D. In "Proc. Symp. on Enzymatic Hydrol. of Cellulose," (Linko, M., ed.), Linko, Aulanko, Finland, p. 485 (1975).
62. Castanon, M. and Wilke, C.R. *Biotech. Bioeng. 22,* 1037 (1980).
63. Klyosav, A.A. and Ribinowitch, M.L. In "Enzyme Engineering," (Klyosav, A.A., ed.), Plenum Press, New York, p. 83 (1978).
64. Kim, C. In "ARO Report 74-2, Proceedings of the 1974 Army Numerican Analysis Conference," Office of the Chief of Research, Development and Acquisition, p. 507 (1974).
65. Reese, E.T. *Process Biochemistry 17*(3), 2 (1982).
66. Lee, Y.H., Fan, L.T., and Fan, L.S. In "Adv. in Biochem. Eng.," (Fiechter, A., ed.), Springer-Verlag, Berlin, p. 131 (1980).
67. Amenura, A. and Terui, G. *J. Ferment. Technol. 43,* 281 (1965).
68. McLaren, A.D. and Packer, L. In "Advances in Enzymology," (Nord, F.F., ed.), Interscience, New York, p. 245 (1970).
69. Ghose, T.K. and Das, K. In "Adv. in Biochem. Eng.," (Fiechter, A., ed.), Springer-Verlag, Berlin, p. 55 (1971).
70. Ross, L.W. and Updegraff, D.M. *Biotechnol. Bioeng. 13,* 99 (1971).
71. Suga, K., van Dedem, G., and Moo-Young, M. *Biotechnol. Bioeng. 17,* 185 (1975).
72. Suga, K., van Dedem, G., and Moo-Young, M. *Biotechnol. Bioeng. 17,* 433 (1975).
73. Huang, A.A. *Biotechnol. Bioeng. 17,* 1421 (1975).
74. Howell, J.A. and Stuck, J.D. *Biotechnol. Bioeng. 17,* 873 (1975).

75. Humphrey, A.E. *Adv. Chem. Ser. 181,* 25 (1979).
76. Howell, J.A. and Mangat, M. *Biotechnol. Bioeng. 20,* 847 (1978).
77. Okazaki, M. and Moo-Young, M. *Biotechnol. Bioeng. 20,* 637 (1978).
78. Lee, S.E., Arminger, W.B., Watteeuw, C.M., and Humphrey, A.E. *Biotechnol. Bioeng. 20,* 141 (1978).
79. Peitersen, N. and Ross, E.W. *Biotechnol. Bioeng. 21,* 997 (1979).
80. Ryu, D.D.Y., Lee, S.B., Tassinari, T., and Macy, C. *Biotechnol. Bioeng. 24,* 1047 (1982).

CHAPTER 13

AUTOMATION AND OPTIMIZATION OF FERMENTATION PROCESSES

Daniel N. Bull

New Brunswick Scientific Co., Inc.
Edison, New Jersey

There has been a number of conferences, symposia, and publications on automation and optimization of fermentation processes since the subject was last reviewed in volume 2 of "Annual Reports on Fermentation Processes" by Weigand (1). His report, and the preceding one by Dobry and Jost (2), were concerned primarily with computer applications to fermentations. We can take it for granted that most fermentation companies routinely use computers for data acquisition (3), and we have reached a point where computer control of fermentations is also no longer novel. A publication is no longer significant because another configuration of computer and interfacing was described, and a method of control of, say, the respiratory quotient was demonstrated. This is an indication of healthy progress, and we are at the point where it can be asked what advance a particular configuration represents.

With this in mind, this review will survey some of the developments in automation and optimization of fermentations over the last several years. Certainly no claim of completeness can be made, and perhaps some arguably less important papers will be mentioned, while seemingly significant contributions might be neglected. The intent is to provide a broad overview of the main trends, and to point out the current problems and road blocks.

Progress can be arbitrarily categorized into a few principal areas. These are developments in 1) new sensors and devices, 2) automation, computerization, and control, and

ISBN 0-12-040306-4

3) optimization. A closely related fourth topic, modeling, is not strictly speaking an automation technique and will be considered only peripherally here.

I. SENSORS, DEVICES, AND MEASUREMENT TECHNIQUES

The most sophisticated automation and control processes are of no use if information on the status of a fermentation is not available. Accordingly, progress in methods of sensing and measuring fermentation state variables is of cardinal importance. In fact, it is probably true that the lack of practical means to make on-line determinations of all but a few variables is the single largest obstacle to wide-spread use of computer control of fermentation processes.

Mor (4) has recently reviewed the major classes of instrumental analysis in fermentation technology. Some progress has been made in the design of on-line, real-time (i.e., response time in minutes or seconds) sensors. Most of the advances of the past few years have utilized membrane technology in some way. Various membrane separators have been reported for the detection of gases and volatile compounds. These are all simply extensions of the tubing method for measuring dissolved oxygen, reported by Philips and Johnson in 1961 (5), to the detection of other materials. Lee, et al. (6), used a semipermeable membrane to separate an ethanol-containing gas stream from a broth, and then detected and measured the alcohol with a semiconductor gas sensor. A response time of one minute was claimed. The device worked better with some fermentations than with others. Heinzle and co-workers (7), have similarly reported on use of a carrier gas with a porous membrane to measure gases and volatiles.

Steam sterilizable carbon dioxide sensors have been reported in the literature and are now commercially available. A CO_2 electrode which can be recalibrated without being removed from the fermentor was reported by Puhar, et al. (8). Another steam sterilizable CO_2 electrode was described by Shoda and Ishikawa (9). Both electrodes are based on the same principle. A pH electrode is immersed in a bicarbonate solution in an electrode holder. The bicarbonate solution is isolated from the fermentation broth by means of a gas permeable membrane. Carbon dioxide permeates the membrane and alters the pH of the bicarbonate solution in accordance with the Henderson-Hasselbach equation. If there is equilibrium on both sides of the membrane, then the pH of the bicarbonate solution is an accurate measure of the activity of CO_2 in the fermentation broth. It must be emphasized that these electrodes do not measure CO_2

concentration, although it could be calculated from the Henry's Law constant if it were known. Rather it is the activity or the partial pressure of CO_2 which is measured. In this respect the electrodes are similar to the various dissolved oxygen electrodes which have been available for many years, and share many of the same deficiencies.

One of the major deficiencies of the dissolved oxygen electrodes has been the lack of a method to calibrate the electrode *in situ* during the course of a fermentation. Puhar, et al., have made some progress in this matter. Their design allows the flushing of the electrode chamber with buffer solutions, which makes it possible to recalibrate the pH electrode. In principle, the electrode described by Shoda and Ishikawa can be removed and similarly recalibrated, so long as the transmembrane pressure is not great enough to cause failure while the pH electrode is not in place. However, drift of the pH electrode is unlikely to be the source of inaccuracies of the CO_2 measurements. Changes in diffusion characteristics of the membrane, e.g., due to formation of a film of fermentation solids on the external side of the membrane, are far more likely to be a source of trouble. However, the availability of any sterilizable-in-place CO_2 electrode is a welcome advance in fermentation technology.

Another development has been the widespread trend to the use of mass spectroscopy for analysis of the gas stream issuing from the fermentor. Pungor and his co-workers (10,11) for example, described the use of a computer coupled mass spectrometer-fermentor system for continuous on-line monitoring of a fermentation outlet gas. Buckland and Fastert (12) described the use of an MS with a computer coupled switching system for sequentially analyzing the exhaust gas from several fermentors. The switching method is very similar to a system reported earlier by Bull (13). The latter system employed a paramagnetic O_2 analyzer and an infrared CO_2 analyzer. The principle of using a mass spectrometer for analysis of fermentor exhaust gases is not new. For instance, this author (with a system developed by C.-S. Chen) used a residual gas MS for determination of CO_2 and O_2 in a fermentor exhaust gas in 1969. The significance of more recent developments lies in the use of microprocessor technology and in the availability of MS instruments specially designed for such applications. Heinzle, et al. (14), described the use of a quadrupole MS, with either a capillary gas inlet for gases or a membrane inlet for volatile compounds, coupled to a desk calculator. The latter was used for resolution of complex peaks.

Yet another area of widespread activity has been in the development of biologically active sensors, using either enzymes or whole cells. The enzyme thermistor as developed at the University of Lund is interesting. Danielsson and Mosbach (15) recently reviewed the potential for these devices. The principle depends upon the detection of the exothermic heat of reaction of a substrate catalyzed by an immobilized enzyme or cell. Detection is by a thermistor in close proximity to the enzyme. Mandenius, et al. (16), employed enzyme thermistors as sensors (for sucrose or glucose) with a PID controller to maintain a constant substrate concentration in an alcohol fermentation. In a more recent report, Mattiasson, et al. (17), linked an enzyme thermistor for sugar and a semiconductor gas sensor for ethanol to a computer to achieve constant product concentration in the effluent stream of a continuous fermentation system. This follows an earlier paper by Mandenius, Danielsson, and Mattiasson (18) in which an enzyme thermistor was used to control substrate concentration in a continuous fermentor.

Enfors (19,20) described a novel technique intended to eliminate problems due to slow transfer of oxygen into glucose oxidase or catalase based glucose electrodes at high glucose levels. Enfors generated oxygen at the enzyme site by electrolysis in a controlled manner so that the electrode was always in a region of high oxygen concentration.

Obana, Hikuma, and Yasuda (21) used a single glucose oxidase electrode combined with chemical hydrolysis and chemical isomerization for on-line measurement of the three principal sugars of molasses fermentations, *viz.*, sucrose, glucose, and fructose. Such a scheme of course necessitates sampling the fermentor and feeding the sample through an automatic wet chemistry apparatus.

Gondo, Morishita, and Koya (22) made a bilayer enzyme electrode with a layer of immobilized glucose isomerase and a layer of immobilized glucose oxidase. By this elegant and clever means these workers were able to significantly increase the sensitivity of the glucose electrode compared to the use of glucose oxidase alone. Once again, analyses were performed on-line using automatic analysis of a sample stream.

Schmidt, Huck, and Schelter-Graf (23), succeeded in modifying carbon electrodes with organic catalysts for electrochemical monitoring of NADH. Coating the electrodes with an immobilized dehydrogenase and NAD+ results in an electrode suitable for determination of the substrate for the dehydrogenase. The same workers also constructed electrodes for the same class of substrates by combining a dehydrogenase and an NADH-oxidase as a coating on an oxygen electrode. The output of the oxygen electrode is the measure of substrate concentration. They used both systems to measure continuously compounds such as ethanol, lactate, and amino acids.

Several reports have appeared on new developments in sensors using immobilized microbes. Hikuma, et al. (24), immobilized several different species of microorganisms and measured the respiratory activity of the organisms with an oxygen electrode. This configuration was used as a detector for methanol and ethanol.

In another study, Hikuma and co-workers (25) described the use of immobilized cell electrodes and a microcomputer to measure ammonia, acetic acid, and glutamate in a fermentation.

Production of carbon dioxide by an immobilized *E. coli* was used by Yasuda (26) as a method of detection of glutamate. The immobilized *E. coli* was placed over a CO_2 electrode, and the output of the CO_2 electrode was used as the actual detector. The probe suffered from serious drift problems, as well as from slow decrease of enzyme activity in the immobilized cells. A microcomputer was employed with a smoothing algorithm and an automated recalibration procedure using standard solutions. Use of smoothing rather than filtering means that estimates are produced for a past time, so it may be difficult to use such a system with a fast-changing fermentation. However, drift in the system was reduced about six-fold.

Suzuki and Karube (27) summarized work on these electrodes and gave a number of references on other work up to 1980 (*q.v.*).

More recently Galindo, Bautista, and Quintero (28) presented results on the use of genetically engineered organisms to improve the performance of microbial electrodes. These workers transformed an *E. coli* strain with plasmid pBR-327 containing a β-lactamase gene. The resulting organism, with high penicillinase activity, was used in conjunction with a pH electrode to make a penicillin electrode. In the same presentation a BOD sensor using immobilized cells was also described. A good estimate of the 5 day BOD was claimed to take only 15 minutes.

It is probable that further progress in electrodes utilizing immobilized cells and enzymes will parallel developments in immobilized cell and enzyme bioreactors in general.

Other methods of sensing the chemical environment are also the subject of much research. Beyeler, Einsele, and Fiechter (29) described a method and application of use of fluorescence of a culture for on-line fermentation measurements. Biryukov, et al. (30), used ion selective electrodes for control of ammonium nitrogen in antibiotic fermentations.

On-line analysis of sample streams using automatic wet chemistry analyzers is also receiving increased attention. For example, Lupin, et al. (31), measured the proteolytic activity of a culture filtrate using dialysis techniques. Leisola and Ojamo (32) successfully used automatic analyzers for on-line analysis for cellulase, α-amylase, glucoamylase, β-galactosidase, and protease in various fermentations. The system could be connected either to a laboratory fermentor or a computer coupled pilot plant. Volesky and Emond (33) devised a system for continuous flow monitoring of lactase using immobilized lactase and glucose oxidase, but they apparently did not connect their system to a fermentor.

Two very interesting papers have appeared in which the heat production of a fermentation was considered as a process varible. Ruocco, Coe, and Hahn (34) measured the heat of fermentation during full scale production of lager beer. Yeast activity was monitored in this way during fermentation with the help of an on-line computer. Control was discussed but apparently not attempted. Eriksson (35) studied heat evolution in several fermentations, with the objective of determining the net efficiency of the biochemical system.

Finally, Spriet and vanSteenkiste (36) and Blachere, et al. (37), have addressed in a general way the fact that the lack of specific sensors is currently a serious drawback in the engineering study of fermentation systems.

It is true, of course, that none of the techniques described can be used to measure or detect events or the status inside the cell. Such determinations cannot yet be made in real time or even on-line except in very special cases. For example, compounds which can be removed from cells by solvent extraction, cell disruption, or other means, can be measured by means of on-line automatic analyzers.

II. STATE ESTIMATION AND DIGITAL SIGNAL PROCESSING

For the most part the number of state variables which can be directly measured is still very limited. This has led to a number of attempts to deduce the values of state variables from models. A notoriously difficult variable to measure directly (except in clear, frequently laboratory media) is biomass. Yamashito and co-workers (38) first proposed making such measurements indirectly from material balances. Later workers developed this theme (39,40,41,42,43, 44). It soon became apparent, however, that this approach was limited by random noise both in the process and in the secondary variable measurements. Stephanopoulos and San (45) proposed applying filters to these stochastic processes for estimating cell biomass and growth rates in a continuous reactor. Basically, they applied various extensions of Kalman filters, including an adaptive filter developed by Jazwinski (46), to computer simulations of continuous fermentations. Endo, Nagamune, and Inoue (47) utilized a Kalman filter for the estimation of several physiological variables using a minicomputer coupled to a 2 liter fermentor. In this case the filter was applied to a batch process, *viz.*, cultivation of *Saccharomyces cerevisiae*. The algorithm was self adaptive. Moilanen (48) applied a nonlinear filter to a fermentation process for both state and parameter estimation. Different combinations of measurements were used in the various experiments. Howell and Jones (49) recognized the desirability of a structured model for parameter estimation. Using a Mean-Value model linearization to obviate difficulties associated with Kalman filters and nonlinear models, they applied a structured model to parameter estimation for the growth of *Pseudomonas putida* on phenol in a chemostat. Various difficulties and suggestions for further work were discussed.

When computers are used for measurement of the fermentation variables, then part of the problem of random noise, at least for input signals, can be alleviated by the use of digital filters. Bernard, Cordonnier, and Lebeault (50) used a simple digital smoothing filter to condition signals in their work with a continuous culture of *Candida tropicalis*. These workers were interested in measurements near the steady state. Batch processes are more problematical, but the theory and application of digital signal processing are well advanced (51), as well as techniques for digital control, especially using the z transform (52).

III. COMPUTER CONTROL AND OPTIMIZATION

We now pass to recent work on computer control and optimization. The most significant event in terms of the number of papers presented and published over the last few years was the Third International Conference on Computer Applications in Fermentation Technology (ICCAFT) held in Manchester, England, August 31-September 3, 1981. Proceedings of this meeting have been published in full (53).

Additional significant papers were presented at the Sixth International Fermentation Symposium, held in London, Canada, July 20-25, 1980, and at the Second European Congress of Biotechnology, held at Eastbourne, England, April 5-10, 1981. A great deal of other work has been published in the general literature. I have selected some of the recent work from all these sources to report in this survey.

Nyeste and Sevella (54) reported on computer evaluation of the results of batch fermentations. This work centered around the use of a generalized logistic equation and a non-linear least squares algorithm. The technique can be used to calculate various kinetic characteristics such as rates, specific rates, yields, etc. It seems probable that a similar technique could be used on-line in control schemes. Tanner (55) discussed problems and techniques in translating batch data to continuous fermentations. A useful discussion of the importance of statistical methods for the design of experiments and analysis of fermentation data is due to Porter (56). Such techniques have been used by fermentation companies for years, but the interesting thing is that in principle such analysis of data can be performed automatically by suitably programmed computers set up for control, data acquisition, and statistical analysis.

Another interesting application of statistical techniques to fermentations was devised by Meiners and Rapmundt (57). These workers used correlation analysis (another time honored technique) for the identification of fermentation process dynamics. A means of selecting amongst candidate models was outlined.

In terms of computer system configurations for data acquisition and control, as I have already mentioned, numerous systems have now been described, with further reports appearing regularly. Fox (58) made a market survey of commercially available computer systems for computer control up to 1981. Thirty-nine manufacturers and suppliers were included in the report, most of whom claimed to be able to supply a range of systems from simple laboratory fermentor control to elaborate systems for control of multiple production fermentors with sophisticated control and optimization. It is problematical as to how many of these systems are actually easily available. Presumably some, at least, are extensions from general process control packages. Others may be designed with fermentation control in mind. Jokela, et al. (59), reviewed the requirements for a modern computer controlled fermentation plant, combining microprocessor based instrumentation with minicomputer systems designed for work with fermentors.

Several industrial users gave us a peak at what goes on in their fermentation plants. Moffett (60) described a computer set-up in a new penicillin plant of Wyeth while Bowski, et al. (61), described a pilot plant data acquisition system at Hoffmann-La Roche. Some of the practices of an industry pioneer, Eli Lilly and Co., were revealed by Alford (62). The practice in the pilot plant of Lederle Laboratories was reported by Lee and Berenbach (63). A direct digital control system with a 64K word computer with floppy disks coupled to 8 pilot fermentors was described by Hjortkjaer and Falch (64) of Novo Research Institute. A computer system for fermentation processes was described by Kuratani and Gota (65) of Ajinomoto Co. A computer system for fermentation research installed at Farmitalia Carlo Erba Research was outlined by Valentini, et al. (66).

A number of academic and research institute investigators also published information about their own computer coupled fermentors. Some of the most fundamental, advanced work is emanating from Purdue University, and that group described the hardware and software configuration they use, along with an example of the system applied to a repeated fed-batch culture (67). Veres, et al. (68), and Nyeste, et al. (69), described a system of computer control with which they were able to maintain minimum generation time as a function of pH and temperature in a turbidostat. Yet another configuration for control of a research fermentor using microcomputers was presented by Clarke, et al. (70).

Two examples of computer monitoring and control in the fermentation laboratory were reported by Unden, et al. (71). One system was used to optimize biomass production and the other to increase amino acid accumulation. Nelligan and Calam (72) discussed requirements for on-line computer control of a fermentor and described their experience with a system for computer control of yeast growth.

Harrison, et al. (73), described another microprocessor system for direct digital control of a 2 liter laboratory fermentor. Some very simple optimization techniques were applied to find the maximum temperature at which a steady state could be sustained in continuous operation. Selection pressure was then possible to select temperature tolerant mutants.

Bungay has been a strong advocate of the use of hobby computers for laboratory fermentation process control. In one recent article (74), he pointed out some problems to avoid in such an application, as well as some guidelines to help achieve success.

An excellent review by Hampel (75) was concerned with various methods for data acquisition and control of bench top bioreactors using various computers and microcomputers. A more recent presentation by Hampel, John, and Roehr (76) demonstrated the use of the system they constructed for experiments on yeast cell wall lytic enzyme formation.

Jefferis (77) suggested the application of a technique which has been used for some time in the process industries, and which is available as a feature of many so-called programmable controllers. This is the technique of fault analysis, whereby a computer can determine a fault in operation (e.g., of a valve) as it occurs, and trigger an alarm condition. Such techniques should find frequent application in the future, especially in view of the high value added by many modern fermentation processes.

Lundell, Karvonen, and Ojamo (78) and Rajala and Lundell (79) presented papers on computer optimization of enzyme fermentations. Models were formed off-line and generated in algorithm form. Fermentations were then run and the models corrected, and so on.

A series of papers by Fawzy and Hinton (80,81,82) described the use of microprocessors in a four-level hierarchical approach for parameter estimation, dynamic optimization of a feedback controller, and optimal control of a penicillin fermentation and a gluconic acid fermentation. Lower computational power and memory capacity are required, it is claimed.

Other reports on optimization are due to Yousefpour and Williams (83) who dealt with real time optimization, and Moresi, et al. (84), who optimized a whey fermentation.

Mattiasson and his co-workers (17) very recently presented a very interesting paper in which a Kalman filter (cf., above) was used for parameter estimation using information from biosensors in a continuous ethanol fermentation, and the resulting information was used for state feedback control. Thus state estimation and stochastic control were coupled, a familiar technique in the process industries (85) but which has not been a practice in fermentation processes.

IV. CONTROL AND OPERATION OF FED-BATCH PROCESSES

A number of interesting papers has appeared on the optimal operation of fed-batch fermentations. Endo and Inoue (86) made an early study aimed at establishing a system for computer coupled fed-batch control.

Boyle and Kalogerakis published a series of papers (87, 88, 89) establishing conditions for existence of a quasi-steady-state (QSS), then devising a control system to force the process to QSS, and finally experimentally evaluating and demonstrating the controller.

Ramirez, et al. (90), reported a simple computer coupled fed-batch controller for yeast using the respiratory quotient and the oxygen uptake rate for control. The concept was demonstrated in a pilot plant fermentor.

Weigand (91) determined the optimal operating policy for a repeated fed-batch fermentation using the continuous maximum principle. The objective was minimum processing time for a fixed final cell concentration. Various operating modes with different specific growth rates were compared. With low residual substrate concentration, cell productivity by repeated fed-batch was greater than for continuous or repeated batch operation.

Hennigan and Lim (92) furthered this work by showing that various modes of repeated fed-batch operation could stimulate continuous stirred tanks, plug flow, or a combination of both, with and without recycle. Once again, under some conditions the productivity of such repeated fed-batch operation was greater than that for continuous or batch operation.

V. CONCLUSIONS AND FUTURE DIRECTIONS

In this review I have intentionally provided few details from the papers cited. Readers should consult the original literature. However, it is still possible to reach some conclusions about where we are and where we are going in automation of fermentation processes. It is certainly true that microprocessors are now firmly established as the most frequently applied device for computer control of fermentations. Microprocessors have made control by computer much easier to do, and consequently new techniques are also easier to apply.

There has been a move to distributed control, but this is not really a new concept (93). In the 1950's control was totally distributed through individual loop controllers. Control was frequently through totally centralized computers in the 1960's and 1970's. Since the advent of the microprocessor, starting in 1970, the trend has been back to distributed systems. The difference is that now we have distributed modular computing units. It should be emphasized that whether the modular computing units are for single loops or for several loops, the basic strategy, that of distributed control with multiple microprocessors, is the same. The decision of whether to use dedicated single loop microprocessors, will depend on the needs of a particular situation.

Fortunately, there are many distributed controllers to choose from. Multi-loop controllers have been available commercially since the mid-1970's. Increasing numbers of single loop microprocessor controllers with bus communication are now becoming available. At least one such set of controllers was designed specifically for fermentation control. Thus cost-effective computer control of fermentors is now available as a routine matter.

An equally important matter in this regard is what is to be controlled and what is the best way of doing so. Smith (94) reviewed these matters, and several points he made, and some further ones are worthy of further elaboration. First, many distributed controllers have very limited sequencing ability. So-called programmable controllers, which were originally strictly sequence controllers, are now being provided with more and more PID control loop capabilities. Also, at least one set of single loop microprocessor controllers designed especially for fermentations is available with a custom designed sequencer for automatic sterilization of fermentors.

The lack of adequate sensors is still a serious obstacle to further advances in fermentation automation. Considerable advances are being made however, so the situation is improving. Given the lack of on-line, real-time sensors, there is some very nice work being done on state estimation with stochastic variables. This information is now being applied to feedback control schemes (17). See (85, Ch. 5) for an introduction to the subject.

Optimization, especially nonlinear programming and use of the continuous maximum principle, is now fairly well in hand. Further progress is anticipated.

Operating aids of a more mundane, but extremely useful and important nature, such as sequencing control and fault analysis are areas for future progress. In terms of actual industrial operation these may turn out to be more useful than much of the sophisticated control and optimization already at hand.

One further area which has received little attention in the fermentation context is that of multi-variable control systems. Microprocessors and other computer devices now allow us to design and choose control algorithms almost at will. Heretofore, we have seen mostly the application of simple PID control algorithms, perhaps with cascade, ratio, adaptive, or other simple extensions. Actually, however, many control loops are interactive. For example, the optimum gain of a vessel pressure controller depends on the gas flow rate through the fermentor. Techniques have existed for some time to accomplish such inter-active multivariable control in the process industries by computer. See (85, Ch. 3) for an introduction. It is to be expected that this will be an important area for future work.

As pointed out by Smith (94), there are probably very few control problems in the fermentation industry which have not been dealt with already in the process industries. Historically, for a number of reasons, fermentation industry practices have lagged those of the process industries by a few years. We can only suspect and hope that the gap is closing.

REFERENCES

1. Weigand, W.A. in "Annual Reports on Fermentation Processes," (Perlman, D., ed.), Vol. 2, Chapter 3, Academic Press, New York (1978).
2. Dobry, D. and Jost, J. In "Annual Reports on Fermentation Processes," (Perlman, D., ed.), Vol. 1, Chapter 5, Academic Press, New York (1977).
3. Armiger, W.B. and Humphrey, A.E. In "Microbiol Technology," 2nd ed., (Peppler, H.J. and Perlman, D., ed.), Vol. II, Chapter 15, Academic Press, New York (1979).
4. Mor, J.-R. In "Computer Applications in Fermentaion Technology," (Proc. 3rd. Int. Conf. on), London, Soc. Chem. Ind., p. 109 (1982).
5. Philips, D.H. and Johnson, M.J. *J. Biochem. Microbiol. Technol. Eng. 3,* 261 (1961).
6. Lee, J.H., Woodward, J.C., Pagan,R.J., and Rogers, P.L. *Biotechnol. Lett. 3*(5), 251 (1981).
7. Heinzle, E., Bolzern,O., Dunn, I.J., and Bourne, J.R. In "Advances in Biotechnology," (Proc. 6th Int. Ferm. Symp.), Vol. 1, Pergamon, Toronto, p. 439 (1981).
8. Puhar, E., Einsele, A., Buehler, H., and Ingold, W. *Biotech. Bioeng. 22,* 2411 (1980).
9. Shoda, M. and Ishikawa, Y. *Biotech. Bioeng. 23,* 461 (1981).
10. Pungor, E., Jr., Perley, C.R., Cooney, C.L., and Weaver, J.C. *Biotechnol. Lett. 2,* 409 (1980).
11. Pungor, E., Jr., Schaefer, E., Weaver, J.C., and Cooney, C.L. In "Advances in Biotechnology," (Proc. 6th Int. Ferm. Symp.), Vol. 1, Pergamon, Toronto, p. 651 (1981).
12. Buckland, R.C. and Fastert, H. In "Computer Applications in Fermentation Technology," (Proc. 3rd. Int. Conf. on), London, Soc. Chem. Ind., p. 119 (1982).
13. Bull, D.N. Paper presented at ASM Ann. Mtg., Las Vegas, NV, May (1978).
14. Heinzle, E., Dunn, I.J., and Bourne, J.R. Abstracts, 2nd Eur. Cong. Biotechnol., London, Soc. Chem. Ind., p. 24 (1981).
15. Danielsson, B. and Mosbach, K. In "Computer Applications in Fermentation Technology," (Proc. 3rd. Int. Conf. on), London, Soc. Chem. Ind., p. 137 (1982).
16. Mandenius, C.-F., Danielsson, B., and Mattiasson, B. J. Abstracts, 2nd Eur. Cong. Biotechnol., London, Soc. Chem. Ind., p. 143 (1981).
17. Mattiasson, B., Mandenius, C.-F., Axelsson, J.-P., Danielsson, B., and Hagander, P. Presented at Biochemical Engineering III, Eng. Found. Conf., Santa Barbara, CA, Sept. 19-24 (1982), to be published by N.Y. Acad. Sci.

18. Mandenius, C.F., Danielsson, B., and Mattiasson, B. *Acta Chem. Scand. Ser. B 34,* 463 (1980).
19. Enfors, S.O. Abstracts, 2nd Eur. Cong. Biotechnol., London, Soc. Chem. Ind., p. 141 (1981).
20. Enfors, S.O. *Appl. Biochem. Biotechnol. 7,* 113 (1982).
21. Obana, H., Hikuma, M., and Yasuda, T. Presented at Biochemical Engineering III, Eng. Found. Conf., Santa Barbara, CA, Sept. 19-24 (1982), to be published by N.Y. Acad. Sci.
22. Gondo, S., Morishita, M., and Koya, H. Presented at Biochemical Engineering III, Eng. Found. Conf., Santa Barbara, CA, Sept. 19-24 (1982), to be published by N.Y. Acad. Sci.
23. Schmidt, H.-L., Huck, H., and Schelter-Graf, A. Presented at Biochemical Engineering III, Eng. Found. Conf., Santa Barbara, CA, Sept. 19-24 (1982), to be published by N.Y. Acad. Sci.
24. Hikuma, M., Kubo, T., Yasuda, T., Karube, I., and Suzuki, S. *Biotech. Bioeng. 21,* 1845 (1979).
25. Hikuma, M., Yasuda, T., Karube, I., and Suzuki, S. "Biochemical Engineering II," *Ann. N.Y. Acad. Sci. 369* (1981).
26. Yasuda, T. In "Computer Applications in Fermentation Technology," (Proc. 3rd. Int. Conf. on), London, Soc. Chem. Ind., p. 167 (1982).
27. Suzuki, S. and Karube, I. In "Advances In Biotechnology," (Proc. 6th Int. Ferm. Symp.), Vol. 3, Pergamon, Toronto, p. 355 (1981).
28. Galindo, E., Bautista, D., and Quintero, R. Presented at Biochemical Engineering III, Eng. Found. Conf., Santa Barbara, CA, Sept. 19-24 (1982), to be published by N.Y. Acad. Sci.
29. Beyeler, W., Einsele, A., and Fiechter, A. *Eur. J. Appl. Microbiol. Biotechnol. 13,* 10 (1981).
30. Biryukov, V.V., Itsygin, S.B., Kantere, V.M., Bashutkin, V.F., Zharov, A.L., and Choban, P.V. *Khim. -Farm Zh. 15,* 71 (1981).
31. Lupin, I.V., Koerner, D., Taeufel, A., and Ruttloff, H. *Enz. Microb. Technol. 4,* 104 (1982).
32. Leisola, M. and Ojamo, H. Lecture abstract, Dept. of Chem., Helsinki Univ. of Tech., Finland (1979).
33. Volesky, B. and Emond, C. *Biotech. Bioeng. 21,* 1251 (1979).
34. Ruocco, J.J., Coe, R.W., and Hahn, C.W. *Tech. Quarterly, Master Brewers Assoc. Amer. 17,* 69 (1980).
35. Eriksson, R.K. Abstracts, 2nd Eur. Cong. Biotechnol., London, Soc. Chem. Ind., p. 138 (1981).

36. Spriet, J.A. and vanSteenkiste, G.C. IMACS Symp. on Simul. of Control Syst., North-Holland Publ. Co., Amsterdam, p. 245 (1978).
37. Blachere, H.T., Peringer, P., and Cheruy, A. 1st. Eur. Cong. Biotechnol., *Dechema Monogr. 82*, 65, Verlag Chemie Weinheim (1978).
38. Yamashito, S., Hishashi, H., and Inagaki, T. *Proc. 4th Int. Ferm. Symp.*, 441 (1969).
39. Jefferis, R.P. and Humphrey, A.E. *Proc. GIAM IV*, Sao Paulo, Brazil, 767 (1979).
40. Cooney, C.L., Wang, H.Y., and Wang, D.I.C. *Biotech. Bioeng. 19*, 55 (1977).
41. Wang, H.Y., Cooney, C.L., and Wang, D.I.C. *Biotech. Bioeng. 19*, 69 (1977).
42. Wang, H.Y., Cooney, C.L., and Wang, D.I.C. *Biotech. Bioeng. 21*, 975 (1979).
43. Ziegler, W.M. and Humphrey, A.E. Paper presented at 178th ACS Mtg., Washington, DC, Sept. 10-13 (1979).
44. Swartz, J.R. and Cooney, C.L. *Biotech. Bioeng. Symp. 9*, 95 (1979).
45. Stephanopoulos, G. and San, K.-Y. In "Advances in Biotechnology," (Proc. 6th Int. Ferm. Symp.), Vol. 1, Pergamon, Toronto, p. 399 (1981).
46. Jazwinski, A.H. "Stochastic Processes and Filtering Theory," Academic Press, New York (1970).
47. Endo, I. and Nagamune, T. Presented at Biochemical Engineering III, Eng. Found. Conf., Santa Barbara, CA, Sept. 19-24 (1982), to be published by N.Y. Acad. Sci.
48. Moilanen, U. *Proc. 19th IEEE Conf. Decis. Control*, Published by IEEE, Piscataway, NJ, 396 (1980).
49. Howell, J.A. and Jones, M.G. In "Computer Applications in Fermentation Technology," (Proc. 3rd. Int. Conf. on), London, Soc. Chem. Ind., p. 57 (1982).
50. Bernard, A., Cordonnier, M., and Lebeault, J.M. In "Advances in Biotechnology," (Proc. 6th Int. Ferm. Symp.), Vol. 1, Pergamon, Toronto, p. 407 (1981).
51. Oppenheim, A.V. and Schafer, R.W. "Digital Signal Processing," Prentice-Hall, Englewood Cliffs, NJ (1975).
52. Franklin, G.F. and Powell, J.D. "Digital Control of Dynamic Systems," Addison-Wesley, Reading, MA (1980).
53. Third Int. Conf. on "Computer Applications in Fermentation Technology," London, Soc. Chem. Ind. (1982).
54. Nyeste, L. and Sevella, B. *Acta Aliment. 9*, 225 (1980).
55. Tanner, R.D. *J. Chin. Inst. Chem. Eng. 9*(2), 99 (1978).
56. Porter, M.A. In "Computer Applications In Fermentation Technology," (Proc. 3rd. Int. Conf. on), London, Soc. Chem. Ind., p. 147 (1982).

57. Meiners, M. and Rapmundt, W. In "Advances in Biotechnology," (Proc. 6th Int. Ferm. Symp.), Vol. 1, Pergamon, Toronto, p. 385 (1981).
58. Fox, R.I. In "Computer Applications in Fermentation Technology," (Proc. 3rd. Int. Conf. on), London, Soc. Chem. Ind., p. 91 (1982).
59. Jokela, J., Oinas, R., and Meskanen, A. In "Computer Applications in Fermentation Technology," (Proc. 3rd. Int. Conf. on), London, Soc. Chem. Ind., p. 85 (1982).
60. Moffett, J.A. Paper presented at ACS 182nd Nat'l. Mtg. New York, Agust 24 (1981).
61. Bowski, L., Perley, C., and West, J.M. Paper presented at ACS 182nd Nat'l. Mtg., New York, August 24 (1981).
62. Alford, J.D. In "Computer Applications in Fermentation Technology," (Proc. 3rd. Int. Conf. on), London, Soc. Chem. Ind., p. 67 (1982).
63. Lee, S.E. and Berenbach, B.A. In "Computer Applications in Fermentation Technology," (Proc. 3rd. Int. Conf. on), London, Soc. Chem. Ind., p. 75 (1982).
64. Hjortkjaer, P. and Falch, E.A. Abstracts, 2nd Eur. Cong. Biotechnol., London, Soc. Chem. Ind., p. 148 (1981).
65. Kuratani, T. and Goto, Y. *Kagaku Kogyo 31*, 900 (1980).
66. Valentini, L., Andreoni, F., Buvoli, M., and Pennella, P. In "Computer Applications in Fermentation Technology," (Proc. 3rd. Int. Conf. on), London, Soc. Chem. Ind., p. 175 (1982).
67. Hennigan, P.J., Rolf, M.J., Weigand, W.A., and Lim, H.C. In "Computer Applications in Fermentation Technology," (Proc. 3rd. Int. Conf. on), London, Soc. Chem. Ind., p. 99 (1982).
68. Veres, A., Nyeste, L., Kuruca, I., Kirchkopf, L., Szigeti, L., and Hollo, J. *Biotech. Bioeng. 23*, 391 (1981).
69. Nyeste, L., Szigeti, L., Veres, A., Pungor, E., Jr., Kurucz, I., and Hollo, J. *Biotech. Bioeng. 23*, 405 (1981).
70. Clarke, D.J., White, N.J.M., Burns, A., Kell, D.B., and Morris, J.G. Abstracts, 2nd Eur. Cong. Biotechnol., London, Soc. Chem. Ind., p. 145 (1981).
71. Unden, A., Rindone, W.P., and Heden, C.G. *Process Biochem. 14*, 8 (1979).
72. Nelligan, I. and Calam, C.T. *Biotechnol. Lett. 2*, 531 (1980).
73. Harrison, D.E.F., Ashby, R.E., Diamond, K.R., Hannah, F.K., Hinton, O., Leung, C., and Wymer, P.E.O. Abstracts, 2nd Eur. Cong. Biotechnol., London, Soc. Chem. Ind., p. 144 (1981).
74. Bungay, H.R. *Chem. Eng. Prog. 76*(4), 53 (1980).

75. Hampel, W.A. In "Advances in Biochemical Engineering," Vol. 13, Springer-Verlag, Berlin, p. 1 (1979).
76. Hampel, W., John, W., and Roehr, M. Abstracts, 2nd Eur. Cong. Biotechnol., London, Soc. Chem. Ind., p. 147 (1981).
77. Jefferis, R.P., III. In "Computer Applications in Fermentation Technology," (Proc. 3rd. Int. Conf. on), London, Soc. Chem. Ind., p. 199 (1982).
78. Lundell, R., Karvonen, E., and Ojamo, H. Abstracts, 2nd Eur. Cong. Biotechnol., London, Soc. Chem. Ind., p. 27 (1981).
79. Rajala, A. and Lundell, R. Abstracts, 2nd Eur. Cong. Biotechnol., London, Soc. Chem. Ind., p. 146 (1981).
80. Fawzy, A.S. and Hinton, O.R. *J. Ferment. Technol. 58,* 61 (1980).
81. Fawzy, A.S. and Hinton, O.R. *IEEE Trans. Syst. Man. Cybern. 10,* 467 (1980).
82. Fawzy, A.S., Hinton, O.R., and Hannah, F.K. *IEEE Trans. Syst. Man. Cybern. 11,* 227 (1981).
83. Yousefpour, P. and Williams, D. *Biotechnol. Lett. 3,* 519 (1981).
84. Moresi, M., Colicchio, A., and Sansovini, F. *Eur. J. Appl. Microbiol. Biotechnol. 9,* 173 (1980).
85. Ray, W.H. "Advanced Process Control," McGraw-Hill, New York (1981).
86. Endo, I. and Inoue, I. *Proc. 2nd Pac. Chem. Eng. Cong. 1,* 583 (1977), AIChE, New York.
87. Boyle, T.J. *Proc. Jt. Autom. Control Conf.,* Philadelphia, PA, Published by ISA, Pittsburgh, p. 213 (1978).
88. Kalogerakis, N. and Boyle, T.J. *Biotech. Bioeng. 23,* 921 (1981).
89. Kalogerakis, N. and Boyle, T.J. *Can. J. Chem. Eng. 59,* 377 (1981).
90. Ramirez, A., Durand, A., and Blachere, H.T. *Biotechnol. Lett. 3,* 555 (1981).
91. Weigand, W.A. *Biotech. Bioeng. 23,* 249 (1981).
92. Henniqan, P.J. and Lim, H. Presented at Biochemical Engineering III, Eng. Found. Conf., Santa Barbara, CA, Sept. 19-24 (1982), to be published by N.Y. Acad. Sci.
93. Morrison, D.L., Snow, R.H., and Lamoureux, J.P. *Science 215,* 813 (1982).
94. Smith, R.J. In "Computer Applications in Fermentation Technology," (Proc. 3rd. Int. Conf. on), London, Soc. Chem. Ind., p. 1 (1982).

Index

T

V

W

X

Y

Made in the USA
Monee, IL
13 September 2026

62977005R10214